OCEAN CURRENTS

Elsevier Oceanography Series

1 J. L. MERO
THE MINERAL RESOURCES OF THE SEA

2 L. M. FOMIN
THE DYNAMIC METHOD IN OCEANOGRAPHY

3 E. J. F. WOOD
MICROBIOLOGY OF OCEANS AND ESTUARIES

4 G. NEUMANN
OCEAN CURRENTS

5 N. G. JERLOV
OPTICAL OCEANOGRAPHY

OCEAN CURRENTS

BY

G. NEUMANN

Professor of Oceanography
Department of Meteorology and Oceanography
School of Engineering and Science
New York University
Bronx, N.Y.

ELSEVIER SCIENTIFIC PUBLISHING COMPANY
AMSTERDAM – LONDON – NEW YORK
1968

ELSEVIER SCIENTIFIC PUBLISHING COMPANY
335 JAN VAN GALENSTRAAT
P.O. BOX 211, AMSTERDAM, THE NETHERLANDS

AMERICAN ELSEVIER PUBLISHING COMPANY, INC.
52 VANDERBILT AVENUE
NEW YORK, NEW YORK 10017

REPRINTED 1973

LIBRARY OF CONGRESS CARD NUMBER: 68-15621

ISBN 0-444-40708-1

WITH 91 ILLUSTRATIONS AND 4 TABLES

COPYRIGHT © 1968 BY ELSEVIER SCIENTIFIC PUBLISHING COMPANY, AMSTERDAM
ALL RIGHTS RESERVED. NO PART OF THIS PUBLICATION MAY BE REPRODUCED,
STORED IN A RETRIEVAL SYSTEM, OR TRANSMITTED IN ANY FORM OR BY ANY
MEANS, ELECTRONIC, MECHANICAL, PHOTOCOPYING, RECORDING, OR OTHERWISE,
WITHOUT THE PRIOR WRITTEN PERMISSION OF THE PUBLISHER,
ELSEVIER SCIENTIFIC PUBLISHING COMPANY, JAN VAN GALENSTRAAT 335,
AMSTERDAM

PRINTED IN THE NETHERLANDS

PREFACE

Knowledge of ocean currents is vital to a better understanding of our natural environment. The physical-chemical stratification and to a great extent, the biological characteristics of the oceans, are closely related to large and small scale horizontal and vertical motions of water masses. One important goal in physical oceanography is to observe, represent and explain the general oceanic circulation, and to predict changes of currents and related characteristics of the physical-chemical and biological environment.

Questions of what ocean currents are, how they can be measured, represented and explained, form the contents of this book. Its purpose is to introduce scientists engaged in the study of the oceans to the basic principles involved in the study of ocean currents.

Navigators, for example, are interested in an adequate description and representation of ocean currents, and their periodic and aperiodic changes. Persons concerned with the drift of pollutants, sewage, radio-active waste products, and other deleterious matter discharged into the sea need to know where a drifting water body finally arrives. For fishery scientists it is important to know, among other things, what the effects of ocean currents are on providing nutrients, oxygen, and other physical-chemical factors to different strata in the sea. Horizontal advection of water masses, with the associated transports of fish larvae and food, is also significant for studies in marine biology and fishery science. Marine geologists and sedimentologists are interested not only in deep and bottom currents, but also in the general circulation of the upper ocean strata and the transport of sedimentary material between surface and bottom. In fact, many valuable contributions to our present knowledge of water movements near the bottom of the deep sea have come from the recent works of marine geologists.

Last but not least, meteorological research and its application to improved weather forecasting on a large scale would not be possible without reference to the general oceanic circulation, its anomalous variations, and its effect on physical conditions at the air–sea interface.

It is hoped that this book will serve the physical oceanographer as an introductory text. For this purpose, an extensive list of references has been added. Moreover, the book will provide the basic material for a study of ocean currents for marine biologists, fishery scientists, chemists, marine geologists, marine engineers, meteorologists, and other researchers interested in general oceanography. Mathematical derivations are kept to a minimum, and when used, they employ simple tools which make it possible for a reader with a background in algebra, trigonometry, and geometry and the basic elements of calculus and differential equations to follow the presentation easily.

The limited space in a book such as this made it necessary to deal briefly with some subjects. For example, observational techniques and the analysis of modern oceanographic data have been developed during the past ten or twenty years at such a rate that these fields may be considered a special science. The same is true for the study of turbulence, mixing and diffusion in the sea. A quantitative explanation of the three-dimensional oceanic circulation requires not only a profound knowledge of higher mathematics and geophysical hydrodynamics, but also a better understanding of the interaction between atmosphere and ocean, as well as friction and diffusion in the sea. We have not yet reached the state of answering these questions adequately in all phases of modern research so that they may help in a summarized view to better comprehend the general circulation of the oceans. The author has made an attempt so as to summarize and integrate established facts and theories.

I wish to acknowledge the support I received from the Office of Naval Research, the National Science Foundation, and the U. S. Naval Oceanographic Office for research in the field of dynamic oceanography and air-sea interaction. My thanks are extended to my colleagues in this country and abroad for their generous permission to quote from their publications and to use reproductions of their figures. I am particularly indebted to

PREFACE

Dr. W. J. Pierson Jr. and Dr. A. D. Kirwan Jr. for helpful suggestions and constructive criticism in preparing the manuscript.

I want to express my greatest appreciation to Mrs. S. Wladaver who not only typed the manuscript but also read the proofs with devotion and reliability. Finally, my sincerest thanks go to my wife for the encouragement she gave me in writing this text in spite of the fact that this writing had to be done in my spare time.

New York, N.Y.
July, 1968

G. NEUMANN

CONTENTS

PREFACE V

CHAPTER I. OBSERVATIONS AND METHODS OF CURRENT MEASUREMENTS 1
  Variability of ocean currents and practical problems in measurement 1
  Direct current measurements—Langrangian and Eulerian methods 9
    Current meters 13
    Drift measurements 24
  Indirect methods of current measurements 34
    The electromagnetic method 34
    The dynamic method in oceanography 37
    TS-Diagrams and the core method 38
    Isentropic analysis 47

CHAPTER II. PRESENTATION OF OCEAN CURRENTS AND WATER MASS TRANSPORTS 49
  The use of current measurements 49
  Singularities in a current field 55
  Current charts 59
  Major ocean currents 73

CHAPTER III. BASIC HYDRODYNAMICAL BACKGROUND . . . 81
  Pressure, density and geopotential 81
  Representation of the field of static pressure—pressure gradients 84
  The equations of motion 88
    Coriolis forces 92
    Frictional forces 95
  The equation of continuity 103
    Kinematic relationships in a current field 105

Use of the equation of continuity for computing vertical motion. 112
Continuity of salt—the Knudsen relationships . . . 115
Turbulence, mixing and diffusion. 120

CHAPTER IV. MAJOR TYPES OF OCEAN CURRENTS 127
Geostrophic currents 127
  Practical application of the geostrophic equilibrium 135
  Practical techniques for transforming relative into absolute dynamic topographies. 145
Inertia currents 149
Circular motion and meandering currents 155
  Gradient currents 158
  Cyclostrophic motion 162
Currents including friction and diffusion. 164
  Decay of a current under the effect of friction . . . 166
  The law of parallel fields and its disturbance by friction and diffusion 168
  Tongue-like distribution of water properties as a result of currents (advection) and diffusion. 174
Wind-driven currents in a homogeneous ocean. . . . 178
  Pure drift currents in deep and shallow water . . . 180
  Slope currents 192
  The elementary current system according to Ekman 195
Currents in a non-homogeneous ocean 197
  Mass stratification in geostrophic and gradient currents 198
  Ekman's relative currents and the elementary current system in a non-homogeneous ocean 212
  Wind-driven currents and slope currents. 216
  Thermohaline circulations 222

CHAPTER V. THE GENERAL CIRCULATION OF THE OCEANS . 227
Horizontal circulation of wind-driven ocean currents . 227
  Ekman's results for a homogeneous ocean 228
  The effect of bottom topography 231
  The planetary curl effect 235
  Modern approaches to the problem of wind-driven currents 238

Wind-driven and thermohaline circulations	261
Special current systems and current branches	271
The equatorial current system	275
Western and eastern boundary currents	284
The Antarctic Circumpolar Current and the Antarctic Ocean	298
Deep-sea circulation and bottom currents	303
REFERENCES	311
INDEX	337

CHAPTER I

OBSERVATIONS AND METHODS OF CURRENT MEASUREMENTS

VARIABILITY OF OCEAN CURRENTS AND PRACTICAL PROBLEMS IN MEASUREMENT

Ocean currents represent a complicated mixture of different types of aperiodic and periodic water movements, ranging over a wide scale of size, velocity and time. Variability of speed and direction of ocean currents is one of their most outstanding characteristics.

Besides seasonal changes of the net motion and periodic changes of shorter duration including tidal motions, a whole spectrum of oscillatory and sporadic movements superimposes the general oceanic circulation and makes the results of current observations in any part of the ocean a complicated record to study. Although a mean motion of oceanic water masses can always be derived from adequate observations, this mean or net motion is defined only for the period of observation. Depending on the time scale that is chosen for deriving the average displacement of the water, the net motion or the residual current can be very different at a given location in the sea when observed at different times. Also, the definition of a mean current is not always the same as the definition of the residual current or of the net displacement observed during a certain period of measurement. This fact implies problems of current measurements and leads to difficulties in the representation of ocean currents in charts. Currents can be shown by velocity vectors, representing speed and direction of water movements at a given place and time. Averages can be taken either as vectorial averages or scalar averages for speed and direction. Both of these averages are not satisfactory for all purposes. For example, a current that fluctuates periodically between two opposite directions with equal speed gives a vectorial mean of zero; that is, no velocity vector at all. The scalar means of speed and direction taken separately for this case would give a resultant current of a certain speed but of no direction.

The accurate observation and analysis of ocean currents and a satisfactory representation of observational results in current charts is by far not an easy task in oceanography. At first glance, the direct measurement of ocean currents may appear very simple, and a great variety of current meters has been developed, particularly during the last twenty years. Reliable results have been obtained with some of these instruments even at depths close to the ocean bottom. However, measurements of currents by means of instruments suspended from ships, buoys, or other observation platforms are subject to unwanted excursions of the current meter as a result of stray motions of the suspension point and the supporting cable.

In the case where these observation platforms drift with wind and surface currents, the current meter will measure only the relative motion between the meter itself and the surrounding water. Thus, in this case the absolute position of the observation platform relative to the ground has to be known very accurately in order to correct relative current measurements into absolute measurements. This is difficult when observations are made far away from shores in the deep ocean. Only in regions where modern navigational aids, such as Shoran, Loran, Decca, and others are available, may ship drifts over ground be adequately controlled. In many parts of the oceans such electronic navigational aids are not yet available and the navigator has to rely completely on less accurate astronomical navigation. Even this may fail completely if sky conditions are unfavorable and neither stars, sun, nor moon are visible for many days.

New navigational aids that permit position location with world-wide accuracy to within about 100 m in any kind of weather are the navigation satellites. The principle is that the frequency of satellite signal with respect to the observer appears to change as the satellite moves to or from the observer (Doppler shift). If this system becomes operational, it will greatly improve position location for oceanographic surveys even in the remotest parts of the world's oceans. (See, e.g., *Ocean Industry*, 2(9) : 32–34, Sept. 1967.)

An approach to the fixed point at sea is an anchored ship or an anchored buoy. Anchoring procedures are time-consuming and expensive. Since most oceanographers have to husband carefully on available ship time and expenses, such anchor stations for current measurements and other observations from fixed points in the deep sea are rare. Successful experiments of this kind are outstanding. In the early past of systematic oceanographic research, remarkable current measurements from anchored ships were obtained. BUCHANAN (1886) anchored in March 1886 the ship "Buccaneer" near the Atlantic equator and successfully completed an anchor station for current measurements at a depth of about 3,400 m. Observations at this station gave first evidence of an Equatorial Undercurrent in the Atlantic Ocean. From 1885 to 1890 the American ship "Blake", a steam schooner only 45 m in length, anchored under the command of PILLSBURY (1891) at different places in the Florida Straits, in the passages of the Windward Islands, and off Cape Hatteras, at depths even as great as 4,000 m. Pillsbury's current measurements with a self-designed current meter and self-designed anchoring gear in one of the strongest ocean currents of the world are considered a great achievement in seamanship and oceanographic planning.

Anchor stations of oceanographic research vessels like those of "Challenger" (BELKNAP, 1885), "Gazelle" (VON SCHLEINITZ, 1889), "Armauer Hansen" (EKMAN and HELLAND-HANSEN, 1931), "Meteor" (SPIESS, 1928; SCHUMACHER, 1930), "Altair" (DEFANT, 1940), and others in modern times are outstanding attempts to measure ocean currents at different depths in the deep sea.

Instead of anchoring a ship, the tendency today is to separate the sensing instrument for recording or telemetering from the ship by use of anchored buoys. For instance, in a recent effort to obtain better information on the variations of currents in deep water, oceanographers from the Woods Hole Oceanographic Institution planted 24 buoyed instrument arrays between Martha's Vineyard, Mass., and Bermuda (RICHARDSON et al., 1963). A continuous line of sensors began recording current speed and direction from surface to bottom. The stations may be left unattended for extended periods of time. Such anchored buoy stations with current meters at different depths are certainly a great advance in the technique of current observations in the deep sea. During recent Equalant Expeditions (since 1963) this technique besides others has also been employed for observations in the Atlantic Equatorial Undercurrent (STALCUP and METCALF, 1966).

Fig.1 shows the design of a buoy station according to RICHARDSON et al. (1963). The doughnut-shaped surface float is 8 ft. in diameter with a 3-ft. hole. It is surmounted by a 10-ft. high tripod tower which supports wind-measuring instrumentation where wind speed and direction are separately recorded. A single transistorized radio transmitter used for location purposes sends signals that can generally be recognized as far as 200 miles from the buoy. To increase the stability and to keep the buoy from capsizing at greater wind force, an apex float of fiberglass-covered foam can be added.

The current measuring devices are attached to the suspension cable. Since they record internally, it is, of course, necessary to recover the cable and to retrieve the instruments and their records. It was found that difficulties and hazards during recovery were greatly diminished if the dead weight of the ground tackle were cast off. For this purpose a release mechanism for the cable just above the ground tackle was employed. A special mooring anchor (STIMSON, 1963) has recently been adopted instead of the previously used 90-lb. Danforth anchor. Details of cable connection between the surface float and the ground tackle are illustrated in Fig.1.

During the last 10–20 years several deep-sea moorings for current measurements by various groups have shown that properly designed stations can be expected to last under all sea state conditions for extended periods of time. Besides rope or wire slack-line moorings like those employed by RICHARDSON et al. (1963), so-called taut-wire moorings utilizing sub-surface floats have also been used. Deep-sea moored buoy instrument stations have been dealt with in greater detail by BASCOM (1957). Deep-sea anchoring and mooring of light devices at depths of 100–3,000 fathoms were more recently discussed by ISAACS (1963).

Certain difficulties in obtaining reliable absolute current measurements in the deep sea are not yet overcome. No completely satisfactory method of measurement exists. This applies

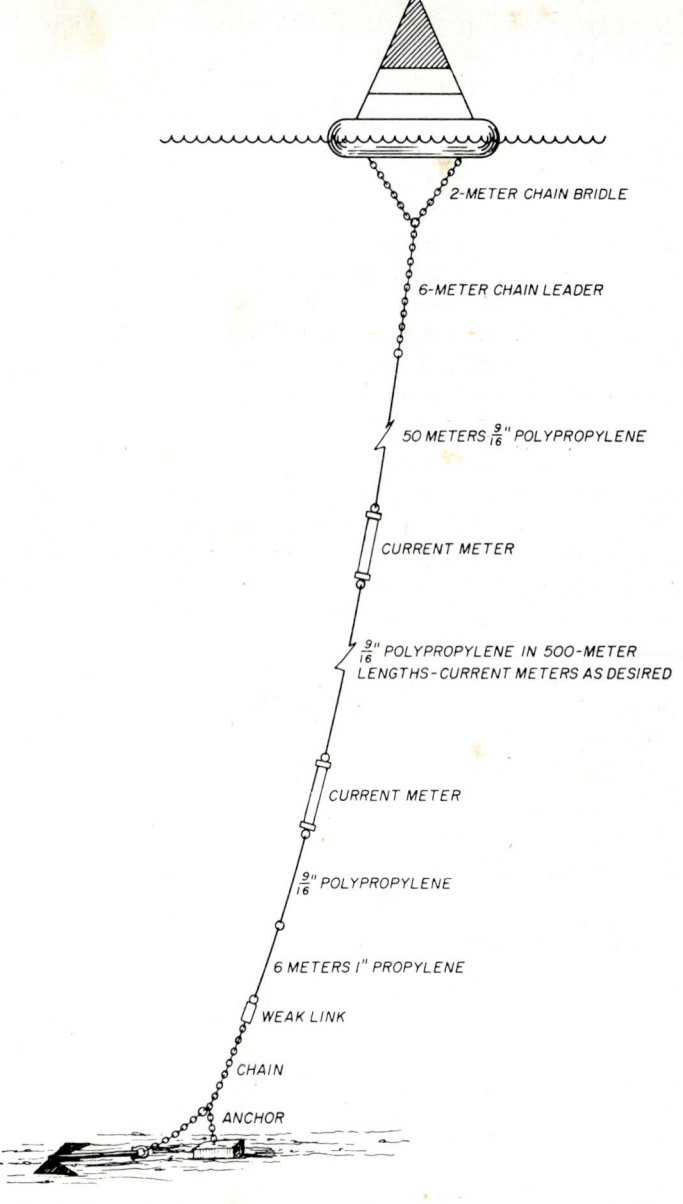

Fig.1. Design of a buoy station according to RICHARDSON et al. (1963).

to current meters as well as to the desired "fixed point" of reference for the recordings of the meter. Even an anchored ship, or a buoy, is far from representing a fixed point at the surface of a restless sea. Observation platforms anchored on long, elastic anchor lines undergo complicated motions of different kinds. A ship may heave, roll, pitch, yaw, sway, and surge. If the ship is at anchor, all of the "degrees of freedom" motions may superpose upon a fore-and-aft "riding" motion due to periodic tightening and slackening of the anchoring gear. Such movements are the result of surface waves, winds and currents. Unwanted motions of this kind are at least partially transferred from the ship or the observation platform to the current meter by the suspension cable. Some of these motions are also found with anchored buoys carrying or supporting current meters. They have to be eliminated from the current meter recordings in the best possible way. It is always better to reduce such motions to a minimum in the first place (CARRUTHERS, 1964).

In the past, current meter recordings from anchored ships have been carefully checked with ship motions that were recorded simultaneously with current measurements (DEFANT, 1961; THORADE, 1933; WITTING, 1930). Correction methods have been employed using observed changes in the position of the ship relative to a buoy anchored by the shortest possible cable to the bottom.[1] THORADE (1933) and more recently DEFANT (1961) and PAQUETTE (1962) have given a summarized review of the difficulties encountered by finding the *fixed point* at sea for current measurements, and the reader may be referred to these discussions.

Even the most elementary demand, the fixed point for current measurements, which is easily provided for on land with wind measurements, is difficult to meet in the ocean. For such reasons alone, absolute and accurate current measurements in the deep sea are hard to obtain. DEFANT's (1940) analysis of current meter

---

[1] In 1965, M. O. Rinkel (University of Miami) used this method successfully for detailed current observations in the immediate vicinity of the Atlantic equator. Rinkel's current observations of the Atlantic Equatorial Undercurrent were combined with detailed observations of temperature and salinity at different depths in the region of this interesting branch in the system of equatorial currents (personal communication, 1965).

recordings obtained at an anchor station of the "Altair" in the North Atlantic Ocean with a Böhnecke current meter is an example of how much additional careful surface recordings, consideration of ship movements and computations are needed to eliminate in the best possible way unwanted components in a current meter record obtained from an anchored ship.

Last but not least, another fact that makes measurement, analysis, and adequate representation of ocean currents difficult is the *variability of the water movements* themselves. Current measurements at any location usually consist of a complex mixture of different components. Superimposed on a "mean motion" (a concept which is difficult to define because it depends on the locality and the time scale of the averaging process) are periodic and non-periodic fluctuations. Currents associated with all kinds of wave motion, particularly with tides, inertia currents, internal waves and surface waves, as well as irregular eddies of different scale, meanders and disturbances caused by rapidly changing winds are registered by a current meter. How much of the shorter-period fluctuations are sensed and recorded, depends on the type of current meter and its sensitivity.

Quite often the variability of the measured currents is large compared to the residual (or mean) current. In this case, the significance of short-term observations is in serious doubt. In the region of strong, well-defined currents, measurements carried out over periods of several hours or a few days are probably meaningful. However, the situation in deep water is considerably more complicated. SWALLOW's (1955, 1957) work with neutrally buoyant floats has shown that the variability of deep-sea currents is the outstanding feature of water movements at depths where not too long ago oceanographers expected a sluggish spreading of water masses in a rather uniform way.

CARRUTHERS (1964) has repeatedly warned against the use of oversensitive meters for measurements of average currents as they are used for many practical purposes; for example, to serve persons interested in the major displacement of water masses. Such persons, like navigators, fishery research scientists, or those who are concerned with the drift of sewage, radioactive waste products, and other deleterious matter discharged into the sea want to know where after some time the drifting water body arrives.

The high sensitivity of many recently developed current meters is not always necessary for general oceanographic purposes. For special work like studies on turbulence, highly sensitive apparatus is needed. However, they then have to meet high quality demands. CARRUTHERS (1964) stated the problem as follows: "In developing and using current meters adequate thought should be given to what observations of current speed and direction are most wanted at sea."

Highly sensitive instruments are mostly of very delicate design, expensive, and often not capable, or at least not very reliable, of operating under heavy sea state conditions. Today, the requirement of reliable operation at all weather conditions is more important than ever, because it becomes more and more desirable to leave current meters unattended for longer periods of time in near-shore and off-shore waters. The need for sufficiently accurate, durable and rapid response current meters is obvious. However, it should always be remembered that unwanted stray motions of the meter, as described before, may make a record of a rapid response current meter an illusion if it is not possible to eliminate the motion of the current meter from the recorded data. Here, the observer must bear the burden of proof that his measurements are meaningful and significant for representing the realistic structure of absolute water movements. Improperly designed sampling methods may even destroy the usefulness of current measurements. Quoting from WEBSTER (1964):

"If the aim of the current measurements is to define long-period variations, such as tidal motions, or even long-term averages, these short-period variations should be filtered out of the data. In order to accomplish this, observations must be made often enough to fully describe the highest frequencies to which the instrument can respond. If the observations are not taken frequently enough, the high-frequency variations will appear in the resulting time series as noise or, even worse, will appear under the apparent guise of low-frequency effects. This effect of high-frequency signal appearing as low-frequency signal has been called *aliasing*. It is a well-known phenomenon that has plagued many programs of time series measurement which use discrete sampling. If strongly aliased data is collected, the results can be very discouraging. Under some conditions, the aliasing produces an extremely high noise level which can mask the low frequency process which is the object of the measurements.

In other cases, very treacherous ones indeed, the aliased high-frequency signal, if systematic and coherent, can be indistinguishable from a low-frequency phenomenon.

The criterion for avoiding aliased data is well-defined. Suppose a measuring device does not respond to frequencies at or above a cutoff frequency, F_c. Then, to avoid aliasing, the time between observations must not exceed $1/(2F_c)$. Therefore unaliased data can, in practice, be collected if either: (*1*) The measuring instrument is damped so that it will not respond to frequencies which cannot be defined by the chosen sampling rate, or (*2*) the rate of sampling is increased so that all the frequencies which the sensor will pass are fully defined."

Although highly sensitive current meters are necessary to solve special problems in connection with ocean currents, it is also necessary to construct and use simpler, less expensive and more robust instruments for more general purposes. CARRUTHERS (1926, 1928, 1954, 1962) has always supported this attitude, and followed it by designing and using reliable and simple current meters to serve many practical purposes.

DIRECT CURRENT MEASUREMENTS—LAGRANGIAN AND EULERIAN METHODS

The direction of ocean currents is given as the direction toward which the current flows, whereas wind directions are recorded as the direction from which the wind comes. The physical unit of speed is one centimeter per sec (cm sec^{-1}, or cm/sec). In navigational practice the speed is often expressed in "knots". This unit is also used by many oceanographers. One knot is one nautical mile per hour. The nautical mile is the length of one minute of geographical latitude and is approximately 1.85325 km. Thus, 1 knot = 51.48 cm/sec.

Observational techniques and methods of measuring ocean currents vary widely. Generally, they can conveniently be classified into two groups: *the Lagrangian method and the Eulerian method*.

The Lagrangian method measures water movements by tracing the path of a water particle over a sufficiently long time interval. It fixes a curve in the current field that is called a trajectory. Combination of many such trajectories in a map leads to a current chart that is called a trajectory chart. Ideal trajectories are three-dimensional. Most often, trajectories in ocean currents are obtained and represented as "horizontal" trajectories, neglecting the vertical motion of the water.

The Eulerian method applies to fixed points where current direction and speed (the current vector) are measured simultaneously by fixed current meters. Also, in this case only the horizontal current is measured by most meters. Current vectors obtained in this way at the same time for different places can be combined in a map showing the synoptic distribution of speed and direction of the current. Continuous lines, following the direction of individual current vectors at the same time are called *streamlines*. Trajectory charts and charts showing streamlines can be basically different. Only for the case for which the water movements are stationary will streamlines and trajectories coincide. Streamlines are obtained from the instantaneous velocity vector distribution ("snapshot picture") and trajectories from a continuous tracking of "tagged" particles ("motion picture").

Both methods of measuring and presenting ocean currents have been used widely and both are of equal value. *The most complete description of oceanic currents is obtained from a combination of both Eulerian and Lagrangian methods.* It is, however, difficult to obtain a simultaneous picture of water movements by either method, covering a large oceanic area. In the past, only a few locally limited attempts were made to approach the simultaneous measurement of currents by either the Eulerian or the Lagrangian methods. An approach to a synoptic survey of parts of the Gulf Stream system was the multiple ship "Operation Cabot" (FUGLISTER and WORTHINGTON, 1951), and the representation of quasi-synoptic current fields in the southern North Sea since 1951 at the suggestion of Carruthers. An outstanding experiment using the Lagrangian method is MOSBY's (1954) measurement of surface floats in the Tromsö Sund (Fig.2). The position of the floats was determined two or three times per minute by means of theodolites from land. The total number of observations used to construct his current charts is about 50,000.

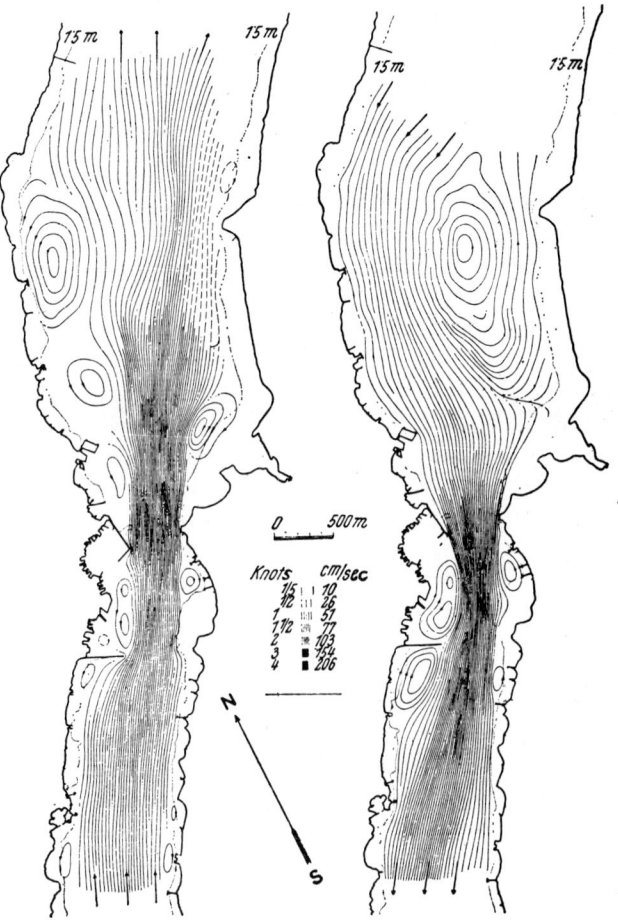

Fig.2. Trajectories for "normal" tidal currents in the Tromösund (no wind; mean vertical tidal range 200 cm). At left: currents to the north; at right: currents to the south. The closer the streamlines, the faster are the currents. The charts also show the effect of the earth's rotation. Currents flowing to the north are concentrated toward the shore in the east; currents flowing to the south are concentrated toward the opposite shore. This is clearly indicated in the northern part of the Sund. (Courtesy of H. Mosby, 1954.)

Results of the quasi-synoptic "Operation Cabot" have shown that the Gulf Stream in the northwestern part of the Atlantic Ocean is by far not a simple continuous stream. A number of disconnected filaments of the current reveals a complicated pattern that changes with time. The concept of a more or less continuous Gulf Stream in this area, as was derived before from averaging individual observations obtained at different times and places has been shattered. Even such majestic flows of great "ocean rivers" as the Gulf Stream in the western North Atlantic and the Kuroshio in the western North Pacific have a much more complicated structure than has hitherto been anticipated. The need for more synoptic surveys of water movements is most obvious.

The oldest, and at first glance, simplest method of determining speed and direction of ocean currents is to follow the position of drifting bodies relative to a fixed point. Basically, this method has provided the enormous amount of data that have served to construct mean sea surface current charts as they are found today in most geographical atlases. Information on worldwide ocean surface currents is almost exclusively derived from *ship drifts* reported in the log books of navigators since about the middle of the nineteenth century. Here, the ship itself is used as a tracer. The currents are determined from the difference between dead reckoning and astronomical fixes, if possible, at least 24 h apart. Although these drifts are inherent to great errors and are partly influenced by direct wind effects, a general picture of average flow conditions can be derived if a large number of observations is available. Since the accuracy of astronomical navigation is usually limited to ± 1 or ± 2 nautical miles, current observations of this type become less accurate where the ship displacements are small, that is in regions of weak currents. Also, ship's log observations refer to a drift of the vessel under way during a longer time interval and, therefore, represent averages over a distance in addition to time averages. All this leads to a presentation of a much smoother picture of ocean currents in atlases than actually exists. In addition, ocean currents can vary considerably not only from month to month and from year to year, but also from week to week or even from day to day. Therefore, ship drift observations collected in many different years and combined in the construction of one map can represent sea surface currents only in a very general way, even if observations are collected for individual months and presented in mean monthly charts.

Nevertheless, such charts showing the general trend of the large systems of oceanic surface water movements are of interest and value. Excellent representations of this kind are the current charts issued by Hydrographic Offices of various countries and SCHUMACHER's (1940, 1943) mean monthly surface current charts for the Atlantic Ocean. However, the user of such charts must always bear in mind that a smooth average picture of the general oceanic surface circulation can be very misleading if details are required. In some parts of the ocean, such details may be even more pronounced than the average circulation pattern itself.

Measurements of ocean currents following the Eulerian method are obtained by a variety of current meters with reference to a fixed point.

Current meters

A great variety of current meters to measure the current velocity in the oceans at different depths from fixed points has been developed in the past. The design of new meters continues, and the market is flooded with new inventions. In many cases, such new devices cannot really be considered an improvement over existing current meters; they often only use more complicated gadgets which may even reduce the reliability of the meter or make its operation in the field more complicated.

An excellent summary of existing methods and of most important types of current meters up to 1955 has been given by BÖHNECKE (1955) which can be considered an updating of THORADE's (1933) outstanding review of this subject. Since the year 1955, other current meters have been developed and used with more or less success. Some of the most promising models of newer design will be mentioned in the following pages. It is, however, difficult to appraise justly every new design, because often adequate information is missing about the details of such instruments, their behavior in oceanographic field work, and results obtained from their records.

The velocity of a current at a fixed point (Eulerian method) is measured by instruments operating on a variety of principles. For measurements of speed, the following principles have been used:

(*a*) Counting the rotations of a free-turning propeller against time. Instead of an ordinary propeller with a horizontal axis of rotation, paddle wheels or rotors with horizontal or vertical axes of rotation (Rauschelbach or Savonius rotors) can be applied. The most efficient and reliable shape of such rotors has not yet been decided upon. Also cup-systems (like cup anemometers) with a vertical axis of rotation have been used.

(*b*) Measuring the slope of wire or cable supporting a known drag, or measuring the ram pressure exerted by flowing water on pendulums, plates, membranes, spheres or pitot orifices.

(*c*) Using ultrasonic devices measuring phase differences of sound which travels in still water and moving water between two fixed points a known distance apart.

(*d*) Measuring the electromotive force which is produced when electrically conducting sea water moves in the earth's magnetic field.

(*e*) Measuring the rate of cooling of hot-wire devices (or thermistors) in moving water. This principle is analogous to measuring wind speed by means of hot-wire anemometers. The development of this method for ocean current meters is still in the beginning. It is not much beyond the state of laboratory work, although some recent measurements for special purposes near the bottom in shallow water have been successfully completed (LUKASIK and GROSCH, 1963). Because of their very small size, thermistors show great promise as a means of measuring small-scale motion (LARSON, 1960; GRANT et al., 1962).

The direction of the current is often determined by means of a compass either mounted inside the instrument or aboard the ship. With bifilar suspension of a current meter, directional readings can be obtained from the recorded angle between the suspension frame of the current meter and the ship's heading. The suspension frame is usually oriented parallel to the ship's keel. This method is of particular value for current measurements in the upper layers of the sea. Disturbances caused by the ship's magnetic field can be avoided when directions relative to the bifilar suspended meter frame are referred to readings of the (compensated) magnetic ship's compass or gyrocompass.

When compared to other single designs, probably the greatest number of useful current meter readings in the past was obtained by the Ekman current meter. Its original design has been modified and improved throughout the years (EKMAN, 1926, 1932; MERZ, 1929), although its simple mechanism and reliability have been preserved. This meter (shown in Fig.3) can be attached to a standard hydrographic wire and lowered to any desired depth with the propeller arrested and protected by a pair of shutters. The meter is not affected by sea pressure because there are no pressure-sealed compartments, and sea water has free access to all parts. When the desired depth of measurement is reached, a messenger weight, clasped around the suspension wire, is dropped. By this means a release mechanism is operated which opens the shutter and releases the propeller. Its revolutions are recorded on a set of dials (Fig.3). After a certain period of time, a second messenger is dropped which locks the propeller and the shutter, and the instrument can be raised to the surface. From the number of propeller revolutions the average flow velocity during the time of measurement can be computed.

The current direction is recorded magnetically by means of a ball-dropping device. With each revolution of one of the metering shafts, a passage is opened which connects to a guide tube above the center of the compass needle. Each time the passage is opened, a small bronze ball is dropped through the guide tube from a magazine filled with such balls which may also carry numbers. A cup mounted on the pivot of the compass needle receives the ball and leads it along a groove into one of the 36 compartments of a box underneath the compass. At the end of an observation period, the ball receiving box can be removed, and the number of balls in individual compartments can be evaluated for water movements towards different directions as indicated by their distribution. The Ekman meter operates accurately even at relatively low velocities. Its only disadvantage is that it is not suitable for continuous operation.

Improvements of the original design are the Ekman repeating current meter (EKMAN, 1926) which allows depth changes between measurements without raising the meter to the surface

16 OBSERVATIONS AND METHODS OF CURRENT MEASUREMENTS

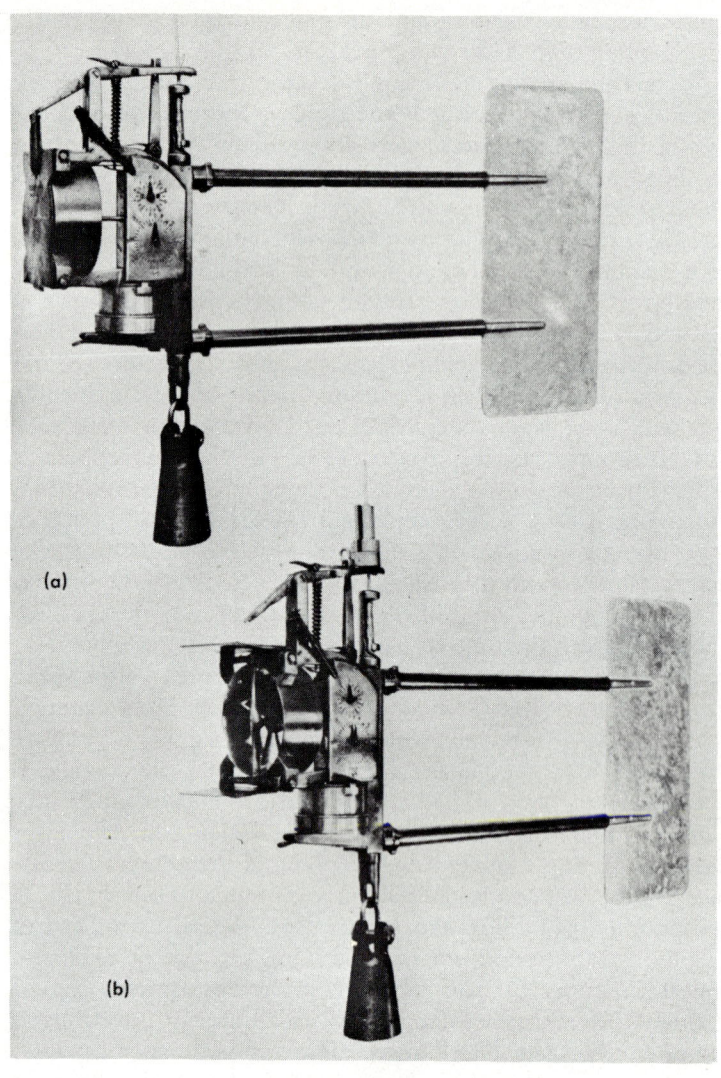

Fig.3. Ekman current meter, (a) cocked, (b) open.
(Courtesy of VON ARX, 1962.)

after each individual recording, and the Ekman–Merz current meter (MERZ, 1929). Also belonging to the great family of Ekman meters is the Ekman-E.M.W. serial current meter (see BÖHN-ECKE, 1955), which can be used for simultaneous measurements at different depths by means of a series of instruments attached to the suspension wire.

A ball-dropping device similar to the Ekman meter has been recently employed by CARRUTHERS (1964) in a totalizing current meter. This instrument registers water movements in all directions over a period of time. Its function is to serve persons who want to find most easily the totality of the varying water movements over one or several tidal periods. A unique part of this instrument is its suspension, designed to amortize in the best possible way unwanted vertical motions due to wave action. A loose attachment of the neutrally buoyant instrument to its suspension rope or wire allows a rise and fall between two stops, usually about one meter apart.

Among other mechanical devices that record speed and direction inside the current meter, the Böhnecke current meter (BÖHNECKE, 1937) deserves to be mentioned here. In this meter, propeller revolutions drive a set of horizontal disc dials with raised numbers around their rims. The compass rose also has such raised numbers around its rim, and is mounted at an even vertical with the speed dials. At certain time intervals, a clockwork transporting tinfoil past the rims of the dial operates a hammer that presses the foil against the numbers on the rim of the discs, thus preserving the instantaneous reading for speed and direction. The time interval between hammer operations can be changed.

Photographic recording has been used in several designs, for example in the D.H.I. paddle wheel current meter (JOSEPH, 1954). This meter is anchored to the bottom and is kept at a desired depth by its own buoyancy. The vanes of the rotating element (paddle wheel) protrude above the streamlined shape of the non-magnetic (iron-free) meter body. The paddle wheel drives a cylindrical water-tight case, and its revolutions are counted with reference to a frame inside the case. This frame contains all necessary recording gear and stays in an upright fixed position by its own weight. Current direction is also indicated inside the paddle wheel case by a magnetic compass. Recording is obtained for speed and direction on 16 mm motion picture film. This meter can operate for about four weeks even under heavy weather conditions before recovery becomes necessary.

Among other photographically recording devices, the IDRAC (1927) current meter is mentioned because of its unique way of recording the direction which is obtained by scanning an illuminated spiral on the compass rose. IDRAC (1933, 1935) has also designed a special current meter for measuring the vertical component of deep sea currents.

The PETTERSSON (1915) current meter is also designed to record photographically both speed and direction. As with all photographic devices, the recording apparatus is enclosed in a water-tight cylinder. The current activates a paddle wheel with vertical axis, and its rotations are transmitted to the recording gear by means of a magnetic drive. PETTERSSON (1929) also developed a special bottom current meter mounted on a tripod which stands on the bottom. Again, the velocity is measured by a paddle wheel mounted on a vertical axis but direction is obtained by projecting the position of a compass needle which carries a radioactive substance on a sensitive film.

Among the more recent current meters using photographic recordings is the Richardson current meter (RICHARDSON et al., 1963). The body of the meter is vertical and cylindrically symmetrical and does not require orientation into the current. Current speed is sensed by a Savonius rotor and direction by a vane in the upper cage. Rotor revolutions and vane orientation are magnetically coupled through end caps of the pressure case to the photographic recording device. Speed is recorded in two channels and direction in fourteen channels running parallel to the length of 100 ft. of photographic film. Two additional channels are occupied by clock pulses (time marks) and by a continuous line. The recording apparatus is enclosed in a pressure case, but the instrument can be used as deep as 6,000 m. The meter (Fig.4) can be employed either suspended from a surface platform or as a link in a mooring system.

A simple method of measuring currents at greatest ocean depths was suggested by EWING et al. (1946) in connection with bottom photography. This method applies several weighted

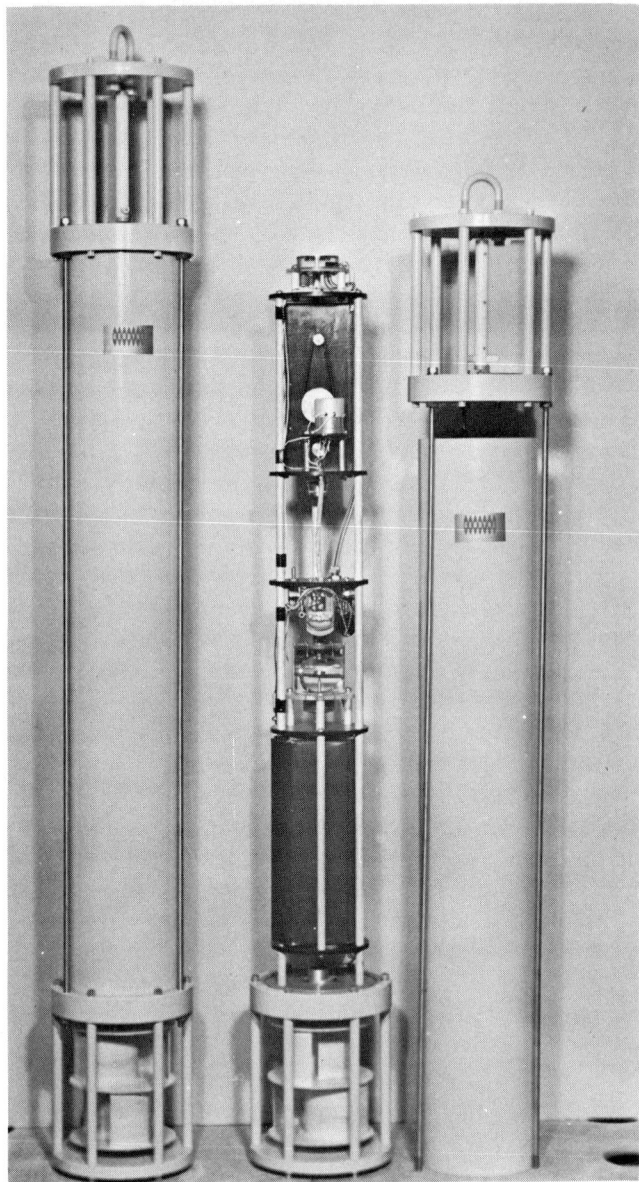

Fig.4. Richardson current meter.
(Courtesy of Geodyne Corporation, Mass., U.S.A.)

spheres mounted on the frame of the underwater camera. The deflection of these spheres under the effect of current drag can be photographed and properly evaluated.

Another simple and efficient instrument, the nylon yarn inclinometer, was used in recent years by LA FOND (1962) to measure the current near the sea floor. LA FOND (1962) has described the measurement of bottom currents by means of this devise from the bathyscaph "Trieste" at a depth of 3,870 ft. in the San Diego Trough off the coast of California. A grid with nylon yarn streamers was suspended between the bathyscaph's gondola and the forward ballast tube such that it could be viewed through the porthole of the gondola. When soaked with water, the nylon yarn streamers are just slightly heavier than water and have a quick response, even to weak currents. Their movements can be photographed with a motion picture camera. A system of mirrors permits the measurements of the movements of the nylon yarn in the viewing plane as well as in the direction of the line of sight. Calibration of the angle of deflection of the nylon yarn in terms of current speed is obtained in a towing tank. The transducer in the instrumentation in this case is quite simple, relatively inexpensive and very efficient, but the vehicle that carried it down to the point of measurement in this special experiment was most complex and expensive. Also, the recording system, and data processing and interpretation are not simple, but the results are rewarding. This method by La Fond is also an example of how a bathyscaph diving into any ocean depths can provide most valuable data in physical oceanography.

Besides paper and ink (which, of course, have to avoid contact with sea water), wax paper has been utilized to preserve records of individual or continuous readings of current meters. Among this group is the BBT-Neypric Currentograph (DUROCHE, 1953). The recording is also done inside a water-tight container, and power to operate the recorder is supplied by batteries. The meter can be suspended from a surface platform or anchored to the bottom. Its buoyancy allows the meter to rise to position at the desired depth. Wax paper recordings are also employed in the bathypitometer by MALKUS (1953). This instrument is designed to make relatively rapid traverses of the water depth. Besides the pitot tube, it contains a Bourdon type pressure sensor for depth determination, and a bimetal strip temperature sensor. Since the pitot tube senses only one component of the flow, it has to be headed into current direction, or the current direction has to be determined by other means.

HERSEY (1952) mentioned the possibility of measuring ocean currents by acoustic means. However, this method is still in the development stage. An ultra-sonic current meter was developed by MIDDLETON (1955) for work in estuaries. A new design by KOCZY et al. (1963), the Doppler current meter, may be mentioned here. The Doppler current meter is a high-frequency sound transmitter. It sends a narrow pulsed beam of sound through a small volume of water. Objects such as small scattering particles in the water through which the sound travels reflect the sound. Since such objects move with the velocity of the water in which they are suspended, the frequency of the reflected sound, sensed by a receiver, is changed according to Doppler's principle in acoustics. The Doppler shift permits the study of changes in velocity with respect to a small volume of water. At present, no report about the use of the Doppler current meter in oceanographic field work is available.

A large group of current meters is designed for telerecording or telemetering. The recording can be aboard a ship or ashore. Tele-transmission does not necessarily have to employ electricity. It can be mechanically effected, and in cases where this is possible, mechanical gear has great advantages over the use of electrical power. Electrical units, cables and their underwater connections have a great dislike of sea water, especially at great depths, and proper operation and maintenance of electronic gear requires much attention in oceanographic deep-sea work.

A sturdy, durable and simple mechanical device for recording and reading the meter aboard a ship was invented by CARRUTHERS (1935) (see also CARRUTHERS et al., 1951) by taking advantage of the most simple mechanical elements such as chains, ropes, wires, poles, ball-bearings, etc. Their use in the Carruthers Vertical Log has shown that mechanical tele-transmission of current recordings at subsurface levels serves adequately its purpose, although the depth range is limited. A great number of light-ships in Great Britain uses this simple device. The speed measuring unit consists of a set of six heavy iron cups suspended vertically in the water. The separate direction-viewing unit is fixed to the ship's rail close to the speed measuring unit. The current direction is given by the deflection of a light chain caused by the drag of the water flow. This deflection is relayed to a metal pointer aboard the ship. A special counting device on shipboard permits an easy reading of the record of how many revolutions are obtained within eight major directions of the compass rose (octants) in the course of, say, a lunar day or longer. This instrument works very well at shallower depths.

With electrical tele-transmission, speed can be counted by closing contacts after a number of propeller revolutions. It can be indicated or recorded by milliammeters or marks on a chronograph. Direction transmission is more difficult and several methods have been used such as in the WITTING (1923) and the SVERDRUP (1926, 1927) meters, the RAUSCHELBACH (1929), the D.H.I.-Bifalar (JOSEPH, 1954), and the VON ARX (1950) meters. The DOODSON (1940) current meter measures the torque of an arrested rotor and records by means of a potentiometer on a milliammeter.

Shore recordings by electric cable connections have been obtained from near-shore, shallow water depths with the PEGRAM (1933) current meter. This meter is mentioned because of its unique way in the electrical transmission of propeller revolutions; it could be operated at even greater depths. MOSBY's (1949) experiments on turbulence and friction near the bottom of the sea, and his bottom current meter using electric multi-recorder equipment are also mentioned at this place.

Although there are many other types of current meters in this and the other groups that deserve to be mentioned, even a fairly complete account would be much beyond the scope of this writing. There is, however, one other current meter that should be briefly discussed: the ROBERTS (1952) current meter which uses radio signals to transmit its readings ashore. It is also commercially available and is reported to record accurately above a speed of 0.3 knots.

The Roberts current meter, shown in Fig.5, can be operated from an anchored buoy or a ship. Fig.6 shows a schematic sketch of the mooring, the suspended meter, and the radio buoy used for transmission. This instrument also makes use of a magnetic drive to transmit the propeller revolutions into the interior of an enclosed recording mechanism, breaking an electric circuit by means of special contacting devices. While one contact operates at each fifth revolution of the propeller, the other operates at each tenth turn. The first contacts indicate the speed and are connected with a magnetic compass. The second contacting device is fixed relative to the meter body. They are so arranged that both contacts occur at the same time when the meter is heading to the south. With other headings,

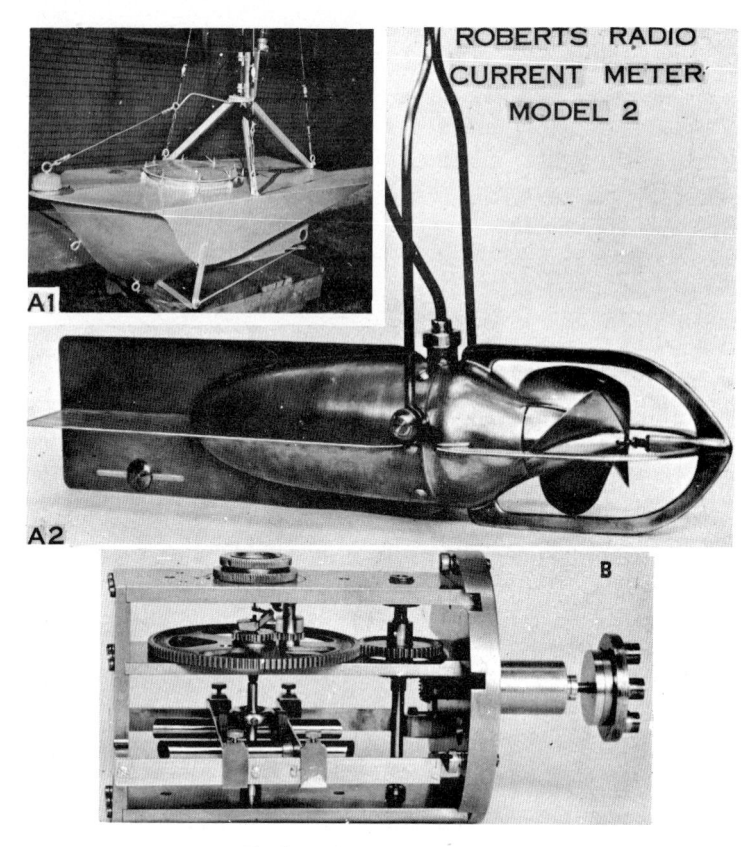

Fig.5. Roberts current meter.
A1. Radio buoy. A2. Meter body. B. Recording mechanism.
(Courtesy of U.S. Naval Oceanographic Office.)

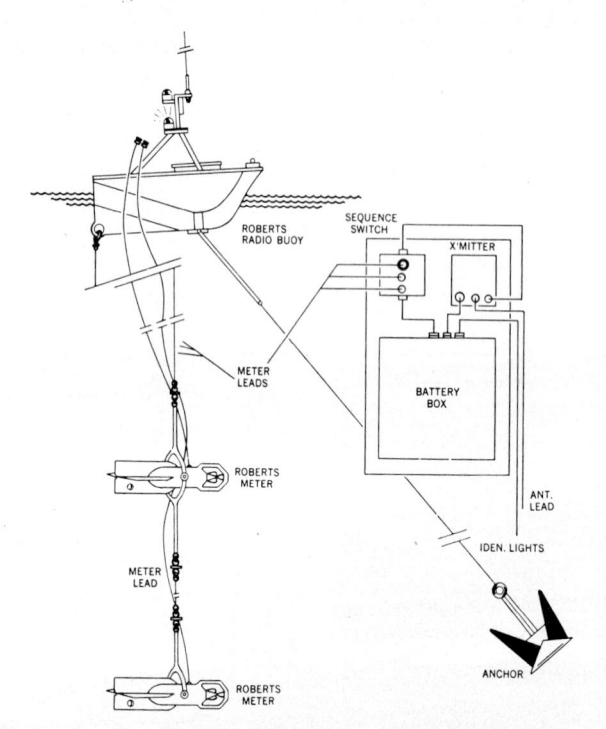

Fig.6. Sketch of mooring for the Roberts current meter.
(Courtesy of U.S. Naval Oceanographic Office.)

the time relation between the two contacts serves as a direction indicator. Several meters can be connected to the suspension cable hanging from a buoy. Sequence switches are used to receive signals from individual meters.

If it becomes necessary to measure accurately the depth at which a meter is located, depth or pressure recorders of different design are commercially available. They can either be built in the meter itself such as in the D.H.I. paddle wheel meter, or can be attached to the current meter body.

Drift measurements

This group of current measuring devices and practices embraces all methods that trace the paths of objects floating with water movements. Although classified under "Lagrangian methods", in most cases these methods can only be considered an approximation to the Lagrangian type of current measurements which actually requires the tracing of "tagged" water particles that are carried along with the current in three dimensions.

All drift measurements require continuous and most accurate positioning of the drifting body (tracer) with reference to a fixed point. In near shore water and in regions where high-precision radio techniques for navigation like Loran-C, Shoran, Lorac and others are available, drift measurements may yield very satisfactory results. Also, anchored buoys in the open ocean are used to mark fixed reference points, and the path of tracers can be followed adequately by a surveying ship. In special cases, or for detailed studies in near-shore or land-locked waters, triangulation methods using theodolites or sextants from shore can be employed. In this way, MOSBY (1954) succeeded in constructing surface trajectory charts in the Tromsö Sund as shown in Fig.2. However, the accuracy of drift measurements is not only limited by available means of navigational position surveying, but also by the type of the tracer.

Different types of floating objects at the sea surface or at subsurface levels have been used as tracers. The old method of *drift bottles* is relatively cheap. Therefore it has been applied in great numbers. Together with *ship drift* observations, this simple method has helped to describe many of the main features of the general surface circulation in the ocean. Drift bottles are properly ballasted and contain a card with necessary information in several languages, explaining their purpose to the finder.

More recently, drift bottles have been replaced by *plastic covered drift cards* to be used in the same manner. Like drift bottles they can be released from ships and also from airplanes. The obvious disadvantage of this method is that in most cases only the end point of their drift can be related to the starting point. Often, these drift cards or bottles may remain undiscovered on a shore for some time, and in this case it is impossible even to determine the magnitude of the average drift velocity between starting and end points.

In coastal waters of more heavily populated areas this problem would probably not be so severe. As old as the method is, the utility of drift bottle and drift card measurements for special purposes has been proven valuable today in the age of much more sophisticated gadgets. The recent atlas by BUMPUS and LANZIER (1965) showing the surface circulation on the continental shelf off eastern North America between Newfoundland and Florida is based on data obtained by drift cards and bottles. Another recent example is the work by BURT and WYATT (1964) on drift bottle observations of the Davidson Current off Oregon State, U.S.A. Remarkable investigations with drift cards to determine the influence of the wind on surface currents were carried out in the North Sea by TOMCZAK (1964). Similar experiments have been conducted in the Atlantic Ocean (LAWFORD, 1956; HUGHES, 1956). Newly designed polythene drift cards have been described by DUNCAN (1964). In tests along the South African coast they have proved to be durable and convenient.

A device developed and later modified by CARRUTHERS (1939, 1954) predetermines the time of drift at the surface. At the end of this time a simple mechanism makes the surface float sink and anchor at the bottom, leaving a buoy at the surface for recovery.

Results obtained from certain types of surface drift bodies must, however, be used with caution. The very surface layer of the water may behave differently from the water layer just a few centimeters below the surface (OLSON, 1951). The drift of different objects depends on the shape and the size of drift materials. Results obtained from drift card measurements in the Mediterranean Sea indicated that plastic drift cards of the thickness of a postcard travelled differently from thicker cards released at the same time and same place (CARRUTHERS, 1957).

The use of parsnip, confetti, dye and other smaller particles or contaminants of sea water as tracers is practical for special measurements on smaller scales; for example, in diffusion, mixing and turbulence studies associated with ocean currents. Dye patches (rhodamine-B, fluorescein, etc.) can serve the purpose of tracing the path of the patch as well as the rate of dispersion. For longer periods of observation, and with small concentration of the contaminant, special instrumentation is needed to detect the dye or contaminant concentration. VINE et al. (1954) have described the problems that are to be faced in using dye patterns as a tracer.

Vertical logs or current poles are buoyant floating devices attached to a log line. This method is similar to the old fashioned chip-log used in measuring the speed of a ship. The direction of the current is obtained from the direction of the log line, and the speed is found by timing the travel time of the vertical log. More sophisticated modern alterations of this method have been suggested by SHARPEY-SCHAFER (1952).

Radio buoys are released from ships and can be traced by radio direction-finding receivers from shore stations, ships or aircraft. Many buoys can be traced at the same time if means are provided to distinguish individual floats. The use of recording and telemetering buoys in deep-sea research was discussed by FRANTZ (1957).

Deep drogues can be of various design using canvas, wooden or metal crosses, sheets, parachutes and other devices that offer the largest possible drag at the level of current measurement. All of these devices have to be properly weighted to reduce the wire angle of the suspension wire and to keep the drogue at the desired level. The drogues are connected either by fine wire or nylon filaments to a small surface float, or buoy, equipped with identification signs (e.g., flags), radar reflector and lights for nighttime observations.

Most commonly used for deep drogues in recent years are aviator parachutes. They have to be designed to open fully at a prescribed depth determined by the length of wire connected to the supporting surface raft or buoy and the ballast weight attached to the parachute. It is advisable to attach a pressure recording device to the drogue in order to control its depth. The principle of the parachute drogue system is explained in Fig.7. Most important is that the drogue moves with the same velocity as the surrounding water. This, first of all, requires that the drag exerted at the drogue be much greater than the drag exerted at the surface float, suspension cable and other parts of the system (CROMWELL et al., 1954; VOLKMANN et al., 1956; KNAUSS, 1963).

More recently, deep drogues were used during Equalant expeditions to track the course of the Equatorial Undercurrent in the Atlantic Ocean. The ship "Explorer" of the U.S. Coast and Geodetic Survey used large truck tire innertubes as floating

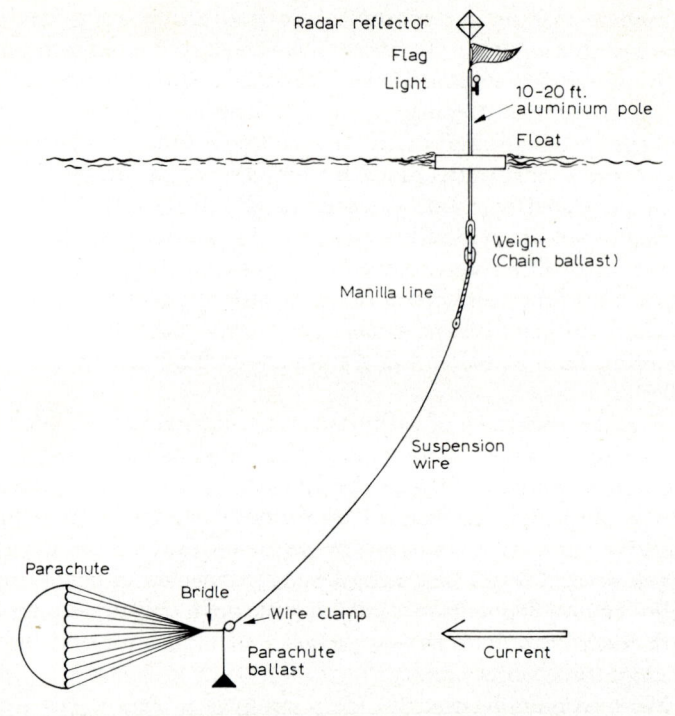

Fig.7. The principle of the parachute drogue.

surface buoys. In the center of each tube was tied a 12-ft. long aluminum pole extending 8 ft. above the sea surface. This pole was surmounted by a radar reflector, flags and lights. From the surface float parachutes with approximately 50 lb. ballast weight were suspended on 1/8 inch stranded wire to a depth of 50–60 m. Radar was used by the attending ship to track the position of the drogues with reference to a set of deeper reference drogues "anchored" at a depth of 500 m. There is, of course, no guarantee that the reference drogues remained at a fixed geographical position. Therefore, a very accurate positioning survey is necessary to control possible drifts of the reference drogues. The "Explorer" surveyed the area between about 15° and 14° West longitude near the Atlantic equator. Since no electronic means of absolute positioning were available in this area, control was based upon astronomical observations. However, favorable sky conditions and the remarkable skill of the navigation officers made this enterprise a success.

Results of deep drogue tracking during phase I of the "Explorer" operation during Equalant I – Expedition are shown in Fig.8. During the period of observation, the course of the Atlantic Equatorial Undercurrent is, on the average, due eastward with an average speed of 80 cm/sec. Minor deviations from a due east direction do occur; their cause is not yet fully understood. The total tracking time of drogue 1 was 2 days, 3 h and 20 min. The trajectories shown in Fig.8 are only approximations to true trajectories of water parcels at the point where the parachute drogues opened because of the fact that the drogues are "tied" to a surface float and may have some "slippage" through the water. Also the elasticity of the suspension cable with variable drags at the drogue may cause some up and down motion of the tracer. This is one of the disadvantages of every deep drogue system. The ideal deep drogue is one that is freely floating at a prescribed level and carried along with the water movement with no attachments to the sea surface.

The first successful subsurface current tracer of this kind, the neutral buoyancy float, was developed and used by SWALLOW (1955). Neutrally buoyant floats will move more or less with the current at a certain depth, or, saying it more accurately, along the level where the weight of the displaced water equals the weight of the float. The tubelike floats, about 8 ft. long, obtain their buoyancy by being less compressible than water. The housing of the floats is made of aluminum. It is of course necessary to detect the position of the float and to survey accurately its path during a sufficiently long period of time. The need for a fixed reference point still exists as it does with all absolute current measurements.

The problem of locating the subsurface neutrally buoyant float was solved by using the float as an acoustic source. An attending ship is necessary to follow the path of the "pings", that is, the acoustic signal given by the float. These pings are received by a hydrophone suspended in the water by the attending ship. Therefore, neutrally buoyant floats are often called "pin-

Panel 1

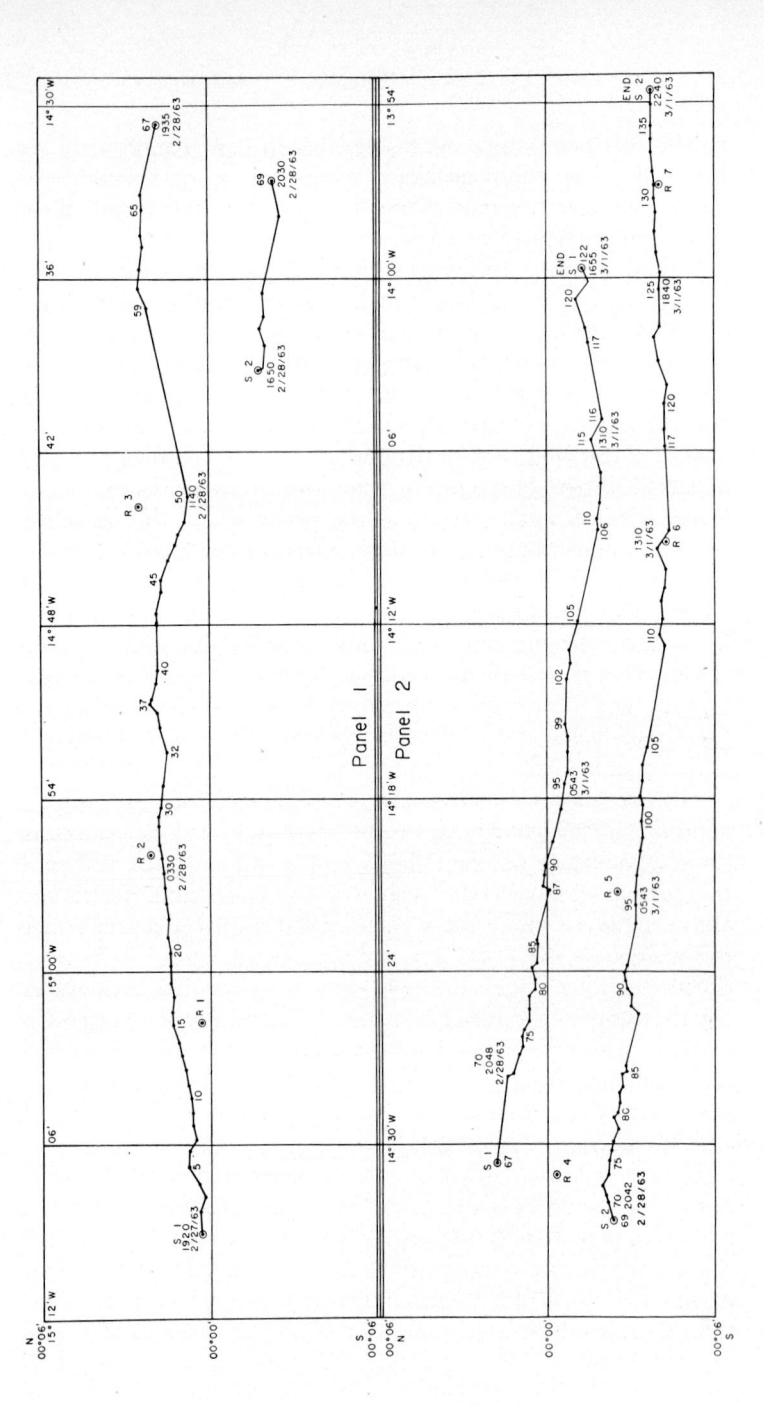

Panel 2

gers". A more recent summary of deep current measurements using neutrally buoyant floats has been given by VOLKMANN (1963). Measurements of currents by this highly developed drift method can be made at greatest ocean depths. Series of measurements over the abyssal plain off Spain have shown that currents did not decrease uniformly with depth. Other measurements using the Swallow floats revealed the existence of a deep countercurrent under the Gulf Stream (SWALLOW and WORTHINGTON, 1957).

The use of neutrally buoyant floats, or "pingers", has been further developed by POCHAPSKY (1961), who studied the movement of water in the ocean by following the motions of pairs of neutrally buoyant floats (POCHAPSKY, 1962, 1963). Measurements were made at various depths in the Caribbean Sea and in the Atlantic Ocean several hundred miles east of Bermuda and near the equator. The floats used with these experiments were equipped with pressure gauges and special transponding features. Two types of floats were employed. One type periodically emits a rapid double ping, and the time interval between pulses in a pair of floats is such as to specify the pressure or depth of the float. The second type of floats "hears" these pulses and responds by transmitting its depth on its own pair of pings. In turn, the first type of float "hears" the response and retransmits it at its own frequency. The first float is called the "master" float, and the second the "slave" float. This ingenious arrangement specifies the depth of the two floats relative to each other and also permits the determination of the separation of the floats. These techniques have been expanded to include a larger number of floats in a cluster. Recently (1963) five floats were used which also transmitted temperature as well as pressure. Floats can be

---

Fig.8. Results of drogue tracking during Equalant I expedition by "Explorer" of the U.S. Coast and Geodetic Survey at the equator between about 15°W and 14°W. The shallow drogues S1 and S2 were at a depth of 50 m in the high salinity core of the Equatorial Undercurrent. Radar fixes of the shallow drogues with reference to the corrected position of the deep drogues (R1 through R7) at 500 m depth are shown by dots along the track. The lower panel 2 is the continuation of the upper panel 1. Date and time are indicated at the beginning and end of each track as well as at selected intervals along the track. (After NEUMANN and WILLIAMS, 1965.)

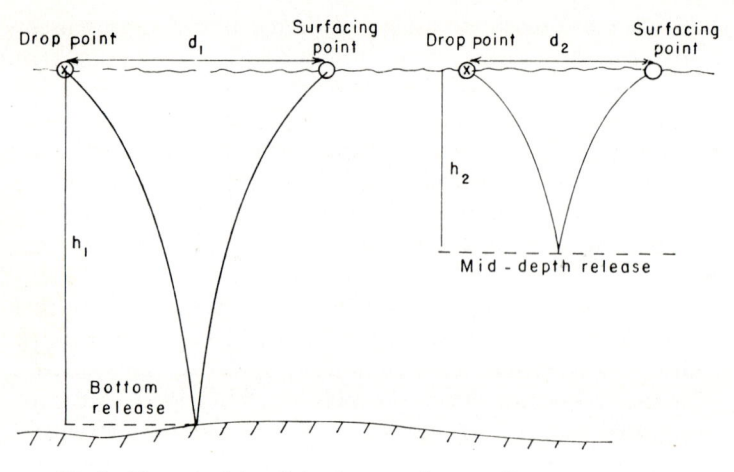

Fig.9. The principle of the drop-sonde according to Richardson.

recalled to the surface at will by means of a coded signal. According to POCHAPSKY (1963): "a noisy ship's screw or heavy rain found the code and put an unwished end to an experiment." Location of such floats at the sea surface was facilitated by a small radio transmitter on each float.

A simple method and instrument to measure directly the average current velocity in a vertical column of water was recently developed by RICHARDSON (1965). The instrument is called the *drop-sonde*. It does not require any suspension; instead, it is dropped into the sea at a known position. The device descends at a constant rate of 2 m/sec either to a prescribed depth or to the bottom, and then rises to the surface. The depth to which the instrument is pre-determined to sink can be selected by means of a pressure release mechanism, or in the case where the instrument is free to sink to the bottom, weights are released on bottom contact. The reappearance of the instrument at the surface and its location has to be precisely noted. The distance between the drop point and the point of reappearance at the surface represents the total (vertically accumulated) drift during the period of descent and ascent (Fig.9). From a series of drop-sonde measurements at a given position a complete picture of water volume transports between surface and bottom can be obtained (Fig.10). Experiments of this kind across the Straits of Florida between Miami and Bimini have been successful. Since the equipment is light and easy to handle, a small, fast vessel equipped with high precision navigational gear can be employed.

Very little is known about the currents near the ocean bottom in the deep sea. Instrumental techniques and devices to measure the bottom currents even at greatest depths are, however, steadily improving. The suspended-drop current meter (THORNDIKE, 1963), designed to measure the velocity of currents near the floor of the ocean, can also be classified under the Lagrangian method. Thorndike's bottom current meter is fixed to a tripod standing on the ocean floor, and a camera takes photographs of colored drops ejected by the device. Knowledge of the time

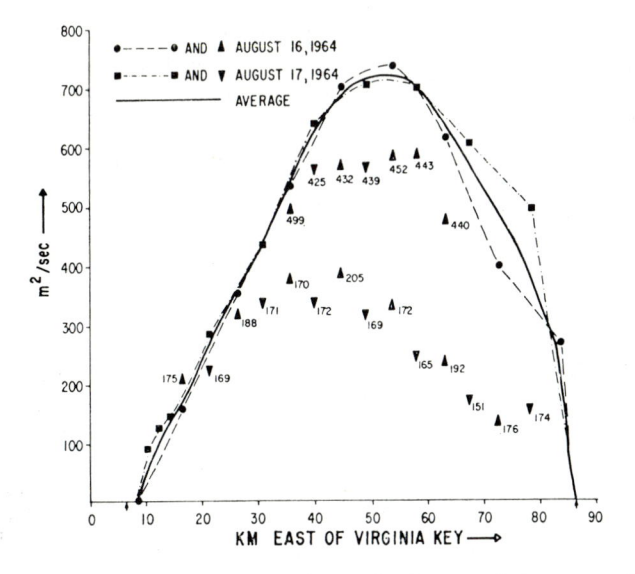

Fig.10. Some of the results obtained with the drop-sonde in an experiment during August 16 and 17, 1964, show the north component of water transport versus distance across the Straits of Florida. Measurements denoted by circles and squares are to the bottom. Triangles show the transports obtained from mid-depth release. The release depth in meters is shown for each triangle. (After RICHARDSON and SCHMITZ, 1965.)

intervals between photographs is provided by a motor-driven timer or watch while a magnetic compass in the field of view indicates the direction. Preliminary results obtained by this device are promising, and further developments of the basic idea may prove very helpful in solving special problems of bottom currents in the deep sea.

A technique for measuring deep ocean currents close to the bottom with a current meter anchored to the bottom has been described by KNAUSS (1965). A special release mechanism permits the self-recording current meter (RICHARDSON, 1963) to rise to the surface after a selected time of operation.

These recent developments in deep-sea instrumentation, and their use in exploring the deeper strata of the oceans, particularly the details of water movements, are the result of modern technical achievement. Both highly developed technical devices and simpler, less expensive means like drift bottles, will be needed in future oceanographic research in answering many questions about the oceanic circulation. The kind of method or technique that is to be employed in the most economic and efficient way depends on the problem or question at hand.

INDIRECT METHODS OF CURRENT MEASUREMENTS

The electromagnetic method

Sea water is an electric conductor, because sea water contains an abundance of highly dissociated salts. If this water moves relative to the earth's magnetic field, an electric field is induced in sea water. The intensity of the electric field can be measured by the gradient of electric potential between two electrodes. These electrodes can be fixed points or be towed behind a ship.

FARADAY (1832) was the first to mention the fact that sea water motions could be detected by electromagnetic induction. However, attempts to prove his theory failed for purely technical reasons. After the laying of the first submarine telegraph cables across the oceans, shortly after the middle of the nineteenth century, certain electrolytic effects were noticed between the ends of broken cables and the ground (LONGUET-HIGGINS, 1949).

It is remarkable that the first direct experiments of this kind to measure ocean currents made in 1918 in Dartmouth Harbor, England, by YOUNG et al. (1920) remained for a long time unnoticed by oceanographers. Only recently, a more comprehensive theory of the phenomenon of electric fields induced by ocean currents moving in the geomagnetic field was developed (LONGUET-HIGGINS, 1947, 1949; STOMMEL, 1948; MALKUS and STERN, 1952; and others). At the same time, VON ARX (1950) developed an instrument to measure ocean currents by means of the observed gradient of electric potential between two electrodes towed behind a ship. This instrument is called the "Geomagnetic Electro-Kinetograph" (GEK).

The GEK electrodes of special design are towed about 100 m apart astern of the ship. They are mounted on a two-conductor cable. In order to have the electrodes far enough away from the magnetic influence of the ship, it is necessary to tow the electrodes at a distance of about two or three times the length of the ship. The electric potential difference between the two electrodes, e, is recorded on a potentiometer. The velocity, v, of the moving water is proportional to e, and $v = ke$, where k is a proportionality factor that depends on the vertical intensity of the geomagnetic field, the distance between the electrodes, the specific electric resistance of sea water and the specific electric resistance of the bottom underneath the water. In deep water, the effect of electric resistance (or conductance) of the bottom becomes less important, and the method is best applied and most reliable for measuring the currents in the upper strata of a deep ocean. Thus, the proportionality factor k depends on the locality and has, therefore, to be determined for different oceanic regions by comparison with current measurements obtained by other means.

Electro-chemical effects may gradually change the characteristics of the electrodes. Therefore, it is necessary to check the zero point of the potential difference from time to time. This is done by changing the course of the towing ship by 180°. Since the GEK measures the water motion at right angles to the ship's direction of progress, it is a regular practice to tow the electrodes for a sufficiently long time along one course, and then turn the ship's course through ninety degrees. This will provide the necessary data for evaluations of current speed and direction.

The method of towing GEK electrodes horizontally below the water surface is shown in Fig.11. Ordinarily, neutrally buoyant cable is used to tow the electrodes at subsurface levels behind a down-pulling "fish" of variable weight. A "drogue", as shown in Fig.11 by a 25 ft. long manila line, is added at the end of the towing section to stabilize the tow of the electrodes at a prescribed depth. Towing at the sea surface is relatively easy (in fair weather) if a buoyant cable and buoyant electrode housings are used.

The use of the GEK since 1950 has provided most valuable data on ocean currents, not only in the Gulf Stream region, but in many other parts of the oceans. For a detailed description of the GEK, its theory and practical use, and the chief difficulties of this method, the reader is referred to VON ARX (1950, 1962).

Stationary electrodes have been used for measuring velocities or total mass transports of ocean current in special locations; for example in the English Channel (LONGUET-HIGGINS and BARBER, 1948; BOWDEN, 1956), in the Straits of Florida (WERTHEIM, 1954), in the Puget Sound area (MORSE, 1957), and other regions. Measurements of turbulence near the sea bed in a tidal current were made by BOWDEN (1962). A series of observations between widely separated points in the North Atlantic Ocean were obtained by STOMMEL (1954), making use of submarine telegraph relay stations at Bermuda and Azores (Horta), and continental stations.

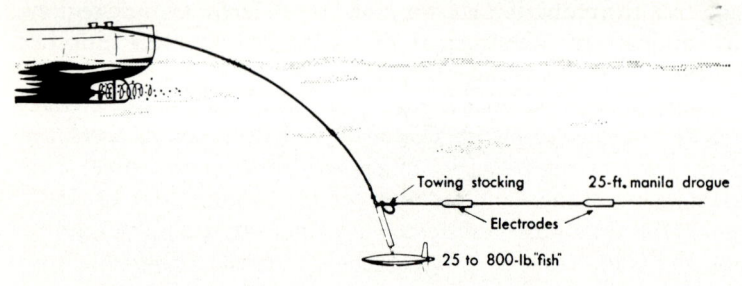

Fig.11. Method of towing the Geomagnetic Electro-Kinetograph according to VON ARX (1962).

The dynamic method in oceanography

Measurements of temperature and salinity[1] of sea water at oceanographic stations provide the necessary data for determining the distribution of water density in the oceanic space. From the field of density, the field of pressure is obtained by means of the hydrostatic equation which expresses the fact that in an ocean at rest the pressure, p, at a given depth h equals $p = g\bar{\rho}h$, where $\bar{\rho}$ is the average sea water density in the vertical column between the sea surface and depth h, and g is the acceleration of gravity. Since the ocean is not at rest, the applicability of the hydrostatic equation in computing the pressure field from the density, or field of mass, may be questioned. However, it can be shown that for the use in the classical dynamic method of current computation, the pressure field can be determined with sufficient accuracy by means of the hydrostatic equation.

If the pressure field in the oceans is known, certain hydrodynamic relationships permit the computation of ocean currents under certain assumptions (SANDSTRÖM and HELLAND-HANSEN, 1903). These assumptions can be more or less rigorous, depending on the hydrodynamical character of the currents. Relationships between the field of pressure and the field of currents will be developed and explained in the following chapters of this book when dealing with the dynamics of ocean currents in a more general way. Also, certain difficulties in applying the classical method of dynamic computations, and its limitation will be discussed in more detail.

Recently, RICHARDSON's (1965) direct measurements in the Florida Current have shown that great discrepancies can exist between results obtained by the classical method and the true current field. S. Broida of the Institute of Marine Sciences, Miami, Fla., directed a hydrographic survey along the same section where Dr. W. S. Richardson applied his drop-sondes.

---

[1] The classical definition of salinity is "the total amount of solid material in grams contained in one kilogram of sea water when all the carbonate has been converted to oxide, the bromine and iodine replaced by chlorine, and all organic matter completely oxidized" (FORCH et al., 1902). The salinity is expressed in parts per thousand (‰).

A disagreement between the two kinds of measurements, direct and indirect, was noticed. The reader will be referred to such discrepancies in subsequent parts of this book. An explanation requires knowledge of the basic hydrodynamic relationships between ocean currents on a rotating globe, current accelerations and friction in relation to the driving pressure differences as computed from the field of mass.

In many parts of the oceans, the classical dynamic method has been employed successfully for computing the relative vertical distribution of ocean currents. More recently, WÜST (1957) has given examples for a successful application in deep-sea layers of the Atlantic Ocean. Since the classical method provides *absolute* current velocities only in the case where a reference level is provided for transforming *relative* into *absolute* pressure differences, the question of a reference level becomes most vital in the application of the indirect classical method.

Wüst's results on deep-sea current transports are based on a reference level of variable depth in the Atlantic Ocean as was first proposed and used by DEFANT (1941). The problem of a reference level for transforming relative into absolute current fields is closely related to the problem of the three-dimensional oceanic circulation and will be considered in more detail in Chapter IV. The classical dynamic method fails where frictional forces play a decisive role in the dynamics of a current and, therefore, cannot be used to compute the pure wind-driven component of the velocity of a current field. It is also evident that in a homogeneous ocean where the potential density[1] of the sea water is constant, horizontal pressure differences can only be the result of the slope of the sea surface as compared to a level surface if frictional forces can be disregarded.

TS-Diagrams and the core method

The vertical distribution of temperature and salinity in the oceans

---

[1] The potential density of sea water is the density of a water parcel raised from its depth in situ to the sea surface without gains or losses of heat and salinity. At subsurface levels, the water parcel is under the effect of the sea pressure and its density in situ is higher than the potential density.

varies from region to region and it seems possible that any temperature may occur with any salinity. However, the structure of oceanic water masses shows that certain temperature–salinity associations are preferred, depending on the origin of the water and the region of observation. If simultaneous measurements of temperature, T, and salinity, S, at different depths for a fixed

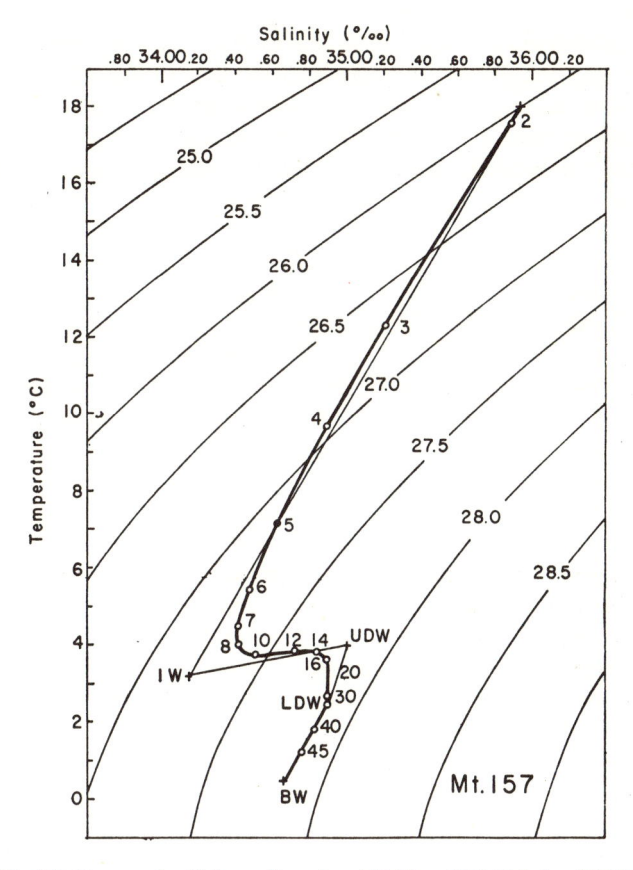

Fig.12. TS-diagram for "Meteor" station 157 ($\phi = 15°36'N; \lambda = 26°57'W$). IW = Antarctic Intermediate Water; UDW = Upper Deep Water; LDW = Lower Deep Water; BW = Antarctic Bottom Water. The family of curved lines labelled from 25.0 through 28.5 are lines of equal σ_t values, or isopycnals (see p.82). Numbers along the TS-curve indicate depths per 100 m.

oceanographic station are plotted in a T–S coordinate system, corresponding temperature and salinity values arrange along lines or curves as shown in the example of Fig.12. Such diagrams are called TS-diagrams. The detailed features of the TS-curve change gradually in the oceanic space, but their major characteristics can often be followed over larger geographical areas.

The introduction of the TS-diagrams into oceanography by HELLAND-HANSEN (1916) has led to many fruitful applications. It was a major step forward in the analysis of the spreading and mixing of oceanic water masses, provided that the basic physical rule of mixing (Richmann's mixing rule) holds. This rule states that if a water mass m_1 of temperature T_1 is mixed with a water mass m_2 of temperature T_2, the temperature of the mixture is T as shown in eq.1:

$$T = \frac{m_1 T_1 + m_2 T_2}{m_1 + m_2} \qquad (1)$$

The same applies to the salinity, S, of the mixture if the water mass m_1 of salinity S_1 mixes with water mass m_2 of salinity S_2:

$$S = \frac{m_1 S_1 + m_2 S_2}{m_1 + m_2} \qquad (2)$$

DEFANT (1935) has shown that these basic rules of physics lead to a TS-relationship that is indeed linear. The meaning of TS-diagrams is best explained by the introduction of water types and water masses. A *water type* is a body of water with a fixed temperature and a fixed salinity.

Consider two water types that originally comprise the vertical structure of the ocean. In Fig.13, the upper water type (*1*) is characterized by a temperature T_1 and a salinity S_1; the lower water type (*2*) has the corresponding characteristics T_2 and S_2. The vertical distribution of temperature and salinity for this kind of stratification is shown in the left part of Fig.13 by fully drawn lines marked (*a*). The TS-diagram for this structure is presented in the first diagram on the right in Fig.13, showing only two points (*1*) and (*2*) which belong to the two water types.

A *water mass* is formed by mixing of the two water types. If mixing or diffusion takes place between water types (*1*) and (*2*), the result would be a vertical stratification for both temperature and salinity as shown by the dashed lines marked (*b*). As mixing continues, the abrupt change (discontinuity) between the temperature and the salinity of the two layers gradually disappears and is replaced by a continuous variation of T and S in the vertical structure. A more progressed state of mixing is indicated in Fig.13 by the dashed lines (*c*). *TS*-diagrams for the two different stages of mixing are shown in the right hand part of the figure. The *TS*-diagram is characterized by a line connecting points (*1*) and (*2*); it represents a water mass formed by mixing of the two water types.

If the vertical distribution of T and S is given by two functions, $T = \Phi(z)$ and $S = \phi(z)$ respectively, representing the dashed curves in Fig.13, where z is the water depth, the resulting relationship $S = F(T)$ is linear and of the form:

$$S = S_1 + (T - T_1)\left(\frac{S_2 - S_1}{T_2 - T_1}\right) \tag{3}$$

S and T, respectively, are the salinity and temperature of the water mass at a given depth after mixing, while S_1, T_1 and S_2, T_2 are the characteristic properties of the original water types before mixing. It can be shown that eq.3 is correct if eq.1 and 2 are substituted for T and S. It is also seen that the relationship

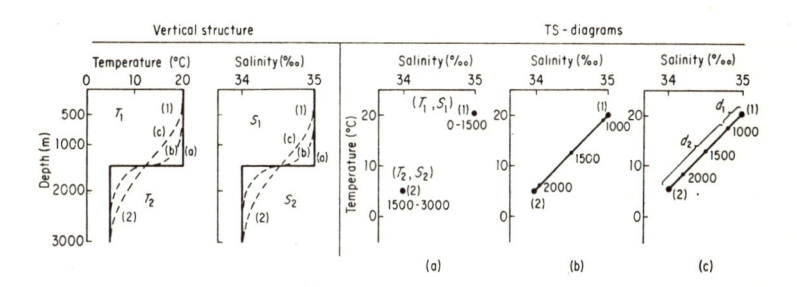

Fig.13. Mixing of two water types, (*1*) and (*2*), respectively. At left, the vertical distribution of temperature and salinity is shown for three stages (*a*), (*b*) and (*c*), respectively. At right, the *TS*-diagrams illustrate the initial stage (*a*) showing the two water types as points (*1*) and (*2*), and the results of progressive mixing for stages (*b*) and (*c*). The depth is indicated in meters at dots on the straight lines connecting points (*1*) and (*2*).

$S = F(T)$ is always a straight line between the points (1) and (2) in the TS-diagram of Fig.13, independent of the form of the functions $T = \Phi(z)$ and $S = \phi(z)$. The only necessary requirement is that the water mass is formed by mixing of the two original water types and that the mixing follows the rules given by eq.1 and 2. This also explains why the very surface layers of the oceans, approximately to a depth of 50–100 m, usually do not show a unique T–S-relationship. Here, heat gains and losses, besides mixing, change the temperature characteristics, periodically and aperiodically, as does evaporation and precipitation in changing the salinity characteristics of the surface layers.

The TS-diagram of a water mass produced by mixing of two water types (1) and (2) is shown in Fig.14. Point (3) on the line connecting points (1) and (2) represents the temperature, T, and salinity, S, of the water mass at a given depth. DEFANT (1935) has shown that the distances, D_a and D_b, from the two end points of any point on the line between water types (1) and (2) are inversely proportional to the ratio of mixing, $D_a/D_b = m_2/m_1$. In Fig.14, D_a represents the distance of point (3) from (1). Since:

$$(D_a)^2 = (S_1 - S)^2 + (T_1 - T)^2$$

it follows after substitution of eq.1 and 2 for T and S, respec-

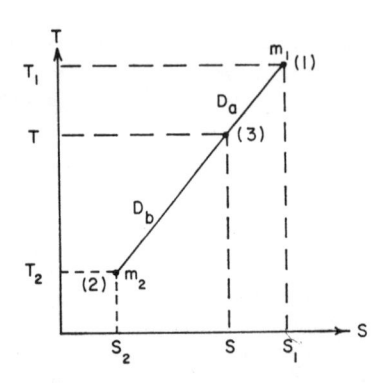

Fig.14. Schematic diagram for TS relationships resulting from mixing of two water types.

tively, that:

$$(D_a)^2 = \left(\frac{m_2}{m_1 + m_2}\right)^2 [(S_1 - S_2)^2 + (T_1 - T_2)^2]$$

or:

$$D_a = \left(\frac{m_2}{m_1 + m_2}\right) D$$

where $D = D_a + D_b$ is the difference between points (*1*) and (*2*). Similarly, it is found that:

$$D_b = \left(\frac{m_1}{m_1 + m_2}\right) D$$

and therefore:

$$\frac{D_a}{D_b} = \frac{m_2}{m_1} \qquad (4)$$

Thus, it is possible to determine for any point on the straight T–S line the percentages of the original water types which led to the formation of the water mass observed at a depth given by the position of the point on the TS-diagram. In the example of Fig.14 at the depth of point (*3*), the water mass is composed of 2/3 parts of water type (*1*) and 1/3 part of water type (*2*), since $D_a/D_b = 1/2$.

The mixing of three or more water types leads to two or more water masses. The resulting TS-diagram is explained in Fig.15 where four water types are involved. The vertical temperature and salinity distribution in different stages of mixing are shown in the left hand part of the figure and the corresponding TS-diagrams in the right hand part of the figure. As a result of continued mixing, the sharp corners in the TS-diagram gradually become more and more rounded. Here, the original TS-characteristics of a water type are lost; however, the vertexes of the TS-diagram still indicate the *core* of the original water type, although this core becomes more and more "diluted" by mixing with the water masses above and below.

In the oceans, TS-diagrams show the essential characteristics of mixing between different water types, and they are often remarkably similar over large oceanic regions. End points and

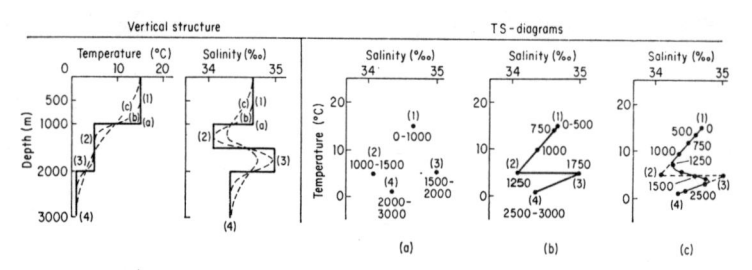

Fig.15. Mixing of four water types and resulting TS-diagrams in three different stages of mixing (a), (b), (c).

vertexes are often connected by nearly straight lines. Such curves permit the precise determination of the *core depth* of individual water types in the deep layers of the ocean as well as the estimation of the amount of the individual components that are involved in the formation of water masses and the gradual change of the temperature and salinity characteristics of water types with progressive mixing. Thus, TS-diagrams can serve as a valuable tool for analyzing the oceanographic stratification as to the origin and spreading of water types. However, the restrictions on the use of TS-relationships for deducing ocean currents should always be kept in mind.

THORADE (1931), CASTENS (1931), and DEFANT (1929, 1936) considered the effects of both mixing (or diffusion) and advection on conservative[1] properties of sea water as, for example, the salinity or the temperature. With the exception of sea surface layers, where neither T nor S can be considered conservative, the use of TS-diagrams for deducing the spreading of water masses is a valuable tool, although it must be applied with caution. A uniform flow, or no water movement at all, as well as a current with a pronounced main current axis, may produce "cores" or "tongues" in the distribution of a property like the salinity, S, or the temperature, T. The appearance of "tongues" or "cores" not only depends on the original distribution of this property in the vaguely known "origin" of a water type, but

---

[1] Conservative properties in ocean water are concentrations that are altered by processes of diffusion and advection only, except at the boundaries. Heat content and salinity are two examples.

also on the characteristics of ocean currents in the region of spreading.

Besides the effects of vertical mixing, lateral mixing has been found to be of great importance in the development of TS-relationships in the oceans and their change from region to region. Probably, the mixing along surfaces of equal potential density is in some parts of the oceans of equal importance as the vertical mixing. This fact was fully discussed by SVERDRUP et al. (1942). It does not restrict the use of TS-diagrams for indirect analyses of the spreading or motion of water masses, but imposes certain caution on its use and on the interpretation of TS data.

ISELIN (1939) has shown that a certain water mass can also be formed by sinking of surface water along σ_t surfaces in the region of surface water convergence in middle latitudes. In this case, a TS-relationship in the vertical direction is obtained at subsurface levels that reflects the horizontal distribution of temperature and salinity along a horizontal distance at the sea surface crossing the region of convergence.

It is difficult to decide whether vertical mixing or lateral mixing processes are of greater importance in the formation of certain TS-relationships as observed in most parts of the ocean. An important point to bear in mind is that the waters of the oceans attain their main original characteristics when under the influence of air–sea interactions (heating, cooling, evaporation, precipitation, changes in gas content, etc.). The interpretation of TS-diagrams, however, has to include lateral mixing of water types as well as vertical mixing, and consideration of the fact that a water type can have different combinations of T and S, depending on the year or season of its formation, the place where it is formed, and, generally, on the fact that there is rarely a place at the ocean's surface where a water type of precise temperature and salinity characteristics can be defined.

The *core method* ("Kernschicht Methode") used by WÜST (1935) in the analysis of the spreading of oceanic water masses has been very successful, particularly in deep-sea water bodies of highly differentiated stratification. The core layer ("Kernschicht") is defined by the water layer within which temperature or salinity, or both, show either maxima or minima when compared to adjacent water masses. Such TS-characteristics are indicated in

TS-diagrams. Often, they can be supplemented and supported by the distribution of oxygen content in sea water. The deep water of the Atlantic Ocean is most clearly differentiated. It is governed by the intrusion of highly saline water which flows from the Mediterranean Sea through the Strait of Gibraltar into the North Atlantic. This well-defined water type—the Mediterranean Water—spreads out from its source region and forms the Upper Deep Water of the Atlantic Ocean (WÜST, 1935). It is characterized by a secondary salinity maximum at depths ranging between about 1,000 and 2,000 m. This water can be traced over large portions of the Atlantic Ocean, although the salinity in its core decreases with increasing distance from the Strait of Gibraltar as a result of mixing processes. Its effects are carried far beyond the boundaries of the Atlantic Ocean in the waters circling the Antarctic Continent, and the three major oceans on the Earth cannot be treated individually. The Antarctic circumpolar currents act as a connecting link between the Atlantic, Indian, and Pacific Oceans.

The TS-diagram in Fig.12 shows the core of the Atlantic Upper Deep Water in 15°36′N 26°57′W at a depth of about 2,000 m with a salinity of 34.91‰.

Higher latitudes in the Southern Hemisphere are the region of origin of Antarctic Intermediate Water. Other water masses in the deep layer of the oceans have been defined by WÜST (1935); however, the Intermediate Water is of special significance. The Antarctic Intermediate Water is formed at the sea surface near the Antarctic Convergence between about 45° and 55°S where the precipitation exceeds evaporation and creates a relatively low surface salinity. From its place of origin in the Atlantic Ocean where the salinity is about 33.8‰ and the temperature about 2.2°C, the Antarctic Intermediate Water sinks to 700–900 m depth while spreading northward, mainly in the western part of the ocean. This water type is characterized by a low salinity core as shown in the TS-diagram of Fig.12 at a depth of 800 m. Since it originates at the surface of the South Atlantic Ocean where the temperature is low, the Antarctic Intermediate Water also carries a relatively high amount of dissolved oxygen along its course of spreading which clearly stands out in contrast to the low oxygen content of the overlying water of the warm water sphere and the underlying water mass of high salinity (Upper Deep Water) which also has a relatively low oxygen content. The oxygen minimum near the lower boundary of the tropical and subtropical warm water sphere and the oxygen minimum below the Antarctic Intermediate Water allowed WÜST (1935) to determine the vertical extent of this water mass in the Atlantic Ocean.

Isentropic analysis

The method of isentropic analysis in the study of the spreading of oceanic water types was introduced by MONTGOMERY (1938) and PARR (1938) in analogy to a similar method applied in meteorology where the distribution of meteorological elements (e.g., moisture) is depicted on surfaces of equal entropy. The main purpose of isentropic charts in meteorology is to study the flow patterns in the free atmosphere and to aid in forecasting atmospheric circulations and weather.

In oceanography, isentropic analysis studies the spreading of water types by depicting notably salinity and temperature on surfaces of equal potential density and by following their changes. Although isentropic analysis requires reference surfaces of constant entropy which are difficult to define for sea water, the use of surfaces of potential density may serve as an approximation as long as the limits of applicability are considered. Since the difference between potential sea water density and the density $\rho_{S,T,0}$, which is often given by $\sigma_t = (\rho_{S,T,0} - 1) \cdot 10^3$, is small in the upper 1,000 m of the sea, it has become practice to use surfaces of equal σ_t as reference surfaces for isentropic analysis. The σ_t values disregard the effect of pressure (adiabatic effects) on the density in situ, that is, on the density of sea water of given temperature and salinity at the depth of observation. The displacement of water masses in such isopycnic surfaces should actually proceed without changes in potential density, that means without changes in salinity and potential temperature. If changes occur, they must be the result of mixing or diffusion, where both lateral and vertical exchange processes can play an equally important part. Thus, as with the core method, isentropic analysis can help in the investigation of currents, water mass transports, and the spreading of different water types. Both methods, if carefully applied, are valuable, qualitative tools in studying ocean currents, especially in conjunction with other indirect or direct means of observing and measuring large-scale ocean currents and their effects on the stratification of oceanic water masses.

In recent years, water properties other than temperature, salinity and oxygen have been either used or suggested as characteristic indicators for the distribution of water masses. Considerable information can be obtained from a knowledge of the distribution of certain radioactive isotopes in sea water. Carbon-14 (^{14}C) has received the most attention as a tracer of water masses. Serious consideration in connection with circulation studies has also been given to the naturally occurring isotopes ^{3}H, ^{32}Si and ^{226}Ra (BROECKER, 1963). Besides these natural isotopes, numerous isotopes produced during nuclear tests have been added to the sea. Most suitable for tracer studies is the isotope ^{137}Cs. This isotope as well as ^{90}Sr were not present in the oceans before atomic bomb testing started.

CRAIG and GORDON (1966) used deuterium and oxygen-18 variations in the oceans as additional variables together with temperature and salinity to characterize water masses and mixing processes. Isotopic-salinity relationships indicate that Pacific and Indian Ocean deep waters are derived for the most part from mixing of North Atlantic deep water and Antarctic bottom water, but that a third component of intermediate water is required. This third component is probably also derived from the South Atlantic. According to Craig and Gordon, the three-component mixture has essentially its Pacific and Indian Ocean characteristics before it leaves the South Atlantic.

CHAPTER II

PRESENTATION OF OCEAN CURRENTS AND WATER MASS TRANSPORTS

THE USE OF CURRENT MEASUREMENTS

Difficulties in finding representative values of ocean currents for the construction of current charts have been discussed in the preceding chapter. Depending on the purpose of such charts, the oceanographer has to apply proper methods and techniques to collect series of measurements that yield a set of data which actually represents those features of ocean currents which he wants to present. Often, current charts are wanted that describe tidal or daily variations, or the pattern of monthly or yearly averages.

Single current measurements obtained in short time intervals, for example from anchored buoys or ships, can vary considerably in speed and direction. In order to describe faithfully longer period variations, measurements should not only be taken for a long enough period of time but also at a frequent enough rate of sampling. With rapid sampling, however, problems in data storage capacity of the current meter may arise, and even if this problem can be solved, the amount of work involved in data processing can become awesome. Nevertheless, properly designed sampling methods are most essential for a meaningful evaluation of current measurements as was recently shown by WEBSTER (1964). An answer to the question of what a "properly designed sampling method" is, and what the bad consequences are if improperly aliased data are collected, cannot easily be given. Besides depending on the motive for the measurements, the answer depends on the characteristics of the currents, the mechanical properties of the current meter, and the behavior of the mooring system.

In general, aliased data collected by the choice of a sampling rate that is too slow will have the effect of introducing a noise level into the resultant time series. This, however, may not be too serious in some applications where a tolerable noise level can be defined and where the use of the aliased data itself prescribes tolerable noise levels. Moreover, the method of data presentation can help to reduce the apparent effects of the aliasing. Fig.16 shows two examples given by WEBSTER (1964). The diagram on the left is a vector summation, or virtual displacement, which describes the path of a water parcel through which it would have moved if the velocities recorded by the current meter are added together by computing the vector sum of displacement at given time intervals. This diagram is based on 10-min averages of data collected every half second by means of a buoy-borne Richardson current meter. Each averaged value of speed, compass and vane direction has been computed from

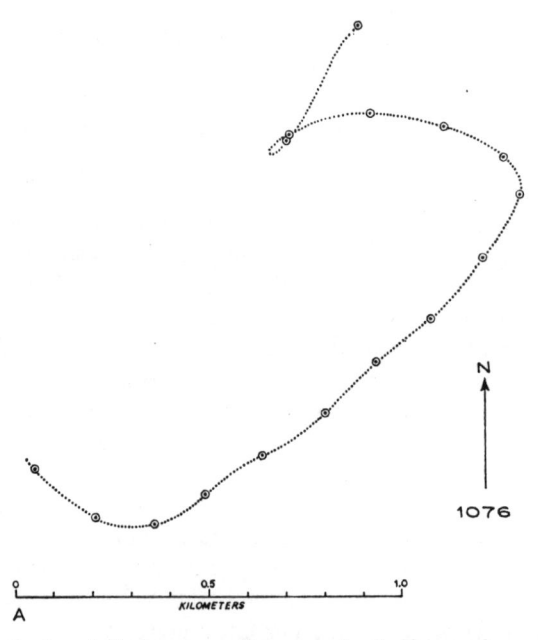

Fig.16A. A virtual displacement diagram using half-second sampling. The points are 4 minutes apart. Hourly positions are shown by circled dots. The current meter was at 200 m depth in 71°00′W, 39°30′N (south of Woods Hole, Mass., a few miles beyond the edge of the continental shelf).

1,200 points of single current meter readings. High frequency motions that were present in the original current meter record are completely smoothed out. The diagram on the right of Fig.16 shows another example of a virtual displacement of a water parcel, where the points are 20 min apart, and each point is based on one observation only. Although the slow sampling rate has produced aliasing noise, a tidal motion with a 12.5 h period is clearly indicated. Even while badly aliased, this plot is useful for a preliminary description of tidal-scale motions.

The computation of a *mean* current defined for a certain period of time meets difficulties, particularly in the case where the digitized data of a current meter record, or other single current observations collected over a period of time, show great variability in speed and direction. To arrive at an average current

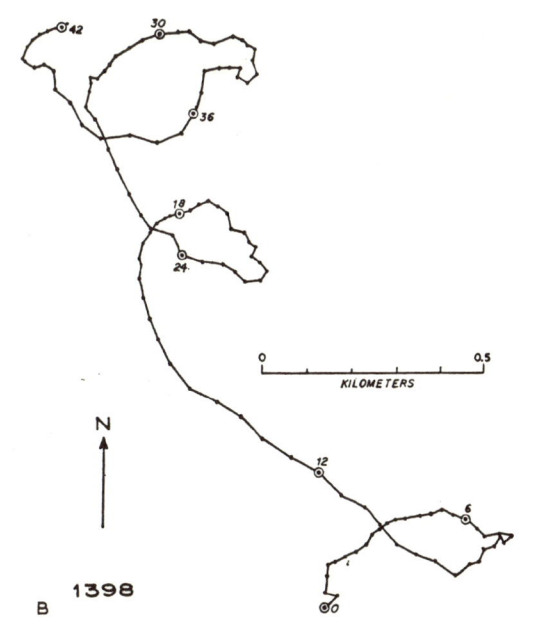

B. A virtual displacement diagram using 20-min sampling. Each point is based on one observation only; 6-h positions are shown by circled dots. Position at 32°10′N, 64°30′W (near Bermuda). The current meter was at 2,000 m depth. Both examples after WEBSTER (1964).

during the period of observation, individual (or instantaneous) current vectors can be averaged either by means of vectorial or arithmetical procedures. Both methods can lead to completely different results, and the question arises as to which result makes more sense. Here, again, the answer depends on what the oceanographer wants to present in his current chart. Actually, the decision on the purpose of his current presentation should have been made before the evaluation of a current meter record starts, and by choosing the right kind of current meter. If the purpose of the current chart is to show net water displacements over a longer period of time, say weeks or months, a current meter that eliminates or filters out short-period fluctuations should have been used in the first place, instead of a complicated meter of high sensitivity and response. Most navigators and fishery scientists are more interested in such long period net water displacements. Some exceptions have to be made in regions of strong tidal currents where net displacements over shorter time periods are desired. The oceanographer studying the general circulation of the oceans is mainly interested in finding long term average movements of water bodies, whereas the oceanographer studying diffusion or mixing processes in the sea needs a more detailed picture of the circulation pattern showing the irregular, up-and-down and lateral motion of water parcels superimposed on the average or net motion during certain time intervals.

In any case, part of the evaluation of current meter recordings is to find an average or mean of vectors recorded over a period

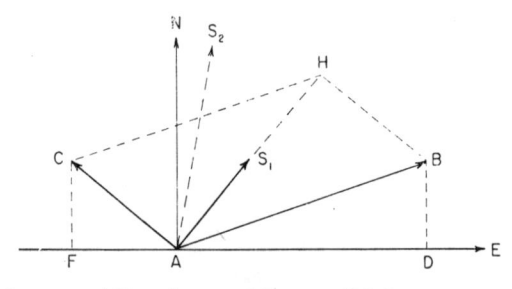

Fig.17. Scalar mean (S_2) and vectorial mean (S_1) for current vectors AC and AB.

of time. Mathematically, there is no problem in computing the average value of vectors. The question, however, is whether a vectorial mean is always wanted for practical purposes. Currents, changing their direction periodically between opposite directions with equal speed in each direction have, certainly, a practical meaning for some purposes, e.g., in navigation. However, a vectorial mean of such currents would give no current at all for longer time intervals. The arithmetical mean would give a current of some speed but of no direction. Presentation of average ocean currents in charts often requires more than pure methodic, mathematical procedures. It requires a thorough study of the nature of currents in different parts of the oceans, particularly of their variability or "stability".

The *stability, B, of ocean currents* is often defined as the ratio of the averaged vectorial velocity c and the averaged arithmetical velocity c, expressed in percent. Thus, $B = 100c/c$ (%). The vectorial mean velocity is obtained by taking the vectorial mean value of single observed current vectors, and the arithmetic mean velocity is obtained by averaging the velocities without regard to the current direction. The stability defined in such a way can vary between zero and 100%. Only in exceptional cases can it be 100%, that is, when the observed single current vectors point in the same direction, and in this case the vectorial mean is the same as the scalar mean. Neither of these averages is satisfactory in areas where the current varies over all directions of the compass rose. The vectors AB and AC in Fig.17 represent two current vectors observed at different times at a fixed place. Their north and east components are DB and AD, and FC and AF, respectively. The vectorial mean of both vectors is AS_1 = 1/2 AH, since the vectorial sum is AH. The *vectorial sum* as well as the *vectorial mean* has a physical meaning in the presentation of current data. The vectorial mean value represents a current that flows with constant speed in a direction as given by the sum of the two vectors, but in order to reach the point of displacement of a water particle following vectors AB and AC it has to flow twice the time of observation (that is, the vectorial sum).

The *scalar mean* of the current observations presented in Fig.17 is the line AS_2 which is dashed in order to distinguish it from a vector. The scalar mean AS_2 is found by averaging the speed and the direction of the original current vectors AB and AC separately. It differs significantly from the vectorial mean AS_1. Its direction is the middle between vectors AB and AC.

Although the vectorial mean is well defined, as far as its physical meaning is concerned, vectorial means do not always represent satisfactory results in a current chart. Proper analysis and complete presentation of current observations require both the computation of vectorial as well as of scalar means. The stability of currents should always be presented in current charts together with speed and direction.

If the variability or stability of average currents is defined in a different way than by the quantity B, the exact meaning of this definition should be stated. For example, different methods were used by SCHUMACHER (1940) in computing the resultant currents for representation in his excellent mean monthly current charts for the Atlantic Ocean. In the case where currents observed at different times do not vary significantly in direction, both scalar and vectorial means do not differ much and no real problem exists. It is only in regions of high variability in the direction of current vectors where the oceanographer, when facing the task of presenting a meaningful current chart, has to spend much thought and effort in the analysis of available data.

Since it is difficult to find useful mean values from a set of current observations that scatter over a wide range, it is even more difficult to establish the relation between the averages of two different vectors. This task has arisen in oceanography, for example, when a functional relationship is sought between the wind and the pure wind-driven ocean currents. Observational data consist of a great number of individual wind and current observations which in most cases are distributed over a great range of directions. The task is to find the average ratio between wind speed and current speed as well as the average angle between the wind and current direction. If the observations scatter over a wide range of directions, both the mean value of wind speed and current speed would be small. Errors and observational chances affect the analysis significantly, and the result is the more unreliable (or insignificant), the greater the variation of the direction of the single observation is. SVERDRUP

(1916) and THORADE (1933) spent much thought on this problem. More recently, ELLISON (1954) has dealt with the special problem of vector correlation. DELAND and LIN (1967) have used such mathematical methods to correlate phases and amplitudes of planetary waves at different levels in the atmosphere.

SINGULARITIES IN A CURRENT FIELD

The field of motion is a vector field and can be represented completely by means of three sets of charts showing the distribution of the three components of a current vector. If the vertical component of ocean currents would be of equal kinematic importance as the two horizontal (the north and east) components, simple current charts could not easily be drawn. The average reader of a book on ocean currents expects to find charts showing the distribution of *horizontal* currents represented by arrows, lines or other means that immediately give a complete picture of "where the water flows". Besides direction and speed, stability can be shown by proper markings. Many of the readers of such charts connect with such presentations the idea that a water particle or any other freely floating object follows the traces as indicated by the graphic presentation. Even for the case of stationary currents for which streamlines and trajectories coincide this is only partly true.

Although the vertical component of large-scale ocean currents is small compared to the horizontal components over most parts of the oceans, in many dynamic studies of ocean currents the vertical component is of greatest importance. Also, in the kinematic consideration of horizontal current charts, the vertical component becomes an essential part of the current system when *singularities* in the presentation of horizontal current fields appear.

Most important singularities are lines and points of divergence and convergence of a horizontal flow. The hydrodynamical meaning of divergence or convergence will be explained in following parts of this book. Here, it can be illustrated by simple examples. If two horizontal currents meet each other "head on", the water particles may continue to flow in horizontal direction and the currents are deflected sideways if ample space in horizontal direction is provided for such a motion. If this space is not available, the colliding water masses have to find a way in vertical direction near the line of "head on collision" unless the fluid is so compressible that it can avoid this vertical motion by becoming more dense. This latter possibility has to be ruled out, because for the speeds of ocean currents, water can be treated as an incompressible fluid.

Another simple example is given for a horizontal surface current that flows away from a coast. This motion can be caused by winds blowing from land toward the sea. To compensate for the near-shore water deficit, water from deeper layers must rise to the surface, thus providing continuity in the flow of ocean water. This region of near-shore horizontal surface water divergence is a region of *upwelling* water which is of greatest importance in oceanography. Such regions are distinguished not only for their physical and chemical properties but also for their outstanding marine biological characteristics and for their effects on the overlying atmosphere. Areas of horizontal divergence of surface currents and upwelling water can also be found in the open sea far away from the coastlines. The correct interpretation of horizontal current charts has to include such lines and points of divergence and convergence where important vertical current components can be present.

Current vector lines, like all vector lines, are continuous and cannot cross each other, except at singular points or lines in a two-dimensional field where the magnitude of the vector is zero. Neither can they begin nor end in a vector field, except at singularities.

The importance of singularities in horizontal current charts was first discussed by SANDSTRÖM (1909) and BJERKNES (1910). Singularities often fix the general outline of the current field and have to be located carefully when current maps are constructed. The most important singularities shown in Fig.18 are the neutral point, the point of convergence, and the line of convergence. If the current directions in the figures are reversed, the neutral point remains a neutral point, whereas the two other singularities become a point of divergence and a line of divergence. In neutral points, vertical motions may, but do not necessarily have to occur. The only requirement is that at neutral points the

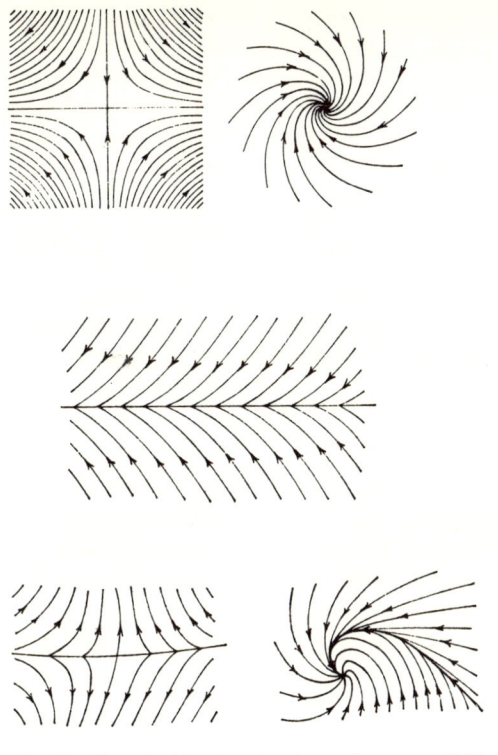

Fig.18. Singularities in a horizontal current field.

horizontal velocity becomes zero. At and along points and lines of convergence or divergence, vertical motions are always involved.

Surface convergences are often caused by heavier water that meets lighter water and spreads in a wedgelike form under the lighter water as shown in Fig.19 where the vertical sections show the transition between the two water masses which move relative to each other (DEFANT, 1929). One- and two-sided convergences and divergences can be formed in this way between water masses of different characteristics. Their explanation is evident if the vertical cross sections in Fig.19 are considered which show the circulation pattern in lateral and vertical directions.

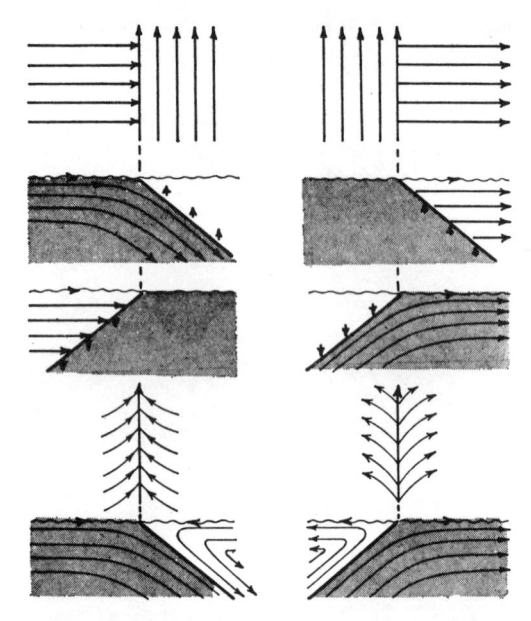

Fig.19. Singularities in a current field and vertical sections showing current conditions on both sides of an inclined gliding surface separating two different water masses. (According to DEFANT, 1929b.)

More complicated current fields can lead to interesting combinations of different points or lines of singularity as shown by two examples in Fig.18.

A convergence line running parallel to a divergence line requires at the sea surface a sinking motion along the convergence and an upwelling motion along the divergence. Horizontal currents in the surface layer and in deeper layers, together with such vertical motions form one closed circulation system. This current pattern is also formed if a propagating wave motion is superimposed on a translatory motion crossing each other. In this case, however, the two lines of singularity are not stationary but propagate with the wave motion. Since streamlines and trajectories coincide only for the case for which the water movements are stationary, a current pattern as in this example has to be cautiously interpreted for practical purposes.

Lines of singularity in the current field are often marked by other phenomena; e.g., by a sudden change in the transparency or color of sea water, temperature and salinity changes, the accumulation of drifting matter like foam or sea weed in the case of converging surface currents, and by changes in the local wind generated wave motion. *Rips*, or rip currents, seem to form frequently in the open ocean between converging or diverging water. They are indicated by strong agitation of the surface water, choppy waves, and sometimes by an increased number of white-caps (breaking waves). Also the opposite change in wave motion can be found at light winds and relatively calm sea when "blanks" or "slicks" appear where the sea surface shows a "glassy calm" or "oily" appearance in contrast to the normal wave motion ("cat-paws" or other small waves) between the slicks.

Remarkable observations of this kind have also been reported by navigators during night hours, when streaks of phosphorescence in surface water appeared in some parts of the ocean, covering many miles of sea surface area (NEUMANN, 1948b). This phosphorescence is caused by the accumulation in streaks of a special kind of plankton.

CURRENT CHARTS

Current charts showing streamlines are based on a large number of direct and indirect current observations at fixed positions. Most of the data used for the construction of surface current charts are based on information obtained from ships' logs. Although these data are not synoptic, and not obtained at fixed points, a reasonable picture of average flow conditions can be derived if a large number of observations is available. Usually, ships' log observations present the drift of the vessel during a 24-h time interval, and therefore, are not only time averages, but also averages over a distance, if the ship is moving through the water.

All of our surface current charts shown in atlases or in pilot charts used in navigation are based on such averages, obtained from observations collected over the years by navigators in international cooperation. In 1853, on the initiative of Matthew Fontaine Maury (1806–1873), an international conference at Brussels, Belgium, was organized at which it was decided that certain observations should be taken at regular intervals and according to a unique scheme by every vessel at sea. In 1873, a congress held in London strengthened this international cooperation in maritime data collection. Since then it has been in force with only minor modifications. This treasure of oceanographic and marine meteorological observations gathered by seamen and entered into their log books for more than 100 years, essentially forms the basis for our present atlases which chart not only the ocean surface currents, but also the average wind direction and force over the oceans, as well as surface water and air temperatures, atmospheric pressure, precipitation and fog, to name only a few variables with which the navigator, as well as the oceanographer and meteorologist, are concerned.

Systematic exploration of the depths of the oceans by research vessels started shortly after Maury's pioneering efforts. Our

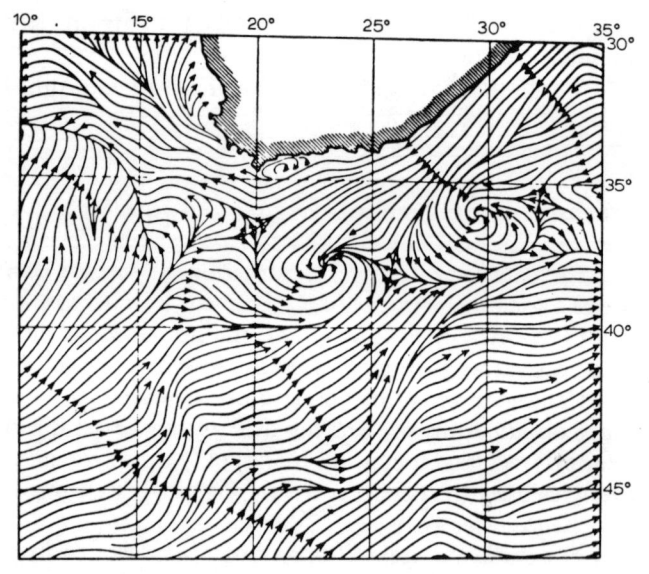

Fig.20. Streamlines of surface currents south of Africa for the month of May. (After MERZ, 1925.)

present knowledge of the waters in the "silent depths of the sea" (ALEXANDER VON HUMBOLDT, 1814) is due to their credit. Of course, very valuable, and often revolutionary, information was also obtained for sea surface conditions from special research that was directed toward a particular goal. The more oceanographers learned about the deep sea, the more was their attention directed toward air–sea interaction processes.

MERZ (1925) and his pupils MICHAELIS (1923), MEYER (1923), PAECH (1926), and others, have to be credited for a most critical and for their time most modern evaluation of ships' log observations, by using BJERKNES' and SANDSTRÖM's (1910, 1912) ideas and suggestions about singularities in the current field and their incorporation in the basic field of horizontal current charts.

MERZ's (1925) chart showing streamlines of ocean currents south of Africa for the month of May is shown in Fig.20. It is seen that most of the current (Agulhas Current) entering the region from the northeast is caught in a number of eddies and other singularity points or lines, returning part of this water into a flow toward the east, such that not much of the Agulhas Current water enters the Atlantic Ocean. Based upon observations obtained during the Antarctic Ob-Expedition, 1956–1958, KORT (1959) estimated $25 \cdot 10^6$ m^3/sec flowing from the Agulhas Current into the Atlantic Ocean.

Besides being continuous, streamlines can also be broken up into arrows. Speed and stability of the currents can be shown either by feathers, the number or shape of arrow heads, the length or thickness of arrows, or by other means like single numbers. Fig.21 is one example showing the currents around the Antarctic Continent (*U.S. Naval Oceanographic Office Publication*, 705).

In order to overcome some difficulties in the presentation of vector data of great variability, current roses have been used. An example of this type is presented in Fig.22 showing the average speed and frequency of currents coming from different directions within sectors of the compass. Eight or sixteen sectors within a circle have been used. Current rose presentations require a great number of single observations and, therefore, the data used for the construction of a current rose are usually collected for larger areas. Often it is even collected for between 9 and 25 "one-degree squares". A "one-degree square" is the surface area of a square of one degree longitude by one degree latitude. Such a combination of observations over large areas requires careful consideration of the general character of the current field in order to avoid combination of observations in different current systems. If the number of single observations is sufficient, and the boundaries of the areas for data collection are properly selected and other precautions in the construction of current

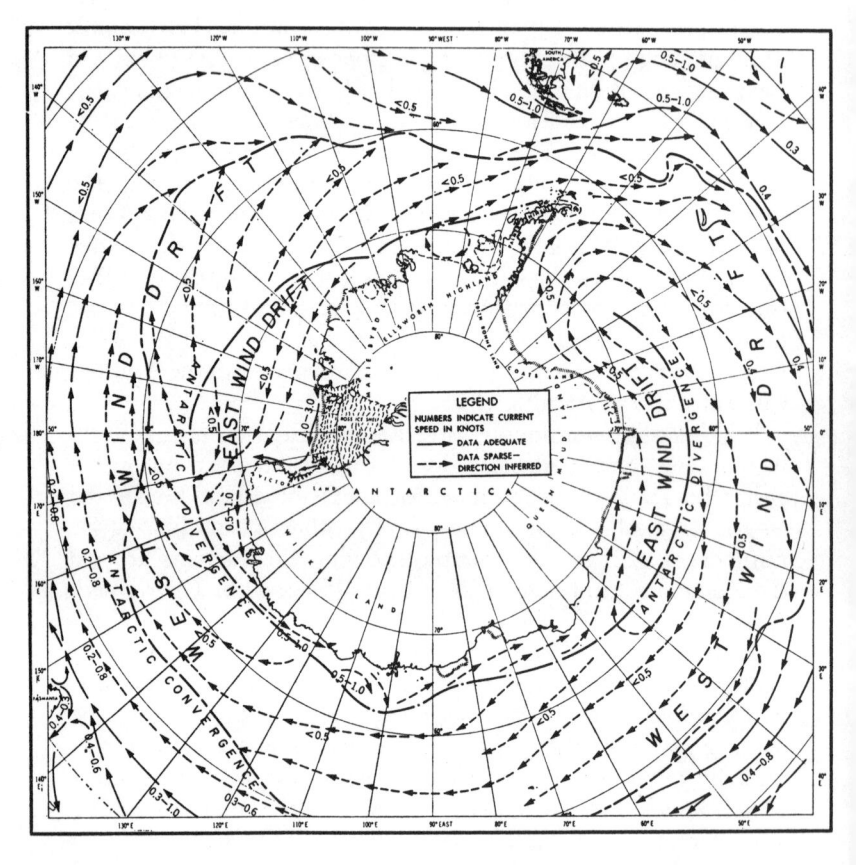

Fig.21. Surface currents around the Antarctic Continent. (From *U.S. Naval Oceanog. Off.*, *Publ.* 705.)

CURRENT CHARTS 63

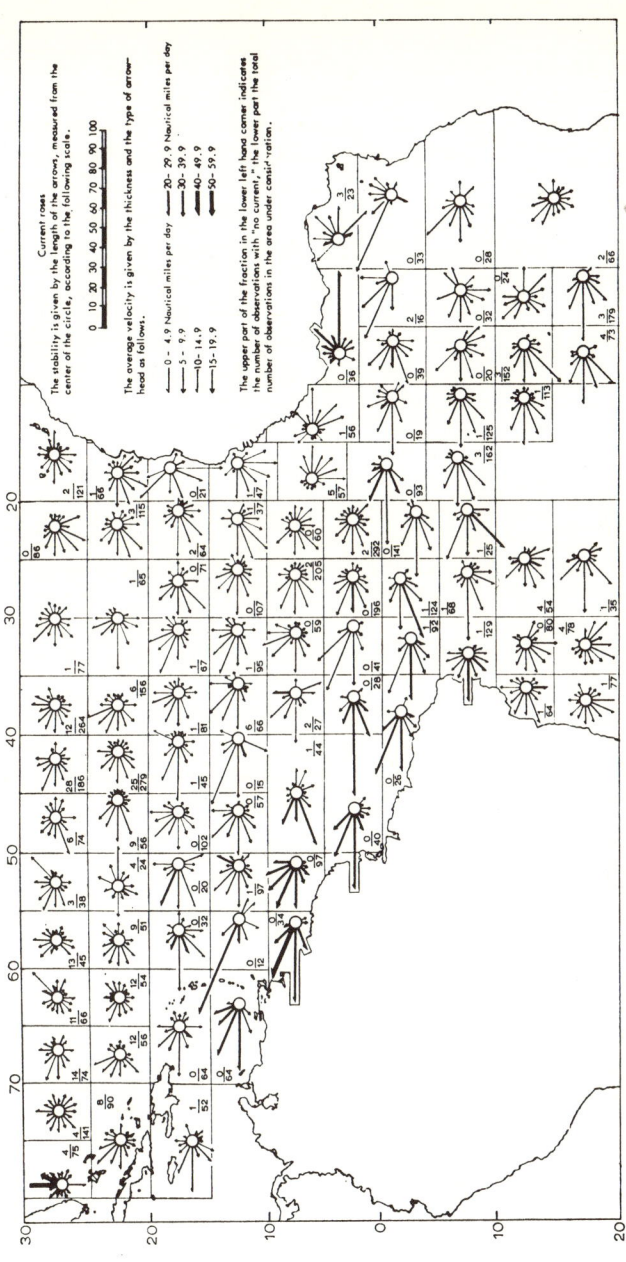

Fig.22. Current roses for the month of February in the tropical Atlantic. From Koninklijk Nederlandsch Meteorologisch Instituut, De Bilt, The Netherlands. *Oceanog. Meteorol. Waarnemingen Atlantische Oceaan*, No. 110, Kaarten, 1918, fotographic reproduction, 1940. (After NEUMANN and PIERSON, 1966.)

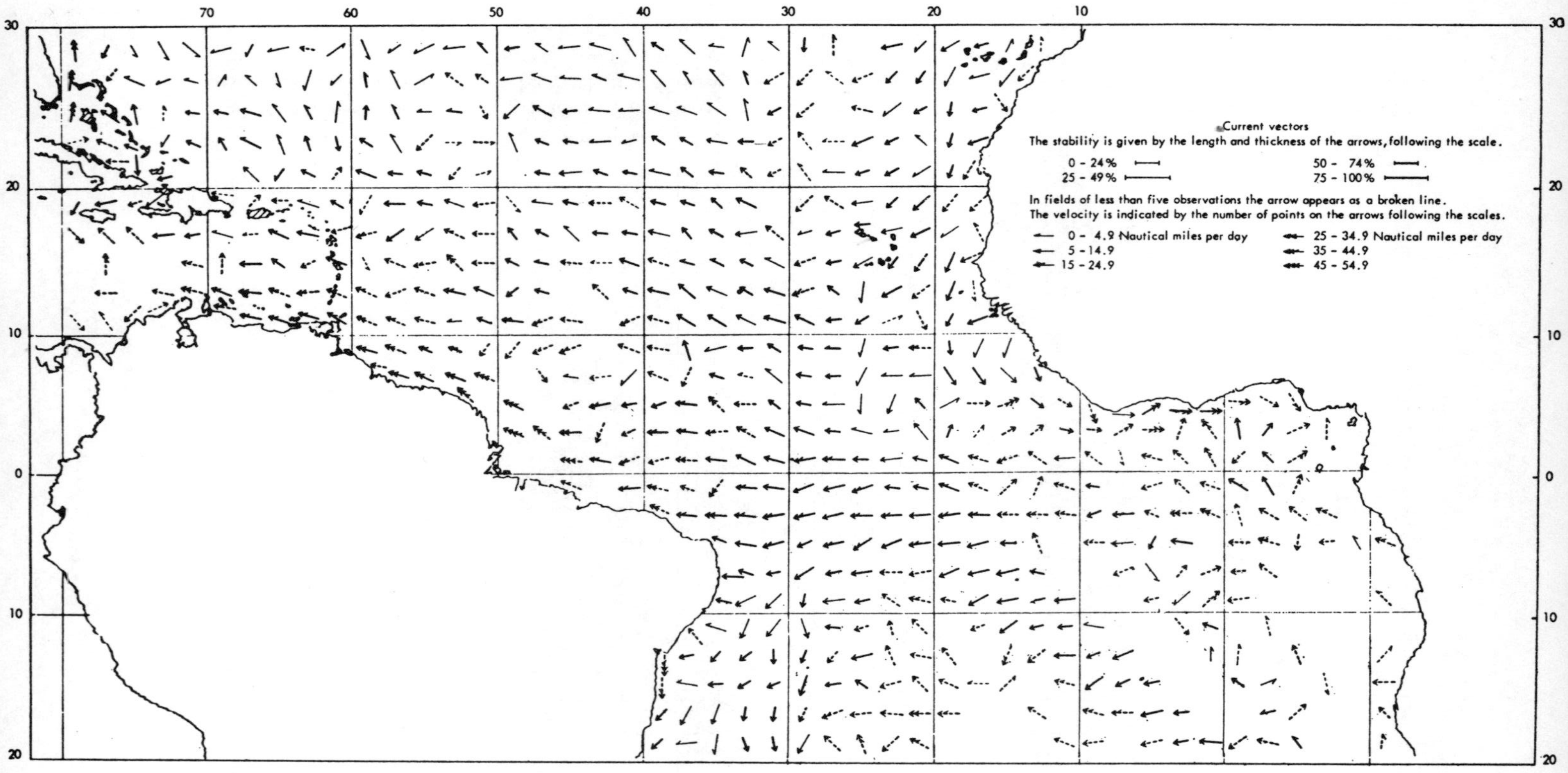

70
60
50
40
30
20
10
30
20
10
0
10
20
Current vectors
The stability is given by the length and thickness of the arrows, following the scale.
0 - 24%
25 - 49%
50 - 74%
75 - 100%
In fields of less than five observations the arrow appears as a broken line.
The velocity is indicated by the number of points on the arrows following the scales.
0 - 4.9 Nautical miles per day
5 - 14.9
15 - 24.9
25 - 34.9 Nautical miles per day
35 - 44.9
45 - 54.9

roses are met, such representations can give objective and significant information not only for practical purposes but also for scientific studies.

Vectorial means of current vectors obtained in two-degree squares are presented in Fig.23. They are based on the same data as used for the construction of current roses in the same region (Fig.22). Explanations are given in the insert of the figure. To arrive at a more comprehensive picture of the average current field, streamlines can be drawn so that the resultant current vectors are everywhere tangential to the streamlines.

WERENSKJÖLD (1922) used current rose presentations to compute charts showing the streamline distribution. Although his work was applied to the wind field, similar techniques can be applied to ocean currents if adequate current roses are available. Lines along which the direction of the current is constant are called isogons, and lines along which the current speed is constant are called isotachs.

A chart of superimposed isogons and isotachs represents uniquely a horizontal current field. According to Werenskjöld, a number of isogons can rather easily be drawn if a chart showing resultant current vectors is available. Since each current vector can be represented by its east and north components (u and v), two charts showing the values of u and v, respectively, can be presented. Lines connecting equal values of v in one chart and lines connecting equal values of u in the other chart usually intersect. At such points of intersection where $u = v$, the angle, α, of current direction is 45° or toward the northeast, since $\tan \alpha = v/u = 1$. This fixes the isogon for currents in northeast direction, and such points can be marked by short arrows pointing in this direction. Along lines where $v = 0$, $\alpha = 0$, and the isogon for currents flowing in east direction is immediately found, as is the isogon for northward flowing currents along the lines of $u = 0$. Intersection of lines $u = 0$ and $v = 0$ obviously gives singular points through which all isogons must pass. Other isogons are easily obtained for intersection points where $u = -v$, $u = \pm 2v$, $v = \pm 2u$, and so on. A line connecting

---

Fig.23. Current vectors for the month of February in the tropical Atlantic. (Same source as Fig.22.)

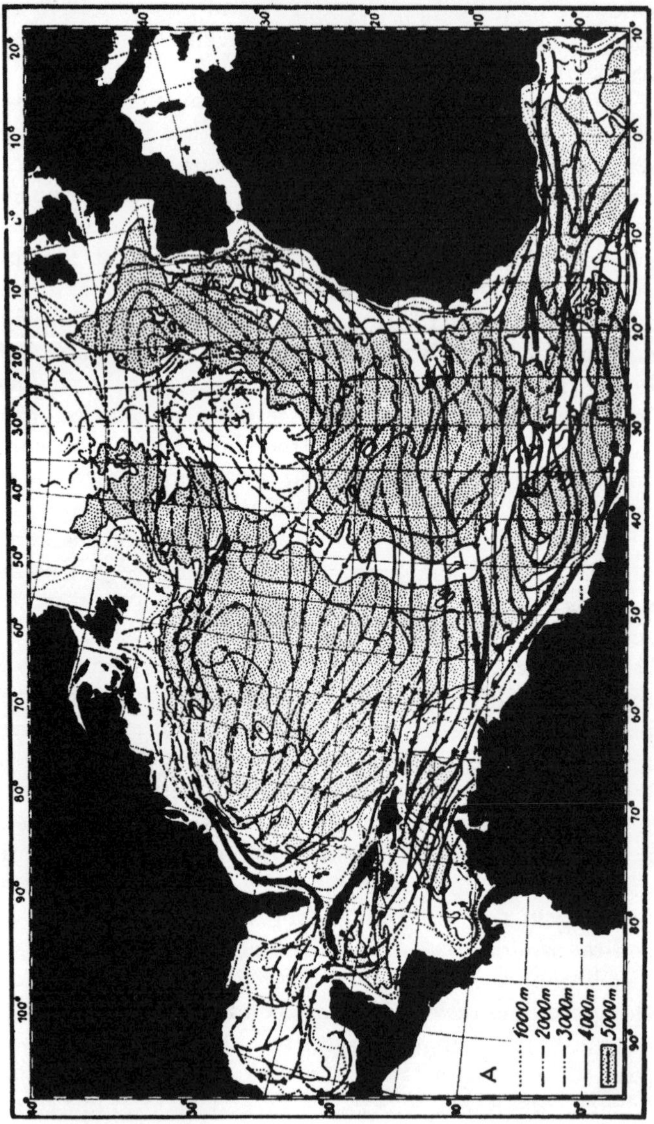
A
1000m
2000m
3000m
4000m
5000m

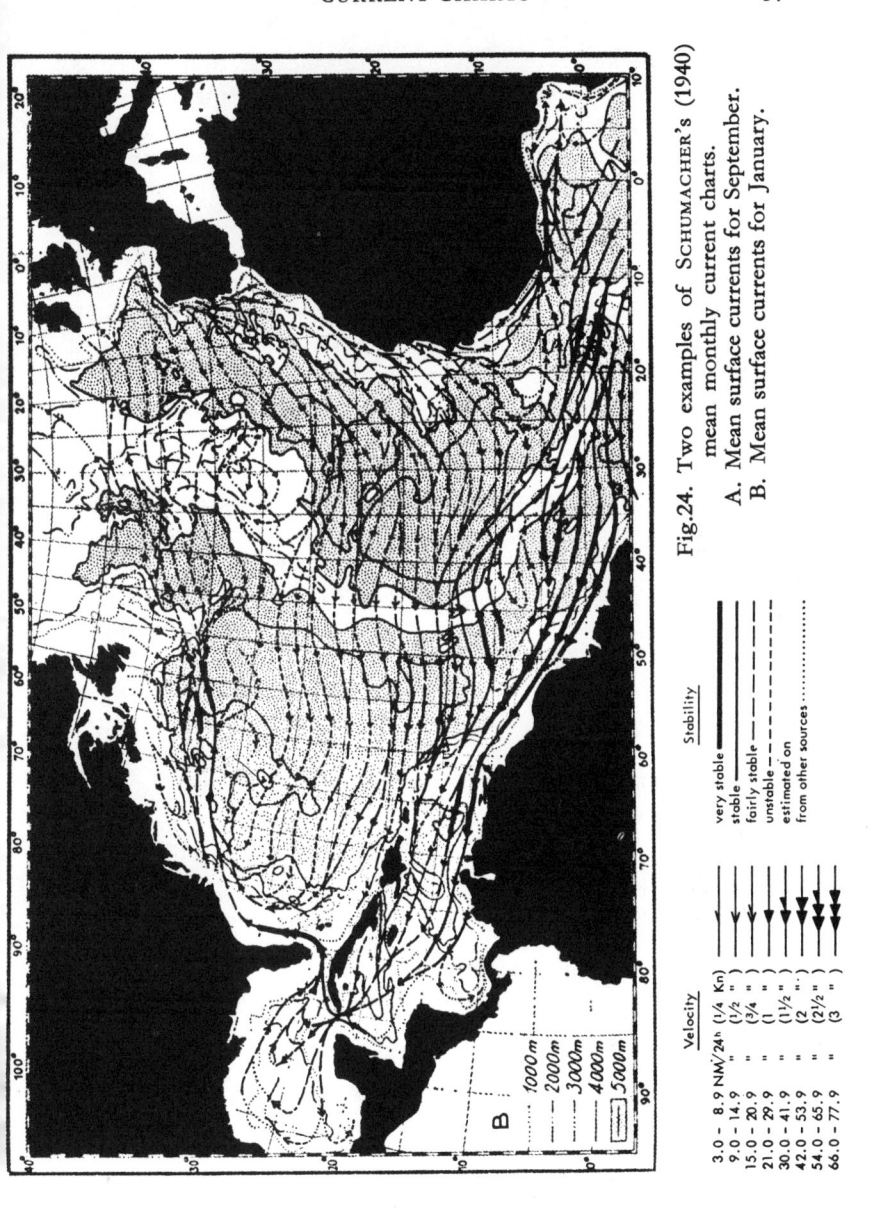

CURRENT CHARTS

Fig. 24. Two examples of Schumacher's (1940) mean monthly current charts.
A. Mean surface currents for September.
B. Mean surface currents for January.

all points for which $u = 2v$ gives the isogonic direction $\alpha = 26.6°$ since $\tan \alpha = 1/2$, and a line connecting all points for which $v = -2u$ gives the isogonic direction $\alpha = 153.4°$. Eight isogonic directions are usually sufficient to fix the current field.

For most practical purposes direct interpolation between the vectors of a current field is sufficient to draw streamlines in a current chart, unless the current fields are extremely complicated.

Mean monthly current charts for the North and South Atlantic Ocean have been prepared by SCHUMACHER (1940, 1943). They represent an excellent source for the study of seasonal variations of many branches of the general oceanic circulation system in the Atlantic Ocean. For the computation of resultant mean monthly currents, Schumacher used the most frequent direction and speed instead of the vectorial mean of *all* observations. Although all available data on ship displacements in a given area were evaluated, for the averaging process only that half circle section (180° sector) that contained most of the observations was chosen. This procedure also required a different definition of "stability" of the current. Stability was defined by the number of observations within the chosen 180° sector, divided by the total number of observations and expressed in percent. The speed in the prevailing direction was computed from scalar means of the ship displacements. Examples of Schumacher's current charts are shown in Fig.24.

Fig.25. World chart of ocean currents during Northern Hemisphere winter. (After NEUMANN and PIERSON, 1966; based on SCHOTT, 1942, and DEFANT, 1961.)

Legend: *1* = North Equatorial Current; *2* = South Equatorial Current; *3* = Equatorial Countercurrent; *4* = Guinea Current; *5* = Antilles Current; *6* = Florida Current; *7* = Gulf Stream; *8* = North Atlantic Current; *9* = Norwegian Current; *10* = Irminger Current; *11* = East Greenland Current; *12* = West Greenland Current; *13* = Labrador Current; *14* = Canary Current; *15* = Guiana Current; *16* = Brazil Current; *17* = Falkland Current; *18* = Antarctic Circumpolar Current; *19* = Agulhas Current; *20* = Benguela Current; *21* = Kuroshio; *22* = North Pacific Current; *23* = California Current; *24* = Aleutian Current; *25* = Oya Shio; *26* = Peru Current; *27* = East Australian Current; *28* = West Australian Current; *29* = Somali Current; *30* = Mozambique Current; *31* = (Indian) Monsoon Current (Northern Hemisphere summer).

pp.71–72

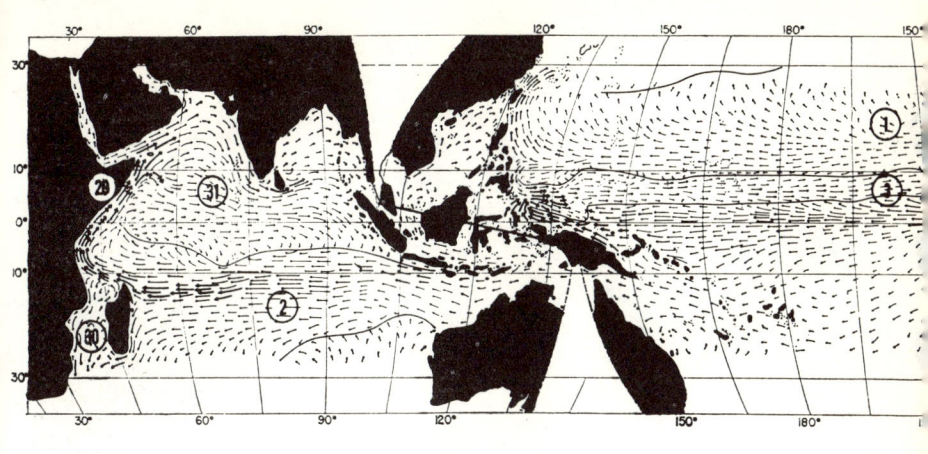

Fig. 26 (lege

pp. 69–70

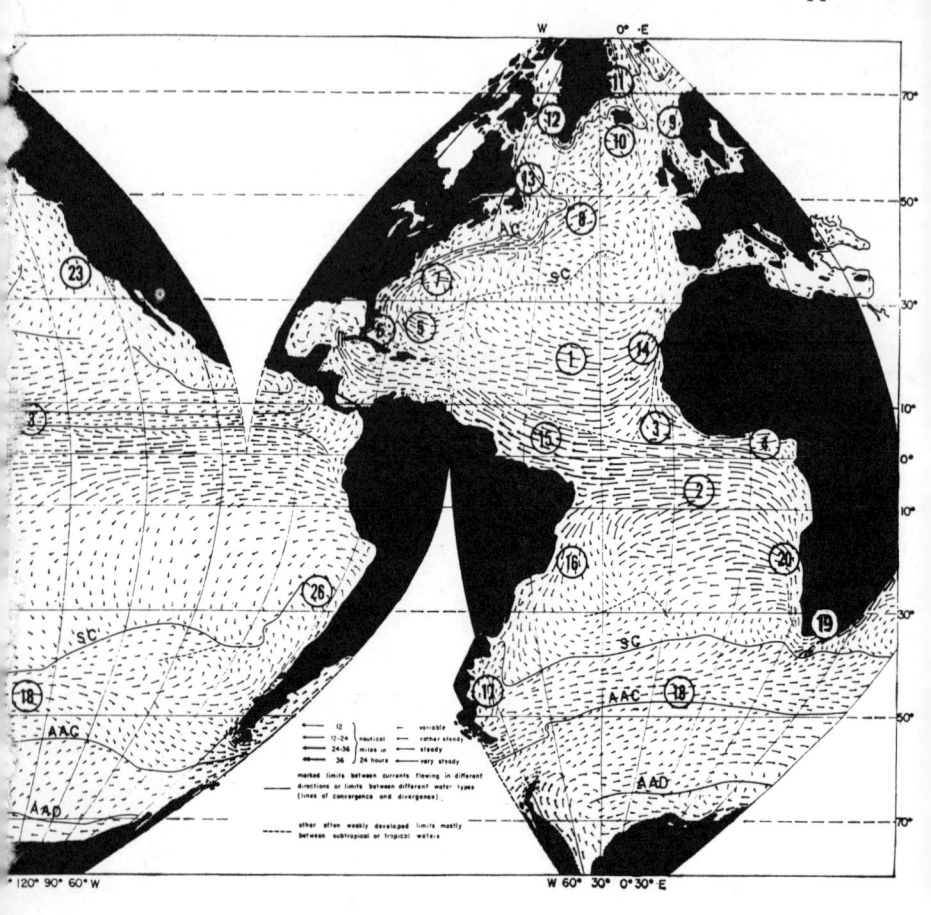

d see p.68).

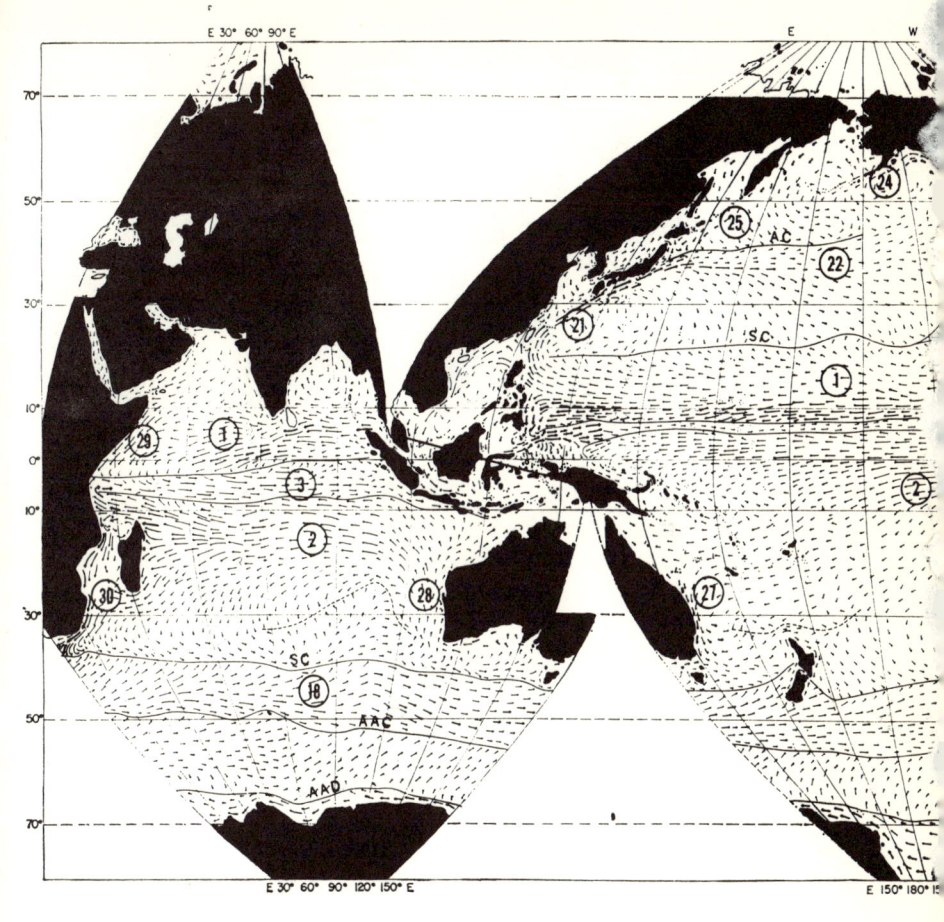

Fig. 25 (lege

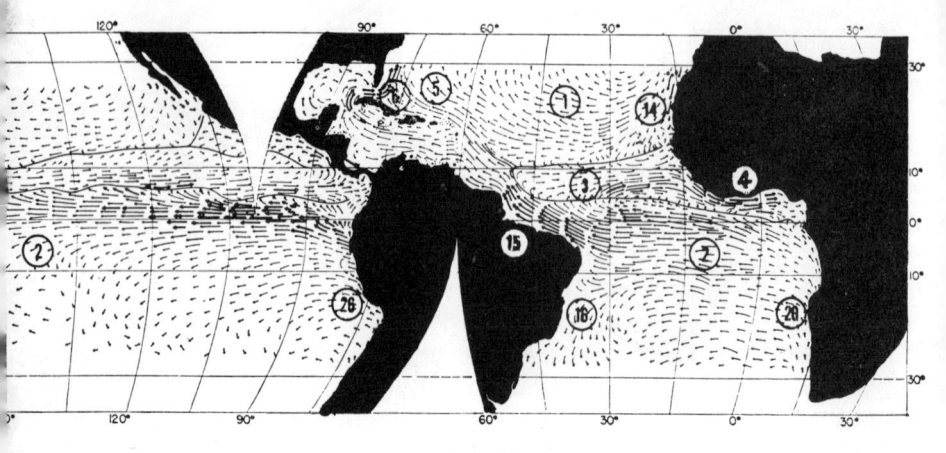

nd see p.73).

Other current presentation may follow more liberally the general streamline distribution by arrows which are partly based on pertinent information referring to the general characteristics of currents in parts of the oceans. Lines and points of convergence or divergence, obtained or suggested by other oceanographic information, can be incorporated in the construction of current charts. Presentations of this kind are very numerous. Since they are based on additional indirect oceanographic information along with the statistical use of a number of direct current observations, such charts often reflect a very personal approach of their originator. The reader will remember the difficulties in obtaining and presenting actual current observations. A thorough study of the literature on the subject, personal experience and knowledge, and subjective decision are often to be combined properly in presenting new current charts.

The current charts used in this book show major ocean currents at the sea surface in Fig.25,26 based on the world map of ocean currents by Schott (1942) and Defant (1961). Other charts could also have been used for the purpose of showing the general circulation pattern of the sea surface layers in the ocean, as long as the reader of such general current charts is interested only in the very general features of surface currents that dominate the oceans.

MAJOR OCEAN CURRENTS

World charts showing ocean surface currents are almost exclusively based on observations collected by navigators. Current charts for all oceans as, for example, those first prepared by Schott (1898) and others, are examples of early data collection and analysis of current observations since Maury's (1849) famous publication of his first nautical atlases. Witte (1879) presented a world chart of ocean currents showing hypothetical regions of upwelling water.

Fig.26. Current chart showing surface currents in tropical and subtropical regions for Northern Hemisphere summer. (After Neumann and Pierson, 1966; based on Schott, 1942, and Defant, 1961.) Legend see Fig.25.

Modern current charts by SCHUMACHER (1940, 1943), the world map of ocean currents by SCHOTT (1942), presentations of ocean currents by the Koninklijk Nederlandsch Meteorologisch Instituut (Royal Dutch Meteorological Institute), the British Admiralty, the United States Naval Oceanographic Office, the Deutsche Seewarte—later Deutsches Hydrographisches Institut (German Hydrographical Institute)—and other sources are available for either parts of the ocean, or for a summarizing view of our present ideas of the general circulation in all of the oceans. Such representations have been continually improved by using the results obtained from general oceanographic research activities. Progress of our knowledge of the structure and the geographical distribution of ocean currents is partly the result of a better theoretical understanding of their dynamics, and partly the consequence of highly improved observational techniques and instruments. The present era of oceanographic research strives for a more detailed survey of the structure and temporal behaviour of the world's water masses and tries to find the interrelationship with the atmosphere on the basis of synoptic observations. To be mentioned here are the more sophisticated experimental techniques like those applied by VON ARX (1952) and FULTZ (1951) who have constructed laboratory models of the circulations in the oceans and in the atmosphere.

The world chart of ocean currents presented in Fig.25 refers to conditions as observed principally in the Northern Hemisphere winter (February–March). In some parts of the oceans, especially in tropical and some subtropical regions, thorough seasonal changes are observed in speed and direction of the currents. Fig.26 represents the average surface currents between 30°N and 30°S during the Northern Hemisphere summer which shows significant changes when compared with the winter map. Details of seasonal variations in current speed and direction, not only in more remote parts of the ocean but also in frequently visited regions (e.g., the region of the tropical western Atlantic off the coast of South America), still await systematic oceanographic studies. Remarkable seasonal changes of the oceanic circulation north and northwest of New Guinea have been studied by WYRTKI (1958), and some recent results obtained in the Somali

Current (SWALLOW, 1965) show unexpected details and fluctuations of one of the strongest ocean currents.

Outstanding features of the surface circulation in all oceans are currents that flow in nearly zonal directions, thus crossing the oceans from east to west in lower latitudes and from west to east in higher latitudes. Near continental coasts, the currents are forced to follow the coastlines, forming eastern and western *boundary currents*. The dominant trend of the general circulation in the upper layers of the oceans is in the form of anticyclonic and cyclonic gyres. In the North Atlantic and North Pacific Oceans the circulation around the anticyclonic gyres is in a clockwise direction. It shows an asymmetric development such that in the western part of each ocean the poleward flow (western boundary currents such as the Gulf Stream and the Kuroshio) tends to become more concentrated and faster than the equatorward flow of the eastern branches of the gyral circulation system. In the Southern Hemisphere, the anticyclonic circulation is in a counterclockwise direction. Stronger, more concentrated boundary currents are the Somali Current and the Mozambique–Agulhas current system in the Indian Ocean, and to a smaller extent, the East Australian Current in the Pacific Ocean. However, taken as a whole, the anticyclonic gyre in the South Pacific Ocean appears to be asymmetrical with the center of its anti-cyclonic circulation displaced rather to the east than to the west.

This general trend of observed ocean currents is mainly the result of the driving forces of the prevailing winds as can be demonstrated in most regions by a comparison of wind and current charts. The prevailing wind system in tropical and subtropical regions over the oceans is governed by the Trade Winds, the doldrum belts and in some areas by monsoon winds. In higher latitudes, approximately north of 30° or 40°N and south of 30° or 40°S latitude winds from westerly direction, the Westerlies, dominate, and poleward of 60° latitude east winds prevail in the climatological average.

The relatively largest surface area over the oceans is occupied by the Trade Winds, then followed by the Westerlies. Both branches of the general atmospheric circulation are separated by a region between about 20° and 30° latitude in both hemispheres where weaker variable winds are observed. This region is often called the "horse latitudes."

The tropical sea surface circulation is dominated by the North and South Equatorial Currents, which are usually separated by an Equatorial Countercurrent flowing in the opposite direction. This countercurrent is not easily explained from the prevailing wind system, although its existence appears to be strongly dependent on the wind distribution in the tropics which includes the possible development of a doldrums region.

On the average, the doldrums belt in the Atlantic Ocean is located in the Northern Hemisphere between about 5° and 10°N. Only in the western Atlantic during late winter or early spring does it move closer to the equator and perhaps extends a little into the Southern Hemisphere. From the current charts it is seen that the Atlantic Equatorial Countercurrent in summer and early fall forms a mighty current of considerable speed which flows all across the ocean from about 50°W, near the American coast, into the inner part of the Gulf of Guinea. Its width occupies the area between about 3°N and 10°N. During winter and spring, the Countercurrent is restricted to the eastern part of the Atlantic Ocean, although signs of a rudimentary eastward motion embedded in the westward trend of the North Equatorial Current can be found locally in some parts of the western Atlantic Ocean. This was shown by SCHUMACHER (1940) in his mean monthly current charts.

Also in the Pacific Ocean, the Equatorial Countercurrent has its greatest development in the Northern Hemisphere summer or early fall when the doldrums belt is located in the Northern Hemisphere all across the Pacific Ocean. During the winter season, the doldrums in the western Pacific shift into the Southern Hemisphere forming a broad belt of calms or light variable winds between about 10°S and 20°S and the equator, west of 180° longitude, approximately. The Equatorial Countercurrent of the Pacific Ocean is much weaker during the winter season, at least in the central part of the ocean and it may even show signs of temporary interruption. However, in the western part of the Pacific Ocean countercurrents of lesser extent are also often observed even during the winter period. Near the western boundaries north of New Guinea, eastward flowing currents can develop great speeds; however, it seems that this eastward flow does not extend far into the central part of the ocean.

In the Indian Ocean, an Equatorial Countercurrent appears only during the Northern Hemisphere winter, and then it is located in the Southern Hemisphere. This is the season when the northeast winds of the Northern Hemisphere reach over the equator into the Southern Hemisphere of the Indian Ocean. Here, a doldrums belt develops at about 5°S–10°S between weak northerly winds and the southeast Trade Winds, south of 10° S. Again, the Indian Ocean Countercurrent develops in the doldrums region, and in contrast to the other two oceans it is found in the Southern Hemisphere.

During northern summer, doldrums or weak variable winds dominate the central and eastern Indian Ocean approximately at the equator, and southeast winds observed in the Southern Hemisphere gradually turn into southerly and southwesterly directions in the Northern Hemisphere after crossing the equator. Southwest winds are especially strong near the Somali coast and in the Arabian Gulf. This is the season of the Indian Southwest Monsoon. Under the influence of the prevailing winds during this period the westward flow of the Indian North Equatorial Current as observed in winter has changed into the eastward flow of the Monsoon Current, and the Southern Hemisphere Countercurrent of the winter season has disappeared. Significant changes in current velocity are also evident in the western boundary currents of the Indian Ocean. The Somali Current, flowing southward in winter, has reversed its direction and flows with extremely high speeds northward. Seasonal changes of ocean currents, associated and directly dependent on the prevailing winds are most strongly pronounced in the Indian Ocean. The concentrated international cooperation during the Indian Ocean Expedition (IOE) to study this ocean in greater detail, particularly the time-dependent features of its circulation, will help in answering many questions about the wind-driven ocean circulation and their effects on the deep-sea circulation and stratification.

Although general relationships between the atmospheric and oceanic circulation are evident, and major changes of speed, direction, and geographical extent of some current branches of the oceanic circulation system, particularly in tropical regions, are clearly related to changes of the driving forces of the wind field, it is often difficult to explain details of seasonal and other time-dependent variations of ocean currents in many parts of the oceans. This is particularly true for currents in middle and higher latitudes like the Gulf Stream and the Kuroshio which are outstanding examples for western boundary currents, and their eastward extensions, the North Atlantic and North Pacific Currents. Also, currents like the California Current, the Canary Current, the Peru Current, and the Benguela Current, which represent eastern boundary currents, are not yet fully explored to explain satisfactorily significant seasonal changes of their structure and dynamics as a result of changes in the atmospheric circulation. Associated with such eastern boundary currents in subtropical regions is the phenomenon of coastal upwelling with its consequences on climate, weather, and life. Not only does the *local* wind field affect the ocean circulation in a particular area, but also changes of the general atmospheric circulation over regions of greater extent, comprising a whole hemisphere or more, seem to affect local currents. As was shown by ISELIN (1940), WERTHEIM (1954), STOMMEL (1958), and others, local changes in the current field of the North Atlantic can be expected as the result of causes far removed from the area of consideration.

Moreover, the wind is not the only cause for the generation of ocean currents, be it by producing pure wind-driven currents or slope currents. Both will be explained in following chapters. Differences in the specific weight, or the density, of ocean water also create horizontal pressure gradient forces and, therefore, accelerations that lead to water movements.

Differences in the heating and cooling of water at the sea surface, together with salinity changes due mainly to differences in evaporation and precipitation over various sections of the oceans located in different climatic zones, lead to the formation of certain water types and water masses. In some distinguished regions, especially in high latitudes, water conditioned at the sea surface tends to sink into deeper layers if its density is greater than the density of the surrounding water body. Such processes can be continuous or intermittent. Most important for the oceanic circulation and stratification is that they produce horizontal inhomogeneities in the density distribution, and, therefore, pressure gradients. Also, the influx, or horizontal advection of water of different density, like the Mediterranean Water that flows through the Strait of Gibraltar into the deeper layers of the North Atlantic Ocean, is of greatest importance. A similar source that affects the circulation in the deeper layers of the Indian Ocean is found in the Red Sea Water flowing through the Strait of Bab el Mandeb into the northwestern Indian Ocean.

The formation and spreading of such water types and water masses is often called *thermohaline circulation*. Mixing and diffusion may obscure the true circulation pattern, particularly in the case of sluggish water movements. Nevertheless, such movements and the spreading of water masses are significant for an explanation of the laminated stratification of the deep sea, which is most obvious in the distribution of temperature, salinity, and oxygen. This superposition of wind-driven and thermohaline components makes the analysis of ocean currents more complicated than the analysis of atmospheric motions, which are essentially thermally driven currents.

In some parts of the ocean, thermohaline circulations may contribute essentially even to the observed pattern of ocean currents in the upper strata. This was demonstrated by STOMMEL (1957) who suggested a qualitative Atlantic model of thermohaline deep-sea circulations interrelated and compatible with the wind-driven circulation in the surface layers of the ocean. Such models will be discussed in Chapter V.

A quantitative explanation of ocean currents, particularly of the three-dimensional general ocean circulation requires not only a profound knowledge of higher mathematics and hydrodynamics, but also a better understanding of air–sea interaction, and friction and diffusion in the sea. However, a great deal can be done without recourse to the advanced aspects or concepts of these sciences. Some basic hydrodynamical background is necessary to understand and explain major features of the general oceanic circulation. This background will be presented in the following chapters. It will be followed by a more detailed discussion of the major types of ocean currents, and the general oceanic circulation.

CHAPTER III

BASIC HYDRODYNAMICAL BACKGROUND

PRESSURE, DENSITY AND GEOPOTENTIAL

The pressure in a fluid at rest is called the *static pressure, p*. Its physical meaning is that of a force, the *pressure force*, **P**, acting in a normal direction on an area **F**. The pressure $p = P/F$ is a scalar quantity in contrast to **P** and **F** which are vectors. Since p represents the force **P** per area **F**, it is measured in units of 1 dyne/cm² or cm⁻¹ g/sec². 1 dyne, the unit of force in the centimeter–gram–second system of physics (cgs system) is the force that accelerates the mass of 1 g by 1 cm/sec². This is a small unit when compared with the acceleration of gravity at the earth surface which is about 980 cm/sec² in middle latitudes. Therefore, BJERKNES (1906) suggested that the unit of pressure for practical purposes in meteorology and oceanography be taken as 1 bar = 10⁶ dyne/cm². Whereas in meteorology, for practical reasons, 1 mbar = 10³ dyne/cm² is used as the pressure unit, the unit of sea pressure in oceanography is practically defined as 1 decibar (dbar) = 10⁵ dyne/cm².

The static sea pressure p at any depth z below the sea surface is given by the weight of a water column of unit cross section between the surface and the depth z. If $\bar{\rho}$ is the mean density of the water column and g the acceleration of gravity, $p = g\bar{\rho}z$. The value of g depends on the geographical latitude and other factors, but for oceanographic purposes it can be taken as 980 cm/sec² in middle latitudes. At the equator, $g = 978$, and at the poles, $g = 983$, as a consequence of the rotation and ellipticity of the earth. The density, ρ, varies from place to place in the oceans; it is most important for computing the field of internal pressure in a stratified ocean.

The density of sea water defined as mass per unit volume (g/cm³) depends on the temperature and the salinity of the water sample, and also on the pressure (depth), since water is slightly compressible. In oceanography, the *specific gravity* rather than the density is generally considered. Specific gravity is the ratio of two densities. If by definition, distilled water at a temperature of 4° C has a density $\rho_m = 1$, then the specific gravity of water (or any other substance) with a density ρ is ρ/ρ_m, and the numerical value of the density is equal to the specific gravity.

The density of ocean water is greater than the density of pure water due to the presence of dissolved material in the form of molecules or ions. These dissolved materials are mostly an ionic dispersion as in a solution of salts, and a minor part consists of colloids and suspensions. At the sea surface, the average density is about 1.025 (g/cm³). In dynamic oceanography dealing with ocean currents, it is necessary to know the sea water density with an accuracy of five decimal places. For example, at a temperature $T = 20\,°C$ and a salinity $S = 35\permil$, the sea-water density at atmospheric pressure (sea surface) is $\rho = 1.02478$. An accuracy of less than five decimal places is insufficient for an accurate computation of the pressure field.

Numerical values of ocean water density always start with 1.0xxxx. Therefore, it has become customary to abbreviate density figures by introducing a quantity called *sigma-t*, defined by $\sigma_t = (\rho_{S,T,0} - 1) \cdot 10^3$. With σ_t values, only the dependence of density on temperature and salinity is considered. The σ_t values always refer to the density at atmospheric pressure, whereas the values $\sigma_{S,T,P} = (\rho_{S,T,P} - 1) \cdot 10^3$ take the additional effect of sea pressure into account. This density is called the *density in situ*. The distribution of density in the oceans determines the field of mass per unit volume.

If the sea surface would be a surface everywhere perpendicular to the direction of the force of gravity, that is, perpendicular to the direction of the plumb line, and in the absence of winds, the internal field of pressure in the oceans would be completely determined by the field of mass. This part of the internal total pressure field is called the *relative* field of pressure. The *total* field of pressure, however, is composed of the relative field and the fields that are caused by external forces such as the atmospheric pressure and the slope of the sea surface caused by wind effects, e.g., by the piling up of water against coasts or in the open sea where wind-driven horizontal water transports either converge or diverge. In an ocean of constant density and of no relative field of pressure, horizontal pressure gradients are possible if the surfaces of equal pressure, called *isobaric surfaces*, are inclined against level surfaces. This requires a tilt of the sea surface.

A level surface is always at right angles to the direction of the gravitational force. If a mass is shifted along level surfaces, and if no forces other than the force of gravity are considered, no work is expended in the displacement, and the potential energy of the mass remains constant. Therefore, level surfaces are surfaces of constant gravitational potential, or of constant *geopotential*.

The amount of work required to move a unit mass a geometric distance h along a plumb line equals gh. Thus, *geopotential surfaces* are defined by different values of gh, and it is seen that not only h but also g enters the value of the gravitational potential. Thus the geometric distance between two equipotential surfaces cannot be constant over the entire earth's surface. At the poles where g is greater, the distance h between equipotential surfaces is less than it is at the equator, where g is smaller.

The numerical value of the geopotential depends on the units used in g and h. If, for practical purposes in oceanography, h is expressed in m and g in m/sec², $gh(\text{m}^2/\text{sec}^2) = 10D$ defines a quantity, D, the *dynamic depth*, in units of a dynamic meter (dyn m). Thus, the geopotential unit of 1 dynamic meter corresponds numerically to a geometric depth of about 1.02 m, if $g = 9.80$ m/sec². At the poles this distance is slightly smaller and at the equator slightly greater. Tables for a conversion of dynamic meters into geometric meters have been given by BJERKNES and SANDSTRÖM (1912).

If the sea pressure, p, is measured in decibars and the dynamic depth, D, in dynamic meters, it follows that:

$$p(\text{dbar}) = \bar{\rho} D \qquad (5)$$

Thus, the pressure of 1 dbar corresponds approximately to the sea pressure exerted by the height of a water column of 1 m. The numerical values of the pressure in dbar, of the dynamic depth in dynamic meters, and of the geometric depth in meters, are, therefore, roughly the same. For example, in a *standard ocean* where the temperature is 0 °C, the salinity 35‰, and $g = 9.80$ m/sec², the pressure of 1,000 dbar is found at a depth of 984.41 m, and the dynamic depth would be 970.40 dynamic m. Although the closeness of these numerical values is quite convenient, the deviations caused by the factor $\bar{\rho}$ in eq.5 are most important in describing the actual relative field of pressure in the sea.

Representation of the field of static pressure—pressure gradients

If the ocean is in the state of rest, the only forces that have to be considered are the gravitational force and internal pressure force. The conditions of static equilibrium of a fluid are fulfilled when all forces acting on a volume, $V = dx\, dy\, dz$ balance each other. Lateral pressure forces acting in level surfaces must balance each other in order to prevent the fluid from being laterally accelerated. In the vertical direction a balance must exist between the gravitational force G, or the weight of the volume, acting downwards, and the vertical pressure difference, that is, the difference between the thrust on the upper surface of the cube, P_z, also acting downwards, and the thrust at the bottom surface of the same cube, $P_{z\,+\,dz}$, acting upwards. The weight of the volume V is $G = mg$ where the mass $m = \rho\, dx\, dy\, dz$. The sum of G and $P_z = p\, dx\, dy$ must be balanced by the pressure force $P_{z\,+\,dz} = (p + [\delta p/\delta z]\, dz)\, dx\, dy$, where δp represents the increase of pressure per vertical distance, δz. If the vertical z-coordinate points downward in positive direction, the balance of forces is:

$$g\rho\, dx\, dy\, dz + p\, dx\, dy - (p + \frac{\delta p}{\delta z}\, dz)\, dx\, dy = 0$$

and therefore:

$$\delta p = g\rho\, \delta z \qquad (6)$$

or, in the limit, as the volume V becomes infinitesimally small, $dp = g\rho\, dz$. Eq.6 in its infinitesimal form is called the hydrostatic equation.

According to the definition of the dynamic depth D, eq.6 can also be written:

$$\mathrm{d}D = \frac{1}{\rho}\,\mathrm{d}p = \alpha\,\mathrm{d}p \tag{7}$$

if p is measured in dbar and D in dynamic m. The reciprocal value of ρ is called the *specific volume*, α. Eq.7 shows that in the case of static equilibrium, the fields of pressure, mass and geopotential must be parallel to each other, or, in other words, isobaric surfaces and surfaces of equal specific volume, called *isosteric surfaces*, must coincide with level surfaces. A field of mass in which the isosteric and isobaric surfaces coincide is called a *barotropic* field. A field of mass in which these surfaces intersect is called a *baroclinic* field. A baroclinic field cannot remain without motion, whereas a barotropic field can be motionless, but does not have to be motionless. Only for the case where isobaric, isosteric, and geopotential surfaces coincide is a barotropic field at relative rest.

The hydrostatic equation was derived with the assumption that the ocean is at relative rest. In fact, the oceans are, generally, in a baroclinic condition and in motion, and the assumptions that led to the derivation of eq.6 are not really fulfilled. The only reason why this equation can be used for practical computations of the pressure field from the observed field of mass is that oceanic water movements are almost horizontal and relatively slow. Deviations from the static equilibrium are neglected. For each vertical column of water in the oceans, hydrostatic equilibrium may be assumed with sufficient accuracy; however, from column to column, the apparent static equilibrium changes stepwise. It should be remembered that we are dealing with a baroclinic ocean, and with a *quasi-static* equilibrium of its water masses. Departures from a true static equilibrium can be significant when dealing with special problems in dynamic oceanography. This fact should be kept in mind when the distribution of the internal field of static pressure is presented in so-called "dynamic charts".

In the oceans, the pressure changes not only in the vertical direction of the plumb line. The changes in the lateral direction are most important when dealing with currents. The rate of change of pressure, p, in a direction, n, normal to the isobaric surfaces is a vector which represents the total pressure gradient force.

If the direction n points from regions of low pressure towards regions of high pressure, the total pressure gradient is $-\partial p/\partial n$. The minus sign indicates that the pressure gradient is directed from high to low pressure. Its magnitude is inversely proportional to the distance between isobaric surfaces which are to be selected for equi-scalar values of the pressure. Fig.27 represents a pressure gradient, $\partial p/\partial n$ (directed from low to high pressure) normal to the isobaric surfaces p_1 and $p_1 + \Delta p$. The magnitudes of its components in the x, y and z directions of the rectangular (left hand) coordinate system are $\partial p/\partial x$, $\partial p/\partial y$, and $\partial p/\partial z$, respectively. If ρ is the water density, the components of the pressure gradient force per unit mass are:

$$-\frac{1}{\rho}\frac{\partial p}{\partial x}, \quad -\frac{1}{\rho}\frac{\partial p}{\partial y}, \quad -\frac{1}{\rho}\frac{\partial p}{\partial z}$$

respectively, if the z-axis points downward in positive direction.

The pressure field in the oceans can be represented either by *isobaric charts* or by charts showing the topography of *isobaric surfaces*. Isobaric charts present the pressure distribution on given level surfaces below the sea surface. Topographic charts show lines of equal dynamic depth, the *dynamic isobaths* of isobaric surfaces. Such charts are called relative dynamic topographies if the dynamic depths refer to and are counted from a sea surface that coincides with a level surface. This method of presenting the pressure field in the oceans is more common in oceanography than showing isobaric charts. A set of relative dynamic topog-

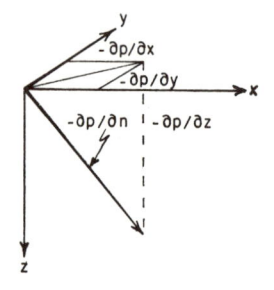

Fig.27. The gradient of sea pressure, $\partial p/\partial n$, and its components in x, y, and z directions, respectively.

raphies for different pressure surfaces provides a fairly complete picture of the quasi-static relative pressure distribution.

The dynamic depth between two isobaric surfaces, with pressures p_0 and p, respectively, is obtained from:

$$D = \int_{p_0}^{p} \alpha \, dp \tag{8}$$

If p_0 refers to the sea surface as the uppermost isobaric surface in the sea, $p_0 = 0$, and D represents the dynamic depth of an isobaric surface with sea pressure p. Of course, D can only be measured relative to the sea surface where $p_0 = 0$. If the sea surface is inclined against a level surface, the dynamic depth, D, as computed from eq.8 can represent only a relative dynamic depth. One of the most important problems in dynamic oceanography is the determination of accurate *absolute* dynamic topographies, because knowledge of the absolute pressure distribution in the oceans is of primary significance in the fundamental equations of motion. If absolute dynamic topographies are known, the oceanographer can apply theoretical relationships between the pressure forces and other forces that cause and affect ocean currents in order to arrive at current speeds and directions at various depths. However, depending on the nature of currents, this procedure can become quite involved if friction as one of the other significant forces has to be considered, and in the case of accelerated currents in which the forces that act upon a mass of water do not balance each other.

To simplify the practical application of eq.8, the concept of a standard ocean of constant salinity, $S = 35\%_0$, and constant temperature, $T = 0\,°C$, has been introduced. The specific volume in situ, $\alpha_{S,T,P}$, can be written:

$$\alpha_{S,T,P} = \alpha_{35,0,P} + \delta$$

where $\alpha_{35,0,P}$ represents the constant contribution of a standard ocean to the specific volume, and the term:

$$\delta = \Delta\sigma_t + \delta_{S,P} + \delta_{T,P} + \cdots$$

represents the *anomaly of the specific volume*. The term $\Delta\sigma_t = \delta_S + \delta_T + \delta_{S,T}$ expresses the dependence of the specific volume anomaly on salinity and temperature, without regard to pressure. This value can be easily obtained from σ_t values, since:

$$\Delta\sigma_t = 0.02736 - \frac{\sigma_t \cdot 10^{-3}}{1 + \sigma_t \cdot 10^{-3}}$$

The other two terms in the expansion of δ represent the effect of pressure at different temperatures and salinities on the specific volume anomaly. Tables for computation of δ were first given by SVERDRUP (1933). More detailed tabulations for the individual terms to facilitate interpolations are also available, for example, the tables by LA FOND (1951).

With the introduction of the anomaly of the specific volume, eq.8 can be written:

$$D = \int_{p_0}^{p} \alpha_{35,0,P}\, \mathrm{d}p + \int_{p_0}^{p} \delta\, \mathrm{d}p = D_{35,0,P} + \Delta D \qquad (8a)$$

Values for $\alpha_{35,0,P}$ were first tabulated by BJERKNES and SANDSTRÖM (1910). For many purposes in oceanography, differences of D between stations are used, and in this case, only the last term in eq.8a needs to be considered. This term is called the anomaly of the dynamic depth, ΔD.

THE EQUATIONS OF MOTION

One of the fundamental laws of mechanics, according to Newton, states that the product of mass and acceleration equals the vector sum of all forces that act on the mass. If this law is applied to a continuous volume of water instead of to a solid object, it forms the basis for the *equations of motion* in hydrodynamics. Since these equations must account for the fact that different parts of the fluid can move in different directions at different times and places, and that two fluid masses cannot occupy the same place at the same time, another equation has to be added, the *equation of continuity*. This equation states that the net flux of a mass into or out of a certain volume must be balanced by a change of density. For moving fluids or gases this simply means that mass can be neither created nor destroyed.

In describing fluid motions, reference must be made to a *system of coordinates*. In most parts of this text, a rectangular x, y, z coordinate system, the Cartesian coordinate system, will be used, where the positive x-axis points toward the east, the y-axis toward the north, and z is taken to be positive upward in the direction normal to a geopotential surface. In oceanography it is often convenient to use a coordinate system in which the z-direction is positive downward. This is called a left-hand coordinate system as opposed to a right hand system where z is positive upward.

Other coordinate systems are used in oceanography for special purposes. In a spherical coordinate system the position of a point is fixed by the distance, R, from the center of the sphere (earth), by the latitude, ϕ, and the longitude, λ. This system is of particular importance and needed if large surface areas are involved and the curvature of the earth becomes important. In dealing with circular motion, it is often convenient to introduce a system of polar coordinates.

Since there are two possible ways to describe the motions of a fluid, there are two ways to formulate the equations of motion. A representation where the fluid motion and the fluid properties and their changes with time are described at fixed points in the coordinate system is called the Eulerian representation. The other way to describe the fluid motion is to "tag" each fluid element as to its position at a given time and to investigate the changes in the state of motion and of the properties of such tagged fluid elements as they move along their tracks. The first approach gives the *Eulerian equations*, and the second approach the *Lagrangian equations* of motion. The basic hydrodynamic equations together with the proper dynamic and kinematic boundary conditions, and initial conditions that prescribe the state of flow at some instant of time, form the essential information for a possible solution to a problem in fluid motion. Sometimes, additional information has to be added in order to describe functional relationships between variables and the equation of state.

The Eulerian equations of motion in their original form

consider only two forces: the total pressure gradient force and an external force. If the only external force that needs to be considered is the gravitational force, and if the rectangular xyz-coordinate system is oriented such that the xy-plane coincides with a level surface, the x and y-components of the external force vanish. There is only a vertical component in the z-direction, and this component of the gravitational force per unit mass is equal to g, the acceleration of gravity. For this case, the Eulerian equation of motion becomes very simple. In a rectangular Cartesian coordinate system with the z-axis pointing down, in positive direction, Newton's basic relationship of mechanics, mass times acceleration equals the sum of forces, is expressed for the three coordinate directions in eq.9:

$$\frac{\mathrm{d}u}{\mathrm{d}t} = -\frac{1}{\rho}\frac{\partial p}{\partial x}, \quad \frac{\mathrm{d}v}{\mathrm{d}t} = -\frac{1}{\rho}\frac{\partial p}{\partial y}, \quad \frac{\mathrm{d}w}{\mathrm{d}t} = g - \frac{1}{\rho}\frac{\partial p}{\partial z} \quad (9)$$

The components of a velocity vector, c, in x, y, and z-directions, are u, v, and w, respectively. The total derivatives $\mathrm{d}u/\mathrm{d}t$, $\mathrm{d}v/\mathrm{d}t$, and $\mathrm{d}w/\mathrm{d}t$ are the components of acceleration of an individual fluid particle moving through an x, y, z-coordinate system fixed with respect to the earth's surface.

In Euler's representation it is necessary to find the acceleration of fluid elements with respect to a fixed point in the coordinate system. Also, the pressure gradient is measured at a fixed point, and this is shown in eq.9 by the differential symbol ∂ which, for example, in $\partial p/\partial x$ denotes the *local* change of pressure in the x-direction. Individual derivatives $(\mathrm{d}\cdots/\mathrm{d}t)$ of a property of a moving particle with respect to time are composed of local changes plus *advective* changes which describe the changing motion as it passes through the fixed point. The notation $\mathrm{d}u/\mathrm{d}t$ is used to denote the acceleration *following* the fluid motion in the x-direction. A similar meaning is connected with $\mathrm{d}v/\mathrm{d}t$ and $\mathrm{d}w/\mathrm{d}t$ for the components of motion in y and z-direction, respectively. Since the velocity components are not only functions of time but also functions of x, y, z, it is possible to write, for example, the total change of u:

$$\mathrm{d}u = \frac{\partial u}{\partial t}\mathrm{d}t + \frac{\partial u}{\partial x}\mathrm{d}x + \frac{\partial u}{\partial y}\mathrm{d}y + \frac{\partial u}{\partial z}\mathrm{d}z$$

and similarly, for dv and dw. Upon division by dt and consideration of dx/d$t = u$, dy/d$t = v$, dz/d$t = w$ it follows that:

$$\frac{du}{dt} = \frac{\partial u}{\partial t} + u\frac{\partial u}{\partial x} + v\frac{\partial u}{\partial y} + w\frac{\partial u}{\partial z}$$

$$\frac{dv}{dt} = \frac{\partial v}{\partial t} + u\frac{\partial v}{\partial x} + v\frac{\partial v}{\partial y} + w\frac{\partial v}{\partial z} \quad (10)$$

$$\frac{dw}{dt} = \frac{\partial w}{\partial t} + u\frac{\partial w}{\partial x} + v\frac{\partial w}{\partial y} + w\frac{\partial w}{\partial z}$$

The advective changes, shown by the last three terms on the right hand side of eq.10 are called the *field accelerations*.

A velocity field is called *stationary* when the local time changes vanish, that is, when:

$$\frac{\partial u}{\partial t} = 0, \quad \frac{\partial v}{\partial t} = 0, \quad \frac{\partial w}{\partial t} = 0$$

In a *non-accelerated* field of motion, the total derivatives vanish, du/d$t = 0$, dv/d$t = 0$, dw/d$t = 0$.

The Eulerian equation of motion as represented in component form in eq.9 with the extensions given in eq.10 have been used extensively for the study of many types of motion. However, one important property of a real fluid, molecular viscosity, has been disregarded. The motions described in eq.9 have been treated as perfect fluids. In many cases, the effects of *viscosity* and *friction* have to be added. Also, the x, y, z-coordinate system is considered to be rigidly connected with the earth and therefore rotates with the earth around her axis. It can be shown that if the equations of motion described in a coordinate system that is fixed to the earth shall also be valid in a system that rotates with the earth, certain *apparent accelerations* have to be added to eq.9. These accelerations are called *geostrophic accelerations* (PROUDMAN, 1953) or *Coriolis accelerations*. Although HADLEY in 1735 used the qualitative effects of geostrophic acceleration in meteorology, and MACLAURIN in 1740 in connection with ocean currents, the formula for the horizontal component of this apparent acceleration was obtained by LAPLACE (1775) in connection with tides. For some reason, the name of CORIOLIS (1835) became associated with this important effect, although the complete explanation was first given in 1837 by POISSON. Since the term Coriolis force (or acceleration if taken per unit mass) has become so familiar in the literature, it will also be used in this text. Quite often it is simply called the *deflecting force of the earth's rotation* for reasons that are explained in the following simple derivation of the horizontal component of the Coriolis force, according to RADAKOVIĆ (1914).

Coriolis forces

Consider in the Northern Hemisphere a point, P, at the earth's surface at a latitude ϕ, and a plane, AT, passing through P tangential to the earth's surface. This plane rotates with the earth around her axis, and its rotation can be thought of as being composed of two components, K_n and K_t, respectively. If $\omega = 7.29 \cdot 10^{-5}$/sec is the angular velocity of the earth's rotation, $K_n = \omega \sin\phi$ and $K_t = \omega \cos\phi$. This is shown in Fig.28A by elementary trigonometric relationships. The meaning of K_t is to keep the tangential plane, AT, tangential to the earth's surface which rotates around OP with the angular velocity, K_n.

Next, consider the tangential plane AT from above, looking down in the direction $P'P$, as shown in the plane view B of

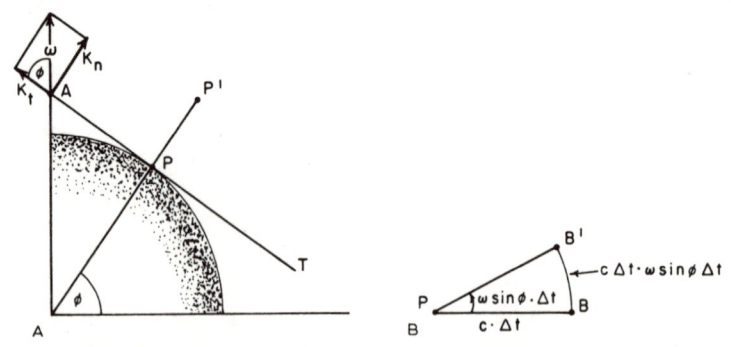

Fig.28. Explanation of the horizontal component of the Coriolis force.

Fig.28. At the time t, an observer at point B observes the motion of a body in the direction from P to B with the constant speed c and this direction will be preserved as long as the body moves unaffected by other forces. During the time interval Δt the body moves the distance $c \cdot \Delta t$ and arrives at point B. However, the observer has moved to B' during the same time interval, since the observer and his coordinate system are fixed to the rotating earth. From the observer's viewpoint, the moving body has experienced a deflection from its original direction with respect to the fixed coordinate system. This deflection is $c\Delta t \cdot \omega \sin\phi$ during the time interval Δt. In a non-rotating coordinate system, this would only be possible if a real force perpendicular to the original motion of the body would have acted. In the rotating system, the deflection is only the result of the earth's rotation, and in the Northern Hemisphere (as in this example) it appears as a deflection to the right when looking in the direction of the moving body. If the force that accelerates and moves the body is K, the distance of movement during the time interval Δt is $\frac{1}{2}(K/m)(\Delta t)^2$, where m is the mass of the body. Thus:

$$\tfrac{1}{2}(K/m)(\Delta t)^2 = c\Delta t \cdot \omega \sin\phi \, \Delta t$$

and the (Coriolis) acceleration in horizontal direction is $2\omega \sin\phi \cdot c$.

Similar simple considerations can be applied to motions in other than north–south directions, and it can be shown also that a vertical component of the Coriolis force exists for motions (or their components) in zonal direction, that is, from west to east or east to west. A simple, but complete derivation of the Coriolis force has been given by BJERKNES et al. (1933).

In a rectangular coordinate system with its x-axis positive to the east, its y-axis positive to the north, and its z-axis positive toward the center of the earth and normal to the xy-plane of a level surface, the three components of the Coriolis acceleration acting on a body moving in any direction with the velocity c (and the components u, v, w) are C_x, C_y, and C_z in x, y, and z directions, respectively:

$$\begin{aligned}C_x &= 2\omega \sin\phi \cdot v - 2\omega \cos\phi \cdot w \\ C_y &= -2\omega \sin\phi \cdot u \\ C_z &= -2\omega \cos\phi \cdot u\end{aligned} \qquad (11)$$

These components of an apparent acceleration have to be added to the Eulerian eq.9 if motions are to be described in a coordinate system that is fixed to the earth's surface. Also, in a real fluid, frictional forces must be added to the equations. In eq.12, they are represented in a formal way by their components F_x, F_y, and F_z per unit mass in x, y, and z directions, respectively. Frictional forces and their effects on ocean currents will be considered later. The equations of motion on a rotating earth in Eulerian representation are given in eq.12:

$$\frac{du}{dt} = -\frac{1}{\rho}\frac{\partial p}{\partial x} + 2\omega \sin\phi \cdot v + F_x$$

$$\frac{dv}{dt} = -\frac{1}{\rho}\frac{\partial p}{\partial y} - 2\omega \sin\phi \cdot u + F_y \qquad (12)$$

$$\frac{dw}{dt} = -\frac{1}{\rho}\frac{\partial p}{\partial z} + g - 2\omega \cos\phi \cdot u + F_z$$

The contribution of the second term in C_x, that is $2\omega \cos\phi \cdot w$ has been omitted in the first of eq.12, because this term is small in most ocean currents where the horizontal velocity components dominate.

The last of the three eq.12 shows that in the case of frictionless motion, non-accelerated currents ($dw/dt = 0$) are described by:

$$\frac{\partial p}{\partial z} = g\rho\left(1 - \frac{2u\omega}{g}\cos\phi\right) \qquad (13)$$

If the second term in parenthesis which has its origin in the Coriolis force is omitted, the hydrostatic equation follows. It can be shown that under ordinary conditions in the ocean the term $2u\omega \cos\phi/g$ in eq.13 can, indeed, be neglected. However, with fast currents near the equator where this term has a maximum, its effect may have to be considered. In the Equatorial Undercurrents of the Pacific and Atlantic Oceans, speeds of zonal currents flowing from west to east of 150 cm/sec can occur. The maximum effect of $2u\omega\cos\phi$ in eq.13 can be estimated either in terms of g or ρ. If the z-component of the Coriolis acceleration is considered to affect the density, an eastward flow of 150 cm/sec would lead to an apparent decrease of density as given by:

$$\frac{\partial p}{\partial z} = g(\rho - 2.3 \cdot 10^{-5})$$

A change of two units in the fifth decimal place of a density value can be significant. With stronger currents in zonal direction, the effect of the vertical component of the Coriolis force will be even more pronounced.

Frictional forces

In moving fluids where the velocity varies in space, frictional stresses are present as a result of momentum transfer between layers of different velocity. In the case of a *laminar flow* (lamina = sheet or layer) the exchange of momentum is the result of molecular motion. However, if the fluid is stirred by some internal or external cause and individual layers are "entangled" by macroscopic, irregular displacements of water parcels, the rates of momentum exchange (as well as of heat exchange, the exchange (diffusion) of dissolved solids, etc.) increase considerably. This is called *turbulent flow*.

Turbulence, stirring, mixing, and diffusion play a very important role in both oceanography and meteorology, but a satisfactory definition of turbulence has not yet been offered. STEWART (1959) has come close to defining turbulence in stating what it does; however, more is needed to understand its physical meaning.

Since many of the motions in the ocean are turbulent rather than laminar, turbulent or "eddy viscosity" effects play a more significant part in controlling stresses or friction between layers of a moving fluid than molecular friction. However, molecular effects are nevertheless most important. Without going into details, it can be stated that the macroscopic effects of eddy viscosity, eddy diffusion and eddy heat transfer would not exist, were it not for the more fundamental molecular effects. Without the molecular effects, small volumes of water with great differences in temperature and other properties could exist adjacent to each other, and the very small turbulent eddy motions in a fluid flow could not be dissipated into heat. The effects of turbulent stirring and mixing of different kinds of water could not produce a relatively homogeneous end product without the basic molecular exchange processes.

In classical hydrodynamics, the effects of molecular viscosity on fluid motion are considered in the *Navier–Stokes equations* for viscous flow of a compressible fluid on a non-rotating earth. Since they have limited practical application in oceanography, they will be omitted in this text and the interested reader can be referred to the derivation and discussion of the Navier–Stokes equations in the book on hydrodynamics by LAMB (1932).

In Newton's formulation, the frictional stress, τ, per unit area, between adjacent layers of water that move relative to each other, is proportional to the velocity shear. If, for example, the horizontal speed in the direction s is c_s, and this speed changes in a direction, n, which is perpendicular to c_s:

$$\tau_{sn} = \mu \; \partial c_s/\partial n \; (\text{dynes}/\text{cm}^2)$$

where μ is the dynamic viscosity coefficient of the fluid and $\partial c_s/\partial n$ is the velocity shear. The first subscript in the notation for the stress, s, denotes the direction of the stress acting on a plane which is shown by the second subscript, n, which gives the direction perpendicular to that plane. A stress component τ_{xz}, for example, denotes a stress acting in the x-direction on the xy-plane.

In oceanography, the molecular dynamic viscosity coefficient, μ, is often simply replaced by a much greater "eddy viscosity coefficient", A, in turbulent motion. Thus, in analogy to Newton's formula:

$$\tau_{sn} = A_{sn} \; \partial c_s/\partial n \qquad (14)$$

This analogy can be quite useful if the values of A are known. However, in a turbulent ocean current, the eddy viscosity coefficients, A, can vary over wide ranges, depending on the scale of the motion (vertical, horizontal, large- or small-scale motions), on the stratification of water masses, and other factors that are not yet clearly understood. Wind stirring in the surface layer of the sea is produced by wind and wave action. With increasing wind speed and more intense breaking of waves, eddy viscosity coefficients increase rapidly.

Turbulent shearing stresses were introduced into hydrodynamics by REYNOLDS (1894), and since then, they are often called *Reynolds stresses*. In a turbulent flow, variations of the instantaneous velocity, c, are observed, and the velocity at any point can be considered as the vector sum of two velocities $\bar{c}$ and c'. The velocity $\bar{c}$ represents an average velocity obtained over a certain time interval, and c' represents the instantaneous departure from the average velocity $\bar{c}$. Thus, the three observed velocity components can be written:

$$u = \bar{u} + u', \quad v = \bar{v} + v', \quad w = \bar{w} + w' \tag{15}$$

The turbulent random fluctuations around the average state are such that (by definition) $\bar{u}' = 0, \bar{v}' = 0, \bar{w}' = 0$. The "bar" over the turbulent velocity component indicates a time average over a sufficiently long time interval. If the actual velocity components (eq.15) are introduced into the equations of motion, for example into eq.9, and continuity requirements (continuity equation) for incompressible flow are properly considered, it can be shown that the equation of motion (eq.9) applied to the *average* flow is valid only if certain terms are added. These additional terms, like those shown in eq.16, result from time averages taken over certain products of the components u', v' and w' of turbulent velocity fluctuations. In the case of an average flow in the x-direction, $\bar{u}$, which is superimposed by u', v', w' fluctuations in x, y and z-directions, respectively, the first of eq.9 gives:

$$\frac{d\bar{u}}{dt} = -\frac{1}{\rho}\frac{\partial p}{\partial x} + \frac{1}{\rho}\left(\frac{\partial}{\partial x}\tau_{xx} + \frac{\partial}{\partial y}\tau_{xy} + \frac{\partial}{\partial z}\tau_{xz}\right) \tag{16}$$

where:

$$\tau_{xx} = \rho\,\overline{u'u'}, \quad \tau_{xy} = \rho\,\overline{u'v'}, \quad \tau_{xz} = \rho\,\overline{u'w'} \tag{17}$$

For arriving at these equations it is also necessary to assume that turbulent fluctuations of both pressure and density, can be neglected.

The terms in eq.17 represent shearing stresses with the dimensions of a force per unit area, and the derivatives of these stresses as they are shown in parentheses on the right side of eq.16 are frictional forces per unit volume, resulting from turbulent velocity fluctuations.

For turbulent flow in the y and z-directions, where the average components are $\bar{v}$ and $\bar{w}$, respectively, similar equations are obtained, and it is seen that in x, y and z directions nine possible stresses result. They can be arranged as follows:

$$\begin{array}{ccc} \tau_{xx} & \tau_{xy} & \tau_{xz} \\ \tau_{yx} & \tau_{yy} & \tau_{yz} \\ \tau_{zx} & \tau_{zy} & \tau_{zz} \end{array} \quad (18)$$

The array as shown in array 18 is called the stress tensor. Fig.29 represents the arrangement of the nine components of the stress tensor acting on three sides of a cube. The stress tensor is symmetrical about a diagonal from the upper left to the lower right in the array 18, and:

$$\tau_{yx} = \tau_{xy}, \quad \tau_{zx} = \tau_{xz}, \quad \tau_{zy} = \tau_{yz}$$

The Reynolds stresses contain fluctuations around the average flow, and in order to determine their numerical values it is

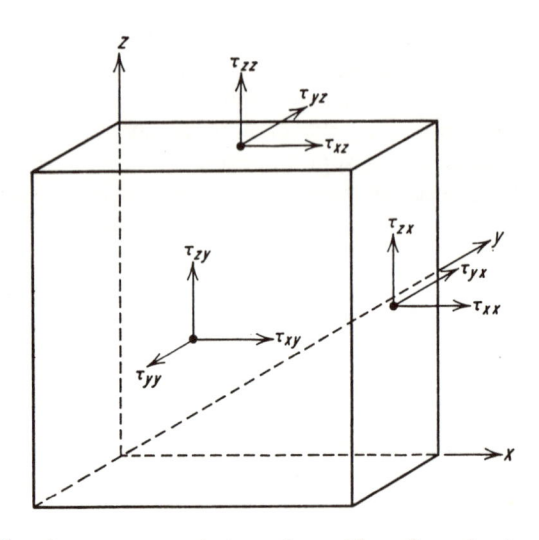

Fig.29. Shearing stresses acting on three sides of a cube in an x, y, z coordinate system.

necessary to measure these fluctuations very accurately and to obtain a representative average of their products. For certain technical and instrumental reasons, it is difficult to obtain this information. For example, instrumentation with sufficiently rapid response has to be developed that is capable of recording adequately the needed turbulent fluctuations, and such recordings have to be free of unwanted stray motions of the sensing element and its suspension (see Chapter I). A start in this direction has been made by BOWDEN (1962) in a tidal current near the sea bottom.

There is, however, the possibility of expressing these stresses by "macroscopic" quantities, although the determination or measurement of representative macroscopic quantities involves other kinds of difficulties and is not always satisfactory. By analogy to the mean free path in molecular motion, PRANDTL (1925) introduced the "mixing length" concept for turbulent motion. After Reynolds' first step, the theory of turbulent fluctuations and their effect on the dynamics of fluids and gases has been further developed, especially for practical applications, by RICHARDSON (1926), VON KÁRMÁN (1930), TAYLOR (1931), and PRANDTL (1925, 1942).

The "mixing length" as defined by PRANDTL (1925) does not have the precise physical meaning as the "mean free path" in molecular motion. Nonetheless, its concept provided the means for a successful hypothetical approach to a definition of eddy viscosity coefficients and eddy viscosity stresses as used in eq.14.

The definition of the mixing length is based on the following hypothetical model: Consider an average flow, $\bar{u}$, in the x-direction only. Superimposed on $\bar{u}$ are turbulent fluctuations, u', v', w' such that:

$$u = \bar{u} + u', \ v = v', \ w = w'$$

If a fluid parcel leaves its original position x, y, z where the average velocity is $\bar{u}(x, y, z)$, it arrives after traveling the distance $l(x, y, z)$ at the place $x + l_x, y + l_y, z + l_z$, where l_x, l_y, l_z are the components of the mixing length, l, in the $x, y,$ and z directions, respectively. At the end of this displacement, the fluid parcel has come into surroundings where the average velocity is:

$$\bar{u}(x + l_x, y + l_y, z + l_z) = \bar{u}(x, y, z) + l_x \frac{\partial \bar{u}}{\partial x} + l_y \frac{\partial \bar{u}}{\partial y} + l_z \frac{\partial \bar{u}}{\partial z}$$

approximately.[1] The velocity difference between the fluid parcel and its new surroundings is, therefore, $-l_x \, \partial \bar{u}/\partial x \; -l_y \, \partial \bar{u}/\partial y \; -l_z \, \partial \bar{u}/\partial z$. Prandtl has shown that the magnitude of these terms is approximately the same as the magnitude of the turbulent velocity components u'. It is also seen that a turbulent u'-fluctuation can be caused in three different ways, that is by original "pushes" in either the $x, y,$ or z-direction. If the manner in which the turbulent u' component is caused is indicated by a subscript in parentheses, $u'_{(x)}$, $u'_{(y)}$, $u'_{(z)}$, the following proportionalities exist:

$$u'_{(x)} \approx - l_x \frac{\partial \bar{u}}{\partial x}$$

$$u'_{(y)} \approx - l_y \frac{\partial \bar{u}}{\partial y} \qquad (19)$$

$$u'_{(z)} \approx - l_z \frac{\partial \bar{u}}{\partial z}$$

With this, the stress components in eq.17 can now be written as:

$$\tau_{xx} = \rho \overline{(u'_{(x)} u')} = - \rho \, \overline{l_x u'} \, \frac{\partial \bar{u}}{\partial x}$$

$$\tau_{xy} = \rho \overline{(u'_{(y)} v')} = - \rho \, \overline{l_y v'} \, \frac{\partial \bar{u}}{\partial y} \qquad (20)$$

$$\tau_{xz} = \rho \overline{(u'_{(z)} w')} = - \rho \, \overline{l_z w'} \, \frac{\partial \bar{u}}{\partial z}$$

According to Prandtl, the remaining turbulent velocity components in eq.20 can also be expressed by products of the gradients of the average flow and the mixing length components. For example, the three stress components in eq.20 become:

---

[1] This is obtained from a Taylor series expansion of $\bar{u}(x + l_x, y + l_y, z + l_z)$ and disregard of higher order terms.

$$\tau_{xx} = \rho l_x^2 \left|\frac{\partial \bar{u}}{\partial x}\right| \frac{\partial \bar{u}}{\partial x} = A_{xx} \frac{\partial \bar{u}}{\partial x}$$

$$\tau_{xy} = \rho l_y^2 \left|\frac{\partial \bar{u}}{\partial y}\right| \frac{\partial \bar{u}}{\partial y} = A_{xy} \frac{\partial \bar{u}}{\partial y} \qquad (21)$$

$$\tau_{xz} = \rho l_z^2 \left|\frac{\partial \bar{u}}{\partial z}\right| \frac{\partial \bar{u}}{\partial z} = A_{xz} \frac{\partial \bar{u}}{\partial z}$$

Comparison with eq.14 shows that the products of density, the square of the mixing length component and the absolute value of the average velocity shear are comparable to eddy viscosity coefficients, A. Thus, Prandtl's mixing length hypothesis leads to a definition of eddy viscosity coefficients and demonstrates the analogy between molecular and turbulent friction. It has provided tools for a successful approach to many oceanographical and meteorological problems of turbulent currents in the ocean and in the atmosphere.

Frictional forces are derived from the stresses as shown in eq.16. This can be further illustrated by considering the stresses in Fig.30, acting on faces perpendicular to the z-axis due to an average flow, $\bar{u}$, in the x-direction. The stress at the bottom of the cube is τ_{xz} and the stress at the top of the cube is $\tau_{xz} + (\partial \tau_{xz}/\partial z)dz$. The total force on the cube is the difference:

$$\left(\tau_{xz} + \frac{\partial \tau_{xz}}{\partial z} dz\right) dx\, dy - \tau_{xz}\, dx\, dy = \frac{\partial \tau_{xz}}{\partial z} dx\, dy\, dz$$

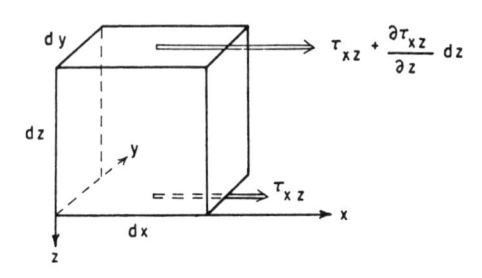

Fig.30. Frictional forces are derived from changes of shearing stresses. In this figure the z-derivative of τ_{xz} is considered.

Thus, per unit mass, the frictional force due to this one kind of stress is:

$$\frac{1}{\rho}\frac{\partial \tau_{xz}}{\partial z} = \frac{1}{\rho}\frac{\partial}{\partial z}\left(A_{xz}\frac{\partial \bar{u}}{\partial z}\right) \qquad (22)$$

In this way, the frictional forces which were symbolically indicated by F_x, F_y, and F_z in eq.12 can be written:

$$F_x = \frac{1}{\rho}\left[\frac{\partial}{\partial x}\left(A_{xx}\frac{\partial \bar{u}}{\partial x}\right) + \frac{\partial}{\partial y}\left(A_{xy}\frac{\partial \bar{u}}{\partial y}\right) + \frac{\partial}{\partial z}\left(A_{xz}\frac{\partial \bar{u}}{\partial z}\right)\right]$$

$$F_y = \frac{1}{\rho}\left[\frac{\partial}{\partial x}\left(A_{yx}\frac{\partial \bar{v}}{\partial x}\right) + \frac{\partial}{\partial y}\left(A_{yy}\frac{\partial \bar{v}}{\partial y}\right) + \frac{\partial}{\partial z}\left(A_{yz}\frac{\partial \bar{v}}{\partial z}\right)\right] \quad (23)$$

$$F_z = \frac{1}{\rho}\left[\frac{\partial}{\partial x}\left(A_{zx}\frac{\partial \bar{w}}{\partial x}\right) + \frac{\partial}{\partial y}\left(A_{zy}\frac{\partial \bar{w}}{\partial y}\right) + \frac{\partial}{\partial z}\left(A_{zz}\frac{\partial \bar{w}}{\partial z}\right)\right]$$

A frictional force acting in one direction can be caused in three different ways, as a result of the effect of the three degrees of freedom in which turbulent fluctuations may act when superimposed on the average flow components, $\bar{u}$, $\bar{v}$, and $\bar{w}$.

The expressions in eq.23 can be considerably simplified if the eddy viscosity coefficients can be taken as constants. Also, in ocean currents, it seems that only shearing stresses resulting from *horizontal* flow components are of importance. The vertical velocity as well as its derivatives are usually small enough to be neglected in the equations of motion. Also, the terms involving A_{xx}, A_{yy}, A_{zz} are often omitted. With these approximations or omissions, the components of the frictional forces in the x, y and z-directions are usually used in the form:

$$F_x = \frac{1}{\rho}\left[\frac{\partial}{\partial y}\left(A_h\frac{\partial \bar{u}}{\partial y}\right) + \frac{\partial}{\partial z}\left(A_z\frac{\partial \bar{u}}{\partial z}\right)\right]$$

$$F_y = \frac{1}{\rho}\left[\frac{\partial}{\partial x}\left(A_h\frac{\partial \bar{v}}{\partial x}\right) + \frac{\partial}{\partial z}\left(A_z\frac{\partial \bar{v}}{\partial z}\right)\right] \quad (24)$$

$F_z = 0$ (with slow vertical motion)

In eq.24, A_h and A_z are simplified expressions for lateral and vertical eddy viscosity coefficients, respectively.

In the past, most of the numerical values for lateral and vertical eddy viscosity coefficients have been determined from observed ocean currents, often under rigorous assumptions about the properties of a current field (THORADE, 1914a,b; DURST, 1924; SVERDRUP, 1926; FJELDSTAD, 1929, 1936; SUDA, 1936; DEFANT, 1936c). From estimates of an energy balance in fully developed wind generated waves NEUMANN (1952b) derived values for vertical eddy viscosity coefficients at different wind speeds that agree fairly well with the results obtained by THORADE (1914a, b) and DURST (1924) for the surface layer of the oceans.

The *Guldberg–Mohn assumption* about *virtual* internal friction was introduced for atmospheric motions as early as in 1876. In their time it was known that in the case of cyclonically curved isobars around a low pressure center the air actually flows into the region of lower pressure. Thus, an acceleration in the direction of the pressure gradient must be expected. The fact that in spite of this no *real* acceleration occurs indicates that in addition to the Coriolis force another "apparent" force is at work. This force can only be friction. GULDBERG and MOHN (1876) made the simple assumption that this frictional force is proportional to the current speed. If this speed is c, they assumed that the frictional force is $F = \rho\, r\, c$ where ρ is the density and $r(\text{sec}^{-1})$ denotes a proportionality factor, the *coefficient of virtual internal* friction. At the time, this assumption marked a great step forward, since it explained in a simple way the angle of departure between the geostrophic flow and the actually observed flow. The Guldberg–Mohn assumption will be used later in connection with some problems.

THE EQUATION OF CONTINUITY

Consider a small cubic volume of fluid whose sides are of length dx, dy, and dz. The mass of water that flows per unit time parallel to the x-axis through the area $dy\, dz$ at x is $\rho u\, dy\, dz$. On the opposite side of the same cube, at $x + dx$ the mass that flows out through the area $dy\, dz$ is $[\rho u + (\partial \rho u/\partial x)dx]\, dy\, dz$, if ρ as well as u are allowed to change in flow direction. The net flow in the x-direction per unit time and per unit volume is, therefore, $\partial \rho u/\partial x$. Similarly, through the dx dy and dx dz faces of this same volume the net flow is $\partial \rho w/\partial z$ and $\partial \rho v/\partial y$, in z and y directions, respectively. If mass is neither created nor destroyed and if the sum of these net fluxes is not zero, it must be balanced by a change of density within the volume. The change of mass per unit time and per unit volume is $-\partial \rho/\partial t$, and the result is:

$$\frac{\partial \rho u}{\partial x} + \frac{\partial \rho v}{\partial y} + \frac{\partial \rho w}{\partial z} = -\frac{\partial \rho}{\partial t} \qquad (25)$$

Eq.25 is called the *equation of continuity*. The quantities ρu, ρv, and ρw are the components of a vector i, which is called the *specific mass flux*, or the *current impulse*, since the product of mass times velocity (per unit volume) represents an impulse.

Application of the differential operator $(\partial/\partial x + \partial/\partial y + \partial/\partial z)$ to a vector (in this case $\rho c = i$) is called the *divergence* of the vector. In vector notation, the divergence is often symbolically abbreviated by div; thus eq.25 can be written in the form:

$$\text{div } i = -\partial \rho/\partial t \qquad (25a)$$

It should be noted that the divergence of a vector is a scalar quantity.

According to basic rules of differential calculus, eq.25 can also be written in the form:

$$\rho\left(\frac{\partial u}{\partial x} + \frac{\partial v}{\partial y} + \frac{\partial w}{\partial z}\right) + u\frac{\partial \rho}{\partial x} + v\frac{\partial \rho}{\partial y} + w\frac{\partial \rho}{\partial z} = -\frac{\partial \rho}{\partial t}$$

or:

$$\rho \text{ div } c = -\frac{\mathrm{d}\rho}{\mathrm{d}t} \qquad (26)$$

On the right hand side of eq.26 stands the total derivative of ρ with respect to time. This includes the local derivative plus the so-called advective terms. On the left hand side the div-operation is applied to the velocity vector c with the components u, v, and w.

From eq.25a it is seen that the divergence of the specific mass flux vanishes if there is no local change of density (stationary density field, $\partial \rho/\partial t = 0$). However, the divergence of a current field vanishes if the fluid is incompressible, and $d\rho/dt = 0$ (eq.26). For most problems in dynamic oceanography, the water can be considered as incompressible, and the *equation of continuity for incompressible fluids* reduces to the equation:

$$\frac{\partial u}{\partial x} + \frac{\partial v}{\partial y} + \frac{\partial w}{\partial z} = 0 \qquad (27)$$

Kinematic relationships in a current field

The basic concepts of *circulation* and *vorticity* (curl), and their relationship to each other are applied in this text. Therefore, an explanation of both quantities will be given by using elementary mathematics. Since ocean currents can often be treated as two-dimensional currents where either the horizontal or the vertical plane is selected for practical application, part of this section will deal, for simplicity, with two-dimensional currents only. The reader who is interested in a more comprehensive treatment of the subject is referred to LAMB (1932).

In a moving fluid, adjacent fluid elements can change their relative positions to each other. Consider a line of adjacent fluid particles as shown in Fig.31A at the time $t = 0$. This line of fluid elements is called a *fluid line*. Fluid elements on this line can be thought of as being divided into a number of equidistant points A, B, C, D, and so on. When the fluid is in motion, the relative positions of the points A, B, C, etc. may have changed after a time interval Δt, since each fluid element has its own velocity. With a very small distance, δs, between points or particles of the fluid, each line element δs can be approximated by a point drifting with the fluid motion. The velocity of this drift, c, has to be neither parallel (c_s) nor normal (c_n) to the fluid line; it can have any direction. However, if only the tangential component of the velocity, c_s, is considered, and each line element δ_s is multiplied by c_s, the *sum* of all products $c_s \cdot \delta s$, or the integral over these products taken around the closed curve of the fluid line is:

$$C = \oint c_s \cdot \delta s \qquad (28)$$

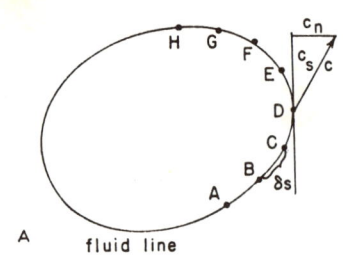

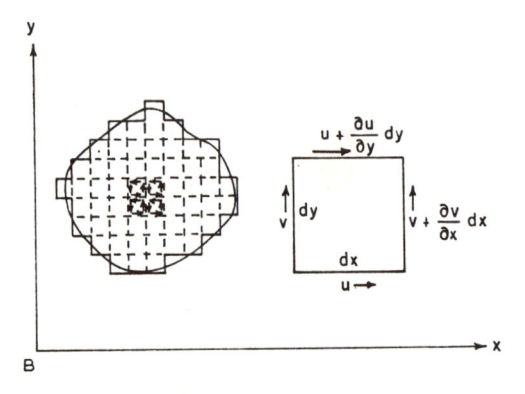

Fig.31A. A fluid line is composed of water particles, $A, B, C, \ldots H$. The velocity c at point D has the components c_s and c_n, tangential and normal to the fluid line, respectively.
B. The area enclosed by a horizontal fluid line is divided into infinitesimally small rectangles. One of these rectangles is enlarged, showing the velocity components along the four sides.

The circle in the integral sign indicates that the integration must be taken around the closed curve.

C as defined by eq.28 is called the *circulation* of that fluid curve. Circulation has the dimensions of cm²/sec and is a measure of the extent to which the fluid exhibits rotary motion. The positive sign of C is defined by fixing the path of integration around a closed curve in such a way that the sense of progression in carrying out the summation and the positive normal on the plane through the curve form a "right hand screw".

In a rectangular x, y, z-coordinate system where u, v, and w are the components of a velocity vector c, and dx, dy, dz are the components of δs, the circulation, C, is:

$$C = \oint (u\,dx + v\,dy + w\,dz) \tag{29}$$

The rate of change of the circulation with time is obtained from:

$$\frac{dC}{dt} = \oint \left(\frac{du}{dt} dx + \frac{dv}{dt} dy + \frac{dw}{dt} dz \right) +$$

$$+ \oint u\,\frac{d}{dt}(dx) + v\,\frac{d}{dt}(dy) + w\,\frac{d}{dt}(dz) \tag{30}$$

The second line integral in eq.30 vanishes, since it represents:

$$\oint (u\,du + v\,dv + w\,dw) = 0$$

The remaining part of eq.30:

$$\frac{dC}{dt} = \oint \left(\frac{du}{dt} dx + \frac{dv}{dt} dy + \frac{dw}{dt} dz \right) = \oint \frac{dc_s}{dt} \delta s \tag{31}$$

is called *circulation acceleration*. It states a relationship between the acceleration of the motion and the time change of the circulation. This relationship is known as Kelvin's theorem.

Next consider a closed curve in the horizontal xy-plane, and let the enclosed area be divided into small infinitesimal rectangles as shown in Fig.31B. The circulation around each elementary rectangle is obtained by a summation of the velocity components tangential to its sides. These velocity components are shown in the enlarged rectangle of Fig.31B. The contribution to circulation around this rectangle is:

$$dC_{dx\,dy} = u\,dx + \left(v + \frac{\partial v}{\partial x} dx \right) dy - \left(u + \frac{\partial u}{\partial y} dy \right) dx - v\,dy$$

or:

$$dC_{dx\,dy} = \left(\frac{\partial v}{\partial x} - \frac{\partial u}{\partial y} \right) dx\,dy \tag{32}$$

If the closed curve in Fig.31B is approximated by the broken, fully drawn line, it is seen that, with a summation carried out over the entire area enclosed by the broken line, contributions from sides common to adjacent rectangles must cancel, and the summation will contain only the contributions along the outer sides of the broken line. If this summation is performed by an integration over the xy-plane enclosed by the curve, s, it follows that:

$$C_{xy} = \int\int \left(\frac{\partial v}{\partial x} - \frac{\partial u}{\partial y}\right) dx\, dy \qquad (33)$$

Thus, for a two-dimensional circulation in the xy-plane ($w = 0$), eq.29 and eq.33 show that:

$$\oint (u\, dx + v\, dy) = \int\int \left(\frac{\partial v}{\partial x} - \frac{\partial u}{\partial y}\right) dx\, dy \qquad (34)$$

This equation represents a special two-dimensional case of Stokes' theorem, which states a general relationship between a line integral and a surface integral. The expression:

$$\frac{\partial v}{\partial x} - \frac{\partial u}{\partial y} = \mathrm{curl}_z\, c \qquad (35)$$

is called the curl or the vorticity of the current vector, c, in the xy-plane (z component of curl c), and eq.34 shows the connection between circulation and vorticity. In a similar way it can be shown that the other two components of the vorticity vector of a current field are:

$$\frac{\partial u}{\partial z} - \frac{\partial w}{\partial x} = \mathrm{curl}_y\, c, \quad \frac{\partial w}{\partial y} - \frac{\partial v}{\partial z} = \mathrm{curl}_x\, c$$

in the xz-plane and yz-plane, respectively.

If the divergence of a horizontal current field in the xy-plane vanishes:

$$\frac{\partial u}{\partial x} + \frac{\partial v}{\partial y} = 0 \qquad (36)$$

and if the vorticity of a horizontal current field in the xy-plane vanishes, $\mathrm{curl}_z\, c$ in eq.35 equals zero. In both cases, the velocity components, u and v, are not independent of each other.

For the case of a non-divergent (incompressible) horizontal current (eq.36), u and v must be related to a function ψ such that:

$$u = \frac{\partial \psi}{\partial y}, \quad v = -\frac{\partial \psi}{\partial x} \qquad (37)$$

The function ψ which satisfies eq.36 is called the *stream function* and in the horizontal xy-plane, the lines of $\psi(x,y) = $ constant are obtained from:

$$d\psi = \frac{\partial \psi}{\partial x} dx + \frac{\partial \psi}{\partial y} dy$$

The inclination of the lines $\psi = $ constant against the x-axis is:

$$\left(\frac{dy}{dx}\right)_\psi = \frac{v}{u} = \tan \alpha \qquad (38)$$

α is the angle between the x-axis and the current vector at each point of the current field. The lines of $\psi = $ constant are called *streamlines*.

If the vorticity of the current field vanishes, the field is called *irrotational*, vorticity-free or curl-free. For this case, in the horizontal xy-plane, eq.35 states that:

$$\frac{\partial v}{\partial x} - \frac{\partial u}{\partial y} = 0 \qquad (39)$$

and the velocity components must be related to another function ϕ, such that:

$$u = \frac{\partial \phi}{\partial x}, \quad v = \frac{\partial \phi}{\partial y} \qquad (40)$$

This function, ϕ, satisfies eq.39 and is called the *velocity potential*.

Both cases are of special importance in the dynamic analysis of ocean currents. In non-divergent currents, a stream function, and in irrotational currents, a velocity potential can be introduced. If this is possible, the mathematical analysis can often be greatly simplified. Actual ocean currents are, in general, neither non-divergent nor irrotational; however, in some problems, approximations to such flow conditions can be made, depending on the character of the currents and the problem to be solved. If friction is included and significant, irrotational motion can hardly be assumed. Also, it can be shown that a straight horizontal motion can be rotational and that certain circular motions are irrotational. Therefore, it is always necessary to study currents carefully before making any assumptions about rotational or irrotational motions.

Kinematic boundary conditions take the place of the equation of continuity along the boundaries of a moving fluid. A particle in the boundary, for example in the sea surface, must move perpendicular to the boundary at the same velocity as the boundary itself. If the boundary is stationary, continuity requires that the velocity component normal to the boundary be zero. The velocity component parallel to the boundary does not have to be necessarily zero under these conditions, but the boundary must be a streamline.

Consider a moving water layer between the sea surface $\zeta(x,y)$, and the bottom, $h(x,y)$, and assume that both boundaries are not level surfaces but slope in x and y directions. For a stationary sea surface, $\partial \zeta/\partial t = 0$, and for the rigid bottom, $\partial h/\partial t = 0$. The vertical velocity component at the sea surface is $w_\zeta = d\zeta/dt$ and at the bottom $w_h = dh/dt$. Since:

$$\frac{d\zeta}{dt} = \frac{\partial \zeta}{\partial t} + u_\zeta \frac{\partial \zeta}{\partial x} + v_\zeta \frac{\partial \zeta}{\partial y}$$

$$\frac{dh}{dt} = \frac{\partial h}{\partial t} + u_h \frac{\partial h}{\partial x} + v_h \frac{\partial h}{\partial y}$$

(41)

it follows for the case of stationary upper and lower boundaries that:

$$u_\zeta \frac{\partial \zeta}{\partial x} + v_\zeta \frac{\partial \zeta}{\partial y} - w_\zeta = 0$$

$$u_h \frac{\partial h}{\partial x} + v_h \frac{\partial h}{\partial y} - w_h = 0$$

(42)

With a stationary mass distribution in the moving layer $\partial \rho/\partial t = 0$, and from eq.25:

$$\frac{\partial \rho u}{\partial x} + \frac{\partial \rho v}{\partial y} + \frac{\partial \rho w}{\partial z} = 0 \qquad (43)$$

EQUATION OF CONTINUITY

Multiplication of eq.43 by dz and subsequent integration over z between the sea surface $\zeta(x,y)$, and the bottom, $h(x,y)$, yields:

$$\frac{\partial S_x}{\partial x} + \frac{\partial S_y}{\partial y} + (\rho u)_\zeta \frac{\partial \zeta}{\partial x} - (\rho u)_h \frac{\partial h}{\partial x} + (\rho v)_\zeta \frac{\partial \zeta}{\partial y}$$

$$- (\rho v)_h \frac{\partial h}{\partial y} - (\rho w)_\zeta + (\rho w)_h = 0 \tag{44}$$

In eq.44, S_x and S_y represent the horizontal components of the total mass transport, $S = \sqrt{(S_x^2 + S_y^2)}$ between the sea surface and the bottom:

$$S_x = \int_\zeta^h (\rho u) dz, \quad S_y = \int_\zeta^h (\rho v) dz \tag{45}$$

The x- and y-components of the specific mass flux at the sea surface and at the bottom are $(\rho u)_\zeta$, $(\rho v)_\zeta$, and $(\rho u)_h$, $(\rho v)_h$, respectively. Thus, with consideration of eq.42 it follows that:

$$\frac{\partial S_x}{\partial x} + \frac{\partial S_y}{\partial y} = 0 \tag{46}$$

Eq.46 states that for stationary boundaries (and in a rectangular coordinate system) the total horizontal mass transport is non-divergent. This, however, does not necessarily mean that the flow is non-divergent in individual layers between the upper and lower boundary of the current field.

A simple vorticity equation

Consider a homogeneous frictionless layer of fluid with density ρ and thickness h in horizontal motion. From eq.12 it follows that:

$$\frac{du}{dt} - fv = -\frac{1}{\rho} \frac{\partial p}{\partial x}$$

$$\frac{dv}{dt} + fu = -\frac{1}{\rho} \frac{\partial p}{\partial y}$$

where $f = 2\omega \sin\phi$ is a function of y. Elimination of the pressure by cross differentiation yields:

$$\frac{d}{dt}(f+\xi) + (f+\xi)\left(\frac{\partial u}{\partial x} + \frac{\partial v}{\partial y}\right) = 0$$

where $\xi = \dfrac{\partial v}{\partial x} - \dfrac{\partial u}{\partial y}$.

The equation of continuity (eq.43) applied to the whole moving layer of thickness h yields:

$$\frac{1}{h}\frac{dh}{dt} = -\left(\frac{\partial u}{\partial x} + \frac{\partial v}{\partial y}\right)$$

and it follows that:

$$\frac{d}{dt}\left(\frac{f+\xi}{h}\right) = 0 \tag{47}$$

In eq.47, the sum $f + \xi$ is called the *absolute vorticity*, $\xi = \partial v/\partial x - \partial u/\partial y$ is the *relative vorticity*[1] and $(f + \xi)/h$ the *potential vorticity*. Eq. 47 states that with the neglect of frictional forces the potential vorticity of a water column within the homogeneous layer cannot change as it moves along the fluid trajectory. In other words, the potential vorticity $(f + \xi)/h$ remains constant.

Use of the equation of continuity for computing vertical motion

The equation of continuity can be applied to compute the field of vertical motion from the horizontal velocity field. The practical application, however, is limited to special conditions which depend mainly on the accuracy of the data obtained by direct measurements or otherwise.

To illustrate the difficulties that are involved in a direct application of the *continuity equation as a prediction equation*, the simple case of homogeneous water of constant depth, h, is considered. With the additional assumption that the horizontal velocity

---

[1] In most oceanographic literature, the symbol ζ is used to denote the relative vorticity. Here, the letter ξ is used to avoid confusion with ζ which, in this text, stands for the free sea surface.

components, u and v, are independent of depth z, it follows from eq.41 and 43 that:

$$(h - \zeta)\left(\frac{\partial u}{\partial x} + \frac{\partial v}{\partial y}\right) = \frac{\partial \zeta}{\partial t} \quad (48a)$$

In reasonably deep water, h is much greater than ζ, and $h - \zeta$ in eq.48a can be replaced by h without impairing the accuracy of the equation. DEFANT (1925) and HANSEN (1938) used the equation of continuity to compute vertical tidal motions from the observed field of horizontal tidal currents.

EXNER (1917) and DEFANT (1925, 1961) have used the divergence of a horizontal current field, div c_H, in a more convenient way expressed in the form:

$$\text{div } c_H = \frac{\partial u}{\partial x} + \frac{\partial v}{\partial y} = \frac{\partial c_H}{\partial l} + \frac{c_H}{\Delta n}\frac{\partial \Delta n}{\partial l} \quad (48)$$

where c_H is the horizontal velocity and ∂l is a length element in the direction of the flow. The distance between two adjacent streamlines is Δn.

Fig.32 represents two adjacent streamlines in the horizontal plane, and two isotachs, $|c_H|$ and $|c'_H|$. The two streamlines are diverging, and the streamline divergence (curve divergence) is given by the angle $\Delta \alpha = \partial \Delta n / \partial l$. If the streamlines are parallel to each other, Δn is constant in flow direction, and div $c_H = \partial c_H / \partial l$. The divergence is the consequence of a velocity change

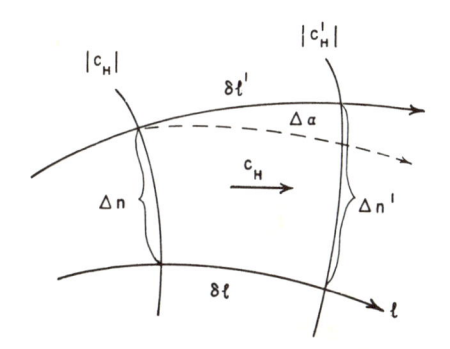

Fig.32. Divergence in a horizontal current field.

only. Increasing velocity in flow direction ($\partial c_H/\partial l > 0$) leads to $\partial \zeta/\partial t > 0$, and the sea surface must sink, since the z-axis points downward in positive direction. With a negative divergence of c_H for the case where the velocity c_H decreases in flow direction, convergence occurs and the sea surface must rise.

With no change of velocity along streamlines, the divergence depends on the variable distance between streamlines, or:

$$\text{div } c_H = (c_H/\Delta n)(\partial \Delta n/\partial l) = c_H \Delta\alpha/\Delta n$$

Thus, with a positive $\Delta\alpha$ in flow direction, divergence occurs, and with a negative $\Delta\alpha$, convergence occurs.

When combined, eq.48a with $h - \zeta \approx h$, and eq.48 show that vertical displacements of the sea surface resulting from changes in the divergence of a horizontal flow can be obtained from eq.49:

$$\frac{\partial \zeta}{\partial t} = h \left(\frac{\partial c_H}{\partial l} + \frac{c_H}{\Delta n} \frac{\partial \Delta n}{\partial l} \right) \tag{49}$$

Eq.49 is a "prognostic" equation, that permits the prediction of a variable quantity, in this case ζ, on the basis of synoptic observations of another variable, in this case c_H as represented by observations over a small area $\Delta n \cdot \partial l$. At first glance, the equation of continuity seems to offer perfect prediction possibilities. In application to meteorology, this possibility was discussed as early as 1904. M. Margules (1904, as quoted in EXNER, 1917, p.71) has shown that for meteorological purposes e.g., for the prediction of air pressure changes with time and vertical motions, the use of the equation of continuity is not practical, although it is theoretically justified. Practical application usually requires a much more accurate knowledge of meteorological factors than obtained from observations.

For oceanographic purposes, DEFANT (1925) has clearly shown the limitations of this method. Defant used eq.49 for computing the amplitudes, ζ_0, of the vertical tide in the North Sea from the observed tidal currents, c_H, at different points in the open sea with consideration of the boundary conditions along the shore lines.

If the water is shallow, the usual accuracy of current meter readings can be sufficient to permit a successful application of eq.49. However, as the water depth, h, increases, the accuracy of determining ζ_o (and of the tidal variations $\zeta(t) = \zeta_o \sin 2\pi t/T$) decreases. If $\zeta_o = 100$ cm along $\partial l = 50$ km and $h = 50$ m, it is necessary to measure accurately the current velocity change $\Delta c_H = 14$ cm/sec along parallel streamlines between points 50 km apart. Similarly, in the case of diverging or converging streamlines, it is necessary to measure $\Delta \alpha/\Delta n$ with the accuracy of about 16° per 50 km distance between observation points. As the distance between observation points and the water depth increases, the method becomes rapidly more and more inaccurate.

Continuity of salt—the Knudsen relationships

The equation of continuity can be applied not only to the field of mass but to any other conservative constituent or property of the moving fluid. Conservative constituents are physical, chemical or biological property concentrations of sea water that are altered locally as a result of the transport by currents (advection) and mixing (diffusion) only, except at the boundaries. If internal processes act to create or destroy the amount of concentration of a constituent or property of sea water, the constituent is non-conservative.

The total salt content in a volume of sea water which is given with sufficient accuracy by the salinity can usually be considered a conservative property if salt is neither added nor lost. The heat content can often be considered a conservative property. However, the oxygen content in sea water is in most cases a non-conservative property, because biological activity and chemical reaction within the water can alter the oxygen content. Photosynthetic processes by plants, especially phytoplankton, produce and tend to increase the oxygen content, whereas respiration by animals and oxidation of decomposing organic matter consume oxygen. Thus, there are "sources" and "sinks" to be considered in addition to advection and diffusion when dealing with non-conservative properties.

The change of concentration of a property, s, per unit volume of water with time along the path taken by the water volume is:

$$\frac{d(\rho s)}{dt} = \frac{\partial(\rho s)}{\partial t} + u\frac{\partial(\rho s)}{\partial x} + v\frac{\partial(\rho s)}{\partial y} + w\frac{\partial(\rho s)}{\partial z} \qquad (50)$$

The property s in eq.50 can represent conservative as well as non-conservative constituents of sea water. If s, for example, represents the salinity given as s grams of total salt per kg of sea water, the concentration per unit volume is ρs. With consideration of eq.26:

$$\frac{d(\rho s)}{dt} = \rho \frac{ds}{dt} + s \frac{d\rho}{dt} = \rho \frac{ds}{dt} - s\rho \, \text{div} \, c \tag{51}$$

Hence, it follows from eq.50 and 51 that:

$$\rho \frac{ds}{dt} = \frac{\partial \rho s}{\partial t} + \frac{\partial(\rho us)}{\partial x} + \frac{\partial(\rho vs)}{\partial y} + \frac{\partial(\rho ws)}{\partial z} \tag{52}$$

For conservative properties, like the salinity, in the absence of diffusion or mixing processes, $ds/dt = 0$. With stationary conditions $\partial \rho s/\partial t = 0$. Thus, for conservative and stationary conditions, eq.52 reduces to:

$$\frac{\partial(\rho us)}{\partial x} + \frac{\partial(\rho vs)}{\partial y} + \frac{\partial(\rho ws)}{\partial z} = 0 \tag{53}$$

This equation states that the vector of the salinity flux is non-divergent. Mixing and diffusion processes in connection with continuity considerations of this kind will be considered in a following section.

Special application of the continuity equation for mass and salt content forms the basis of Knudsen's Hydrographical Theorem. Simple relationships between the water and salt budget in limited oceanic regions often yield valuable information on mean flow conditions across the open boundaries. Knudsen's relationships are of particular value when applied to adjacent seas, bays, straits, river estuaries and other semienclosed water bodies.

Fig.33 shows a longitudinal section through a channel where the upper layer of lighter water flows to the right and the lower layer of denser water flows to the left. The transition between the two layers is indicated by a broken line. Consider a particular portion of the channel bounded by two vertical cross sections located at *1* and *2*. The average velocities and salinities at section *1* are u_1, and s_1, in the upper layer and u_1', s_1' in the lower layer, respectively. At section *2*, the corresponding average values of

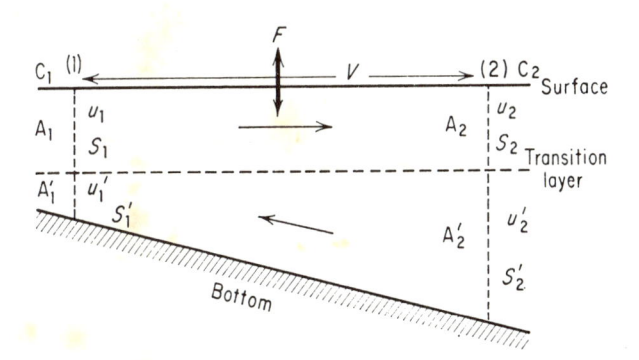

Fig.33. Longitudinal section through a two-layer channel where u denotes velocities, S salinities, and A cross section areas. Indices *1* and *2* apply to sections at *1* or *2*, respectively. Primed values refer to the lower layer, unprimed values to the upper layer. F is the net water exchange through the sea surface per unit area, and V is the surface area of the channel between *1* and *2*.

velocity and salinity are u_2, s_2, and u_2', s_2', respectively. F is the water gain or loss through the sea surface per unit area; for example, the difference of evaporation as against precipitation. If the surface area of the channel between the two cross sections is V, the product FV represents the total water exchange across the sea surface. This term can also be regarded as some kind of inflow or outflow through lateral boundaries in addition to the flow through sections *1* and *2*. For example, it can represent the inflow of river water.

Let A_1, A_1', and A_2, A_2' denote the cross section areas of the upper and lower layers in both sections, respectively. Continuity of water volume flux through the boundaries of the water body between sections *1* and *2* for stationary conditions requires that:

$$A_1 u_1 - A_1' u_1' + FV = A_2 u_2 - A_2' u_2'$$

and continuity of salt flux requires that:

$$A_1 s_1 u_1 - A_1' s_1' u_1' - A_2 s_2 u_2 + A_2' s_2' u_2' = 0$$

If the net rate of transport of mass of salt across each section is zero, then:

$$A_1 s_1 u_1 = A_1' s_1' u_1' \text{ and } A_2 s_2 u_2 = A_2' s_2' u_2'$$

provided that the water between sections *1* and *2* maintains its salinity in both layers.

From these conditions, the Knudsen relationships are derived as shown in eq.54:

$$A_2 u_2 \left(1 - \frac{s_2}{s_2'}\right) = A_1 u_1 \left(1 - \frac{s_1}{s_1'}\right) + FV$$

$$A_2' u_2' \left(\frac{s_2'}{s_2} - 1\right) = A_1' u_1' \left(\frac{s_1'}{s_1} - 1\right) + FV \tag{54}$$

KNUDSEN (1899, 1900) applied his equations to mean conditions in the Baltic Sea and to the waters around the Danish Islands. GEHRKE (1909) used similar relationships for an investigation of the mean current off the north of Scotland, where the North Atlantic Current flows towards the northeast.

More recently, simple continuity considerations of this kind were applied to part of the Irish Sea (PROUDMAN, 1953) with only one layer involved. The Irish Sea was bounded by two sections: section *1* connected Carnsore Point and St. David's Head, and section *2* was drawn from Kingstown to Holyhead. The average salinity in sections *1* and *2* was determined from observations; $s_1 = 34.83\%_0$, and in section 2: $s_2 = 34.33\%_0$. The mean velocities through the cross section areas, A_1 and A_2, of these sections are u_1 and u_2 respectively. Since $A_1 s_1 u_1 = A_2 s_2 u_2$ for the continuity of mass of salt, it follows that $A_1 u_1 / A_2 u_2 = s_2 / s_1 = 0.9856$. With stationary conditions, also $A_1 u_1 - A_2 u_2 + FV = 0$, and therefore $A_2 u_2 = FV/(1 - 0.9856)$. When allowance is made for the influx of fresh water from land, the effective F is, according to BOWDEN (1950), $F = 61$ cm/year and $FV = 11.38$ km³/year. The area, A_2 of cross section *(2)* is $A_2 = 7.135$ km². Thus, it follows that:

$$u_2 = \frac{FV}{(1 - 0.9856) A_2} = 111 \text{ km/year} = 0.35 \text{ cm/sec}$$

Of course, this result implies that the rate of salt transport into the region is equal to that out of the region.

Another example is given for the mean water exchange across the Strait of Gibraltar. For the Mediterranean Sea as a whole, evaporation exceeds the gain of fresh water (precipitation plus fresh water influx from rivers). In the upper layer, A, of a cross section through the Strait of Gibraltar, Atlantic water of mean salinity $S = 36.25\%_0$ flows into the Mediterranean Sea. In the bottom layer across A', Mediterranean Sea water of mean salinity $S = 37.75\%_0$ flows into the Atlantic. SCHOTT (1915) estimated the rate of transport of water volume in the upper layer (into the Mediterranean Sea) as $1.75 \cdot 10^6$ m^3/sec. If the total amount of water in the Mediterranean Sea remains constant it follows that:

$$Au - A'u' + FV = 0$$

If the average salinity of Mediterranean water also remains constant:

$$A S u = A' S' u'$$

This condition gives $A'u' = Au(S/S') = 1.68 \cdot 10^6$ m^3/sec for the outflow from the Mediterranean Sea. The difference, $FV = A'u' - Au = (1.68 - 1.75) \cdot 10^6 = -7 \cdot 10^4$ m^3/sec represents the net rate of loss of fresh water from the Mediterranean Sea. If this loss of fresh water is distributed evenly over the surface area of the Mediterranean Sea ($2,507 \cdot 10^9$ cm^2), it means that a water layer of 88.1 cm is lost every year from the surface. With a stationary sea level, this amount of water is, of course, supplied by the surplus influx from the Atlantic. If the Strait of Gibraltar would be thought to be closed by a dam, the water level of the Mediterranean Sea would lower at a rapid rate. WÜST (1951) based his calculations of the net loss of fresh water in the Mediterranean area on completely different methods and arrived at a figure of 96.5 cm/year. Thus it is seen that in the relatively short period of about 50 years, the surface of the Mediterranean Sea if shut off from the Atlantic would be lowered by as much as 50 m, exposing large areas of the continental shelf to the use of habitable land.

TURBULENCE, MIXING AND DIFFUSION

In motionless water, differences in the concentration of a water property, like the salinity or the heat content are slowly diminished by the process of molecular diffusion. Turbulent motion in natural water bodies causes in a similar manner a transfer of property concentrations by turbulent mixing of water masses. In the case of turbulent momentum exchange and friction, the concept of eddy viscosity has been introduced for practical purposes. Although this concept is not always clearly defined and fully understood, it has served some practical demands. In the case of turbulent diffusion and mixing, a similar concept of *eddy diffusion* and *eddy heat conduction* has been the mainstay of most of the past investigations of the distribution of water masses and currents in the oceans that involve mixing. Many objections have been raised by oceanographers and physicists against a straightforward application of analogies between molecular motion and eddy motion. Modern theories of turbulent diffusion are, probably, on their way to develop a better understanding of what "mixing" actually means and why "eddy" terms have the specific values that they have. It is also unlikely that a better understanding of turbulent diffusion will come from the area of the classical approach, as it was used in oceanography. Despite this fact, it has to be stated that what is known at present about the oceans has been learned by application of classical methods and their quantitative procedures.

The classical theory of diffusion and mixing in oceanography dates back as early as 1909, although EKMAN (1905) was first in pioneering into this subject for turbulent momentum transfer. GEHRKE (1909) realized the importance of turbulent mixing of water masses. He found that the vertical transfer of heat was proportional to the product of the specific heat and the vertical temperature gradient. In the case of molecular heat conduction, a similar relationship exists, with the exception that the proportionality coefficient for heat exchange in turbulent water bodies is much larger. GEHRKE (1909) called this larger proportionality coefficient the *coefficient of turbulent mixing*. Because the transfer of conservative properties by turbulent eddies is connected with a random transport of mass, this transfer is often referred to in scientific literature as *turbulent mass exchange*. A more detailed study of turbulent mass exchange in the oceans followed quickly, and a profound basis for what is called today the *classical theory of turbulence* in the oceans was laid by JACOBSEN (1913, 1915, 1918), TAYLOR (1915, 1918, 1922), and SCHMIDT (1917, 1925). Also, in the case of turbulent diffusion, Prandtl's ideas about a mixing length were adopted for turbulent diffusion and mixing. As the mixing length depends on the rate at which water property concentrations are transferred between turbulent eddies, it can be expected that the exchange coefficients would vary from one property to another. This effect has, indeed, been noticed (JACOBSEN, 1913; TAYLOR, 1915; see also PRANDTL, 1949).

If the turbulent motion is isotropic, the exchange coefficient for a given property can be expected to be the same for the eddy transport in any direction. However, in most cases of eddy diffusion in the oceans, turbulent motion cannot be considered isotropic, as, for example, in the case of a stable stratification of water masses. If the available kinetic energy of turbulent motion is insufficient to overcome the stabilizing effects of gravity and buoyancy, vertical turbulent motion will be suppressed, and mixing in lateral direction, particularly along isopycnal surfaces will dominate the eddy diffusion, or "spreading", of a given property concentration. As the intensity of vertical mixing decreases with increasing stability of water mass stratification, it seems that lateral mixing increases. For the oceans, evidence for this case of non-isotropic mixing was provided by PARR (1936b). It appears that either vertical or lateral mixing dominates. In the wind-stirred surface layers the mixing process is dominated by vertical turbulent fluctuations, whereas in stably stratified subsurface layers vertical turbulent exchange can become insignificant as compared to lateral exchange. An investigation of the space distribution of vertical and lateral mixing coefficients in the Atlantic Antarctic Intermediate Water by KIRWAN (1965) has clearly demonstrated the fact that vertical eddy diffusivity coefficients show a minimum, and lateral diffusivity coefficients a maximum in layers of high stability.

The flux of a property concentration, s, is proportional to the gradient, $\partial s/\partial n$, of this concentration (s represents the mean value of concentration). If "exchange coefficients" are introduced for turbulent motion in analogy to molecular diffusion coefficients, the turbulent flux components in the x, y, and z-directions can formally be expressed in simplified form as:

$$F_{(x)} = -A_{(x)} \frac{\partial s}{\partial x}, F_{(y)} = -A_{(y)} \frac{\partial s}{\partial y}, F_{(z)} = -A_{(z)} \frac{\partial s}{\partial z} \quad (55)$$

The minus sign on the right of eq.55 indicates that the flux, F, is counted positive from areas of high concentration to areas of low concentration. The coefficients, $A_{(x)}$, $A_{(y)}$, $A_{(z)}$ are the exchange coefficients in x, y, and z-directions, respectively, for a given exchangeable property concentration, s.

When dealing with a specific property, it is often necessary to add other factors to eq.55, in order to make the equations dimensionally correct. For example, in the case where F represents the flux of heat amount, the right hand side of eq.55 has to be multiplied by the specific heat. Thus, the heat transport due to vertical turbulent mass exchange per unit area and per unit time is $F_{(z)}$ [heat] $= H_{(z)}$, and:

$$H_{(z)} = -c_p A_{(z)} \frac{\partial \theta}{\partial z} \text{ [cal cm}^{-2}\text{/sec]} \quad (56)$$

where c_p (cal. g^{-1}degree^{-1}) is the specific heat at constant pressure, θ is the potential temperature (degrees) and $A_{(z)}$ (cm^{-1} gsec^{-1}) represents the exchange coefficient for heat, or, the eddy heat conductivity coefficient.

The concentration distribution of a conservative property is determined by eddy diffusion and advection processes. The classical equation that governs turbulent diffusion can easily be obtained from the turbulent fluxes as given by eq.55. However, in addition to turbulent mixing, molecular diffusion becomes important in the presence of strong gradients in producing the end product of continued mixing, that is in producing a homogeneous distribution of the exchangeable property.

Consider a volume with the side elements dx, dy, dz. The turbulent transport of water per unit time through the dy dz plane in the x-direction at x is $F_{(x)}(x) = -A_{(x)} (\partial s/\partial x)$ per unit area. At the opposite dy dx plane, at $x + $ dx, the outflowing transport per unit time and unit area is $F_x(x + $ d$x) = F_{(x)}(x) +$

$(\partial F_{(x)}/\partial x)\,\mathrm{d}x$. The difference between the two fluxes per unit time and unit area is:

$$F_{(x)}(x) - \left[F_{(x)}(x) + \frac{\partial F_{(x)}}{\partial x}\,\mathrm{d}x\right] = -\frac{\partial F_{(x)}}{\partial x}\,\mathrm{d}x \quad (57)$$

Thus, with consideration of the first of eq.55:

$$\frac{\partial}{\partial x}\left(A_{(x)}\frac{\partial s}{\partial x}\right)\mathrm{d}x$$

represents the net transport along the distance dx per unit cross section area dy dz and per unit time. In the y and z-directions the corresponding net transports are given by $-(\partial F_{(y)}/\partial y)\mathrm{d}y$ and $-(\partial F_{(z)}/\partial z)\mathrm{d}z$, respectively. The total change per unit time and per unit volume (dx = dy = dz = 1) is:

$$\left(\frac{\partial s}{\partial t}\right)_{\text{diffusion}} = \frac{\partial}{\partial x}\left(\frac{A_{(x)}}{\rho}\frac{\partial s}{\partial x}\right) +$$
$$+ \frac{\partial}{\partial y}\left(\frac{A_{(y)}}{\rho}\frac{\partial s}{\partial y}\right) + \frac{\partial}{\partial z}\left(\frac{A_{(z)}}{\rho}\frac{\partial s}{\partial z}\right) \quad (58)$$

The coefficient of diffusion enters here in the *kinematic* form (that is A/ρ), because concentrations have been defined as amounts per unit volume.

If in addition to pure eddy diffusion an *average current* is present, a net change of concentration can also be the result of differences in *advection*. The concentration per unit volume carried through a unit area by a current component, $u(x)$, in the x-direction at x is $s(x)u(x)$ per unit time. At the distance $x + \mathrm{d}x$ the transport is $s(x + \mathrm{d}x) \cdot u(x + \mathrm{d}x)$, if both s and u vary along the x-direction. The net change of s, due to a current transport in the x-direction is:

$$(su)_x - (su)_{x + \mathrm{d}x} = -(\partial su/\partial x)\mathrm{d}x$$

Similarly, in the y- and z-directions, the changes of concentration due to advection are $-(\partial sv/\partial y)\mathrm{d}y$ and $-(\partial sw/\partial z)\mathrm{d}z$ respectively. Thus, the local net change of concentration due to eddy diffusion and advection per unit time and per unit volume is:

$$\frac{\partial s}{\partial t} = \frac{\partial}{\partial x}\left(\frac{A_{(x)}}{\rho}\frac{\partial s}{\partial x}\right) + \frac{\partial}{\partial y}\left(\frac{A_{(y)}}{\rho}\frac{\partial s}{\partial y}\right) + \frac{\partial}{\partial z}\left(\frac{A_{(z)}}{\rho}\frac{\partial s}{\partial z}\right) -$$

$$- \frac{\partial}{\partial x}(su) - \frac{\partial}{\partial y}(sv) - \frac{\partial}{\partial z}(sw) \qquad (59)$$

The last three terms of eq.59 can be written:

$$-u\frac{\partial s}{\partial x} - v\frac{\partial s}{\partial y} - w\frac{\partial s}{\partial z} - s\left(\frac{\partial u}{\partial x} + \frac{\partial v}{\partial y} + \frac{\partial w}{\partial z}\right)$$

For incompressible fluids the sum of the terms in parentheses is equal to zero. Since sea water can be considered as an incompressible fluid for the exchange of property concentrations, it follows from eq.59 and:

$$\frac{\mathrm{d}s}{\mathrm{d}t} = \frac{\partial s}{\partial t} + u\frac{\partial s}{\partial x} + v\frac{\partial s}{\partial y} + w\frac{\partial s}{\partial z}$$

that:

$$\frac{\mathrm{d}s}{\mathrm{d}t} = \frac{\partial}{\partial x}\left(\frac{A_{(x)}}{\rho}\frac{\partial s}{\partial x}\right) + \frac{\partial}{\partial y}\left(\frac{A_{(y)}}{\rho}\frac{\partial s}{\partial y}\right) + \frac{\partial}{\partial z}\left(\frac{A_{(z)}}{\rho}\frac{\partial s}{\partial z}\right) \qquad (60)$$

This equation relates time changes of property concentration to changes caused by eddy diffusion when following a water parcel with the current. However, *local* time changes of the concentration of a conservative water property, *s*, are governed by the combined effects of advection and diffusion.

In practice, eq.60 has been used in a greatly simplified manner by assuming that A_x, A_y and A_z are constant. Also, in most cases (DEFANT, 1929a; THORADE, 1931; SVERDRUP, 1940b, and others) a particular plane was selected for the major exchange process. Although this assumption can often be justified, the assumption that the effective exchange coefficients are constant in space is much more serious. For example, in the case of vertical turbulent transfers:

$$\frac{\partial}{\partial z}\left(\frac{A_{(z)}}{\rho}\frac{\partial s}{\partial z}\right) = \frac{A_{(z)}}{\rho}\frac{\partial^2 s}{\partial z^2} + \frac{\partial s}{\partial z}\frac{\partial (A_{(z)}/\rho)}{\partial z}$$

should be considered, because it is evident that in most cases of vertical eddy exchange $A_{(z)}/\rho$ is a function of depth. The same applies to the exchange coefficients in the x- and y-directions and their changes in lateral direction.

The problem of diffusion in the sea becomes more and more important in studying the effects of sewage pollutants in river estuaries and coastal regions, as well as the spreading of radio-active wastes in the open sea. Recent work by PRITCHARD and CARPENTER (1960) has stimulated new experimental approaches to the study of dispersion. They have shown that a certain dye (Rhodamine-B) can be detected in extremely minute concentrations by a fluorometer, and concentrations of this dye in sea water can be measured to considerable accuracy. A thorough discussion of recent results on dispersion in the ocean with the idea that the eddy diffusion can be a function of both space and time, was given by OKUBO (1962). CORRSIN (1959, 1962) has presented summaries of present theories of turbulent dispersion.

CHAPTER IV

MAJOR TYPES OF OCEAN CURRENTS

In the following sections special types of motion in the ocean will be considered. Depending on the character of forces that act upon a water body, and on the state of balance of these forces, the resulting motions can be different. If water movements in a lake or in a small part of the ocean are considered, it may be possible to neglect the effect of Coriolis forces in comparison with the other forces acting on the fluid; however, large scale motions on the earth are affected most significantly by geostrophic acceleration.

Frictional forces are present in any type of current, although in some currents their effect may be less significant and can be ignored when compared to other forces that act upon the motion.

If the balance of forces is such that at any fixed place in the moving water the velocity does not change, the current can be treated as a *stationary* current, and if the sum of all forces that act upon a volume of water is equal to zero, the current will be *non-accelerated*.

The most simple case of ocean currents is a non-accelerated, frictionless motion. This case will be considered first. Following this, some special cases of frictionless water movements, including the case of accelerated currents, will be chosen to explain certain typical features of large and small scale motions. Finally, the important effect of frictional forces will be introduced.

GEOSTROPHIC CURRENTS

When a frictionless current flows horizontally without change of velocity and the only external force is gravity, it follows from eq.12 that for each horizontal coordinate direction the components of the Coriolis force and the pressure gradient force balance each other. In the vertical direction the pressure gradient is balanced by the vertical component of the Coriolis force and the gravitational force:

$$2\omega \sin\phi \cdot \rho v = \frac{\partial p}{\partial x}$$

$$2\omega \sin\phi \cdot \rho u = -\frac{\partial p}{\partial y} \qquad (61)$$

$$2\omega \cos\phi \cdot \rho u = -\frac{\partial p}{\partial z} + g\rho$$

It was shown for eq.13 that the term $2\omega \cos\phi \cdot \rho u$ in most ocean currents can be neglected in the third of eq.61, and hydrostatic equilibrium is approximated by the hydrostatic equation $\partial p = g\rho \, \partial z$. The equation of continuity is satisfied for a stationary mass distribution ($\partial \rho / \partial t = 0$).

If the first and second of eq.61 are squared and added together, the result can be written:

where:
$$\frac{\partial p}{\partial n} = 2\omega \sin\phi \cdot \rho c \qquad (62)$$

$$c = (u^2 + v^2)^{1/2} \text{ and } \frac{\partial p}{\partial n} = [(\partial p/\partial x)^2 + (\partial p/\partial y)^2]^{1/2}$$

The equilibrium of forces as shown in eq.62 expresses the fact that the Coriolis force must be equal and exactly opposite to the horizontal pressure gradient force. This means that the horizontal current vector, c, must be parallel to the isobars, and in such a direction that in the Northern Hemisphere, the higher pressure is to the right when one faces in the direction of the current. In the Southern Hemisphere, where ϕ is counted as a negative quantity, the higher pressure is to the left when looking "downstream". This type of current is called a *geostrophic current*[1] and the equilibrium of forces expressed by eq.62 is called a *geostrophic equilibrium*.

Instead of using the horizontal pressure gradient along level

---

[1] PHILLIPS (1963) has presented a comprehensive monograph on geostrophic motion.

surfaces, the slope of isobaric surfaces can be introduced in eq.62. Fig.34 represents two isobars in a vertical plane, p and $p + \Delta p$, which are inclined against a level surface, n. The vertical nz-plane is perpendicular to the current velocity c. The pressure at point A on the level surface is p, and at point B it is $p + \Delta p = p + g\rho\Delta z$ where ρ is the density of the water column between points B and C. Thus $\Delta p/\Delta n = g\rho(\Delta z/\Delta n)$ and:

$$\frac{\partial p}{\partial n} = g\rho \tan\beta \tag{63}$$

A negative sign on the right side of eq.63 has to be used when the z-axis points downward in a positive direction. In this case, β is positive in the clockwise direction. If eq.63 is introduced into eq.62, it follows that:

$$\tan\beta = \frac{2\omega \sin\phi}{g} c \tag{64}$$

In the Southern Hemisphere, the slope reverses since the Coriolis term changes sign and the isobaric surfaces slope upward to the left when one faces in the direction of the current c.

The slope of isobaric surfaces including the sea surface as the uppermost isobaric surface with sea pressure zero is small when compared to the slope of isothermal, isohaline (surfaces of equal salinity) and isopycnal surfaces. Surfaces of equal temperature, salinity, and density can easily be plotted in sections showing the vertical and horizontal distribution of these factors as obtained by observations. Such vertical sections are shown in Fig.35 for

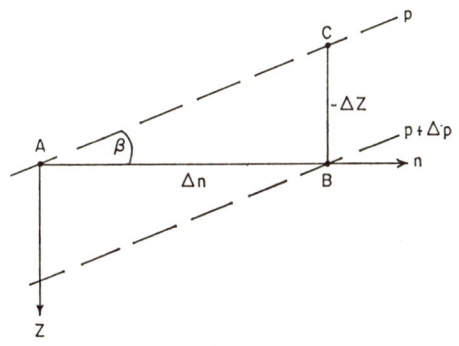

Fig.34. Two isobars, p and $p + \Delta p$, in the vertical nz-plane.

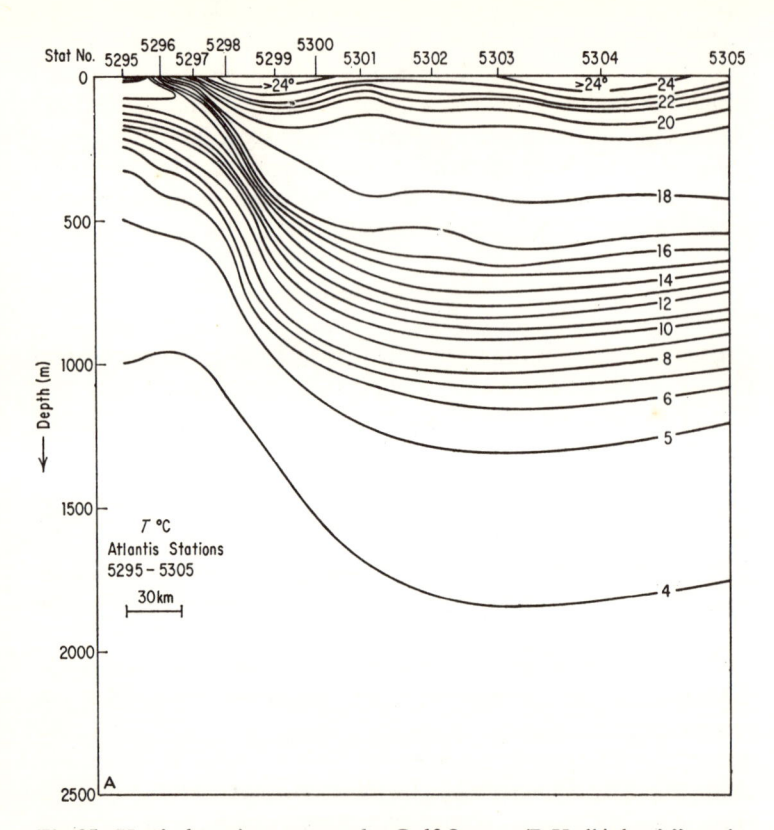

Fig.35. Vertical sections across the Gulf Stream (R.V. "Atlantis" station numbers 5295–5305) showing the distribution of temperature ($T°C$), salinity ($S‰$), σ_t, and anomaly of specific volume (δ).
(After NEUMANN and PIERSON, 1966.)

the observations obtained in June 1955 by the R.V. "Atlantis". The section was taken nearly perpendicularly through the Gulf Stream between 36°35'N, 74°20'W and 35°02'N, 71°20'W. The slopes of isotherms, isohalines, isopycnals, and of the anomalies of the specific volume are strongest in the region of swiftest currents across the section. The slope of the isopycnals, represented in Fig.35 by lines of equal σ_t, is about 700 m per 100 km horizontal distance in the region of the Gulf Stream. However, the slope of isobaric surfaces in the same region is much smaller.

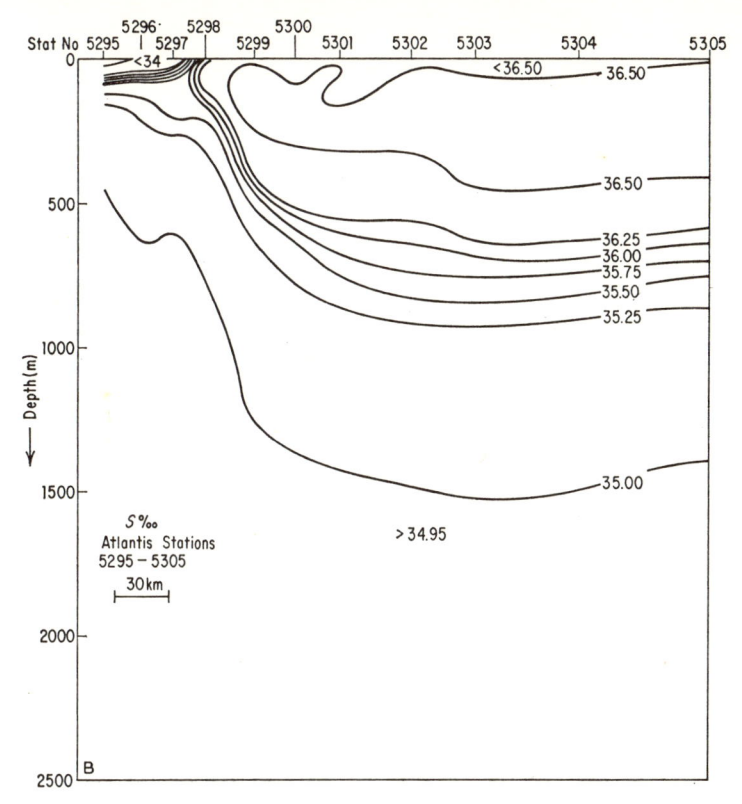

Fig.35B (legend see p.130).

If, for example, $c = 150$ cm/sec, $2\omega \sin\phi \approx 10^{-4}$ in middle latitudes, and $g = 980$ cm/sec², $\tan\beta \approx 1.5 \cdot 10^{-5}$ according to eq.64. This yields a slope for isobaric surfaces of only 1.5 m per 100 km horizontal distance.

In order to test eq.64, it is necessary to measure sea surface slopes and associated currents with a sufficient degree of accuracy. The accuracy of current measurements has been discussed in Chapter I. For the open ocean, there is, at present, no possibility of measuring large scale sea surface inclinations against a level surface. However, across straits, or channels, where the absolute geodetic "zero" point of level surfaces can be established by precise leveling, the "mean sea water level", or the height

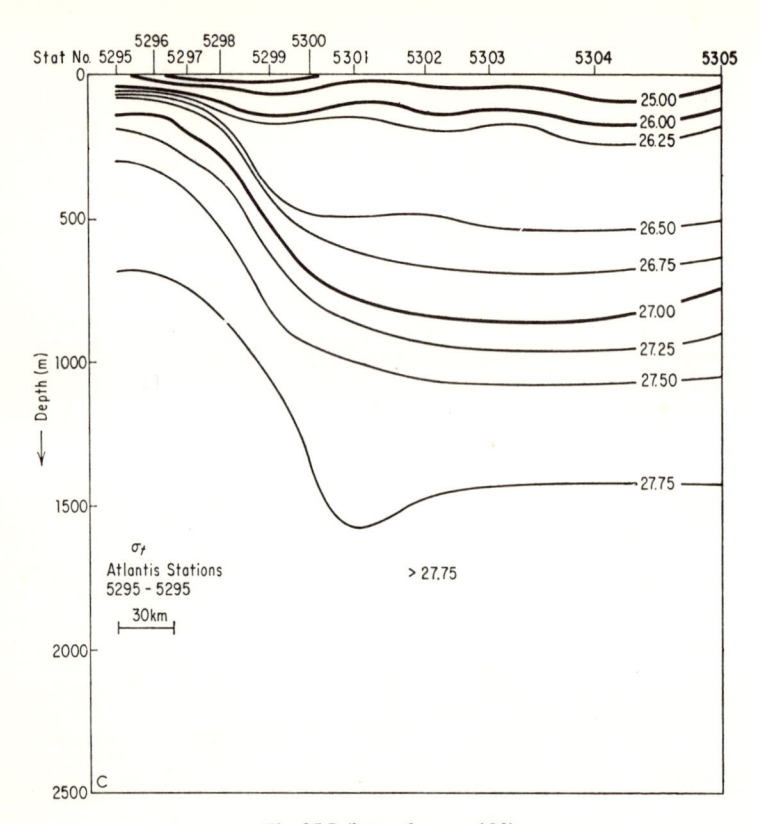

Fig.35C (legend see p.130).

of the actual sea surface above or below the zero point, can be found from tide gage observations. Such observations in conjunction with simultaneous current measurements in a strait or channel have been used to check on the validity of eq.64.

One of the more recent investigations of this kind are those by DIETRICH (1946) who used current observations made at the Danish light ship "Halskov-Rev", and tide gage recordings across the Great Belt between the Danish islands Seeland and Fünen (Korsör–Slipshavn). Fig.36 shows the results obtained by DIETRICH (1946). The dashed line B indicates a slope of the sea surface, $\tan\beta \approx \beta = 1.4 \cdot 10^{-7}c$. The same slope is indicated by observations, approximated by line A. The small

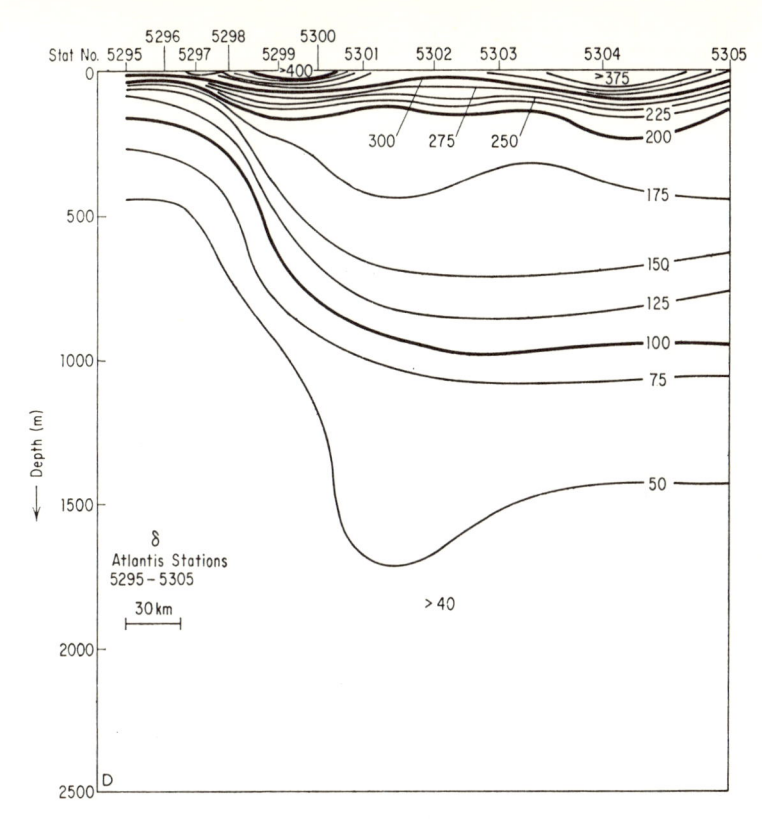

Fig.35D (legend see p.130).

difference between the two lines (1.5 cm) in Fig.36 is within the accuracy of precise leveling on which the "zero" points at both tide gage stations depend. If eq.64 is applied to compute the slope of the sea surface between Seeland and Fünen for the observed mean current speeds, Dietrich finds that across this strait $\beta/c = 1.22 \cdot 10^{-7}$. This result agrees remarkably well with the observations in Fig.36 and is a very good verification of eq.64. Particularly in shallow water and in a relatively narrow channel as in the Great Belt, one would expect a much greater discrepancy between eq.64 and observations of c and β, at least on account of frictional effects.

In another case, eq.64 was applied to the sea surface slope

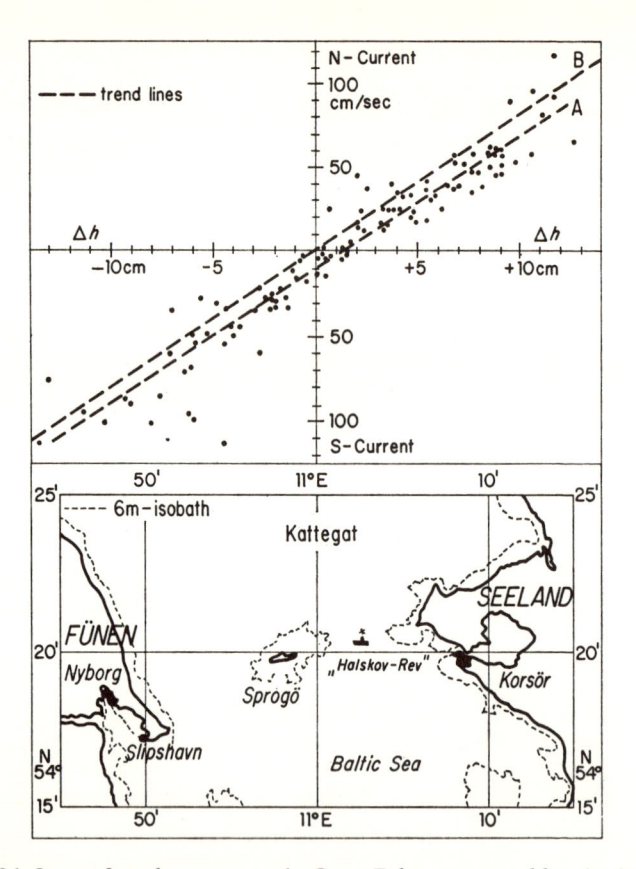

Fig.36. Sea surface slopes across the Great Belt as measured by the height difference, Δh, of sea level between Korsör and Slipshavn, and their relationship to the surface current speed as observed at the lightship "Halskov Rev". (After DIETRICH, 1946).

across the Strait of Dover after elimination of tidal currents (PROUDMAN, 1953). According to CARRUTHERS (1935), the mean speed of currents in a section across the strait is about 6.43 cm/sec, directed from the English Channel towards the North Sea. Although great variations from this mean value occur, it can be taken as a typical speed. In a latitude $\phi = 51°$ for a current speed $c = 6.43$ cm/sec and $g = 981$ cm/sec², $\tan\beta = 0.743 \cdot 10^{-6}$ according to eq.64. The width of the strait is about

35.2 km. Therefore, a mean rise of sea level from the English to the French coast of 2.62 cm can be expected. Continuous observations of the mean water level on both sides of the strait and simultaneous current observations in the strait have helped to determine the relationship between the zeros of the levelling systems in England and France.

Practical application of the geostrophic equilibrium

The classical dynamic method of computing ocean currents is a direct application of eq.62 and of the hydrostatic equation. If p is the pressure at depth z in the ocean, p_a the constant atmospheric pressure at the surface ζ, and $\rho(z)$ the sea water density as a function of depth:

$$p = p_a + g \int_\zeta^z \rho(z) \mathrm{d}z$$

The horizontal pressure gradient, $\partial p/\partial n$, at depth z is obtained from this equation and Leibnitz' rule for differentiating integrals:

$$\frac{\partial p}{\partial n} = g \int_\zeta^z \frac{\partial \rho}{\partial n} \mathrm{d}z - g\rho(\zeta) \frac{\partial \zeta}{\partial n} \qquad (65)$$

In a homogeneous ocean where ρ is constant, the first term on the right side of eq.65 equals zero, and the horizontal pressure gradient at depth z is the same as the gradient resulting from the slope, $\partial \zeta/\partial n$, of the sea surface. $\rho(\zeta)$ represents the water density at the surface. In a stratified ocean where $\partial \rho/\partial n \neq 0$, the horizontal pressure gradient has two components. Besides the contribution of a sloping free sea surface, the first term on the right of eq.65 adds a component due to horizontal density differences in the water. This term depends largely on the vertical coordinate and usually increases its value with depth. This means that with increasing depth the contribution of the sea surface slope to the total horizontal pressure gradient, $\partial p/\partial n$, can gradually be compensated by the integral term of eq.65. In this case, at a certain depth $z = H$, $(\partial p/\partial n)_{z=H} = 0$, and:

$$\int_\zeta^H \frac{\partial \rho}{\partial n} \mathrm{d}z - \rho(\zeta) \frac{\partial \zeta}{\partial n} = 0 \qquad (66)$$

According to eq.62, the absolute geostrophic current should also become zero at the depth $z = H$. This depth is often called the "zero surface" or "layer of no motion". It is evident that there may be none or even more than one layer of no motion in a stratified ocean.

Eq.66 requires that $\partial \rho/\partial n$ be a negative quantity, and the isopycnals $\rho =$ constant must slope in a direction that is opposite to the sea surface slope. This is represented in Fig.37. If this were not the case, and the isopycnals would slope in the same direction as the sea surface, the integral term in eq.65 would have the same sign as the term $\rho(\zeta)\ \partial \zeta/\partial n$ and the pressure gradient as well as the current speed must increase continuously toward the bottom. Experience has shown that this is generally not observed in steady state ocean currents. Normally, the current speed decreases with depth, although important exceptions are known. These exceptions, however, often refer to non-geostrophic currents where either acceleration or friction are significant terms to be considered.

The meaning of eq.66 is illustrated in Fig.37. In a vertical section between two oceanographic stations, A and B, the sea surface rises from A to B over the distance Δn along a level surface by the amount $\Delta \zeta$. The isopycnals, $\rho_1, \rho_2, \rho_3, \rho_4$ are given by temperature and salinity observations obtained at station A

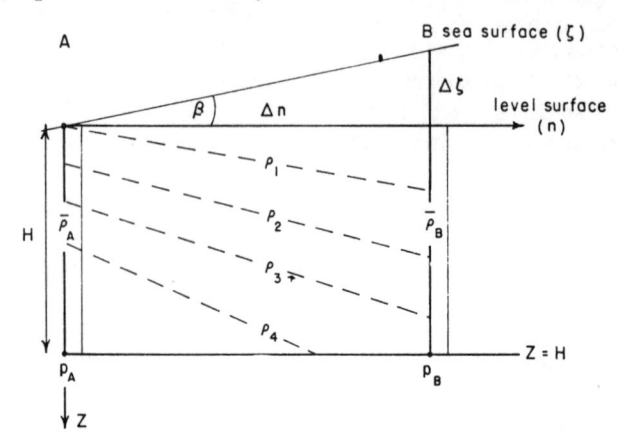

Fig.37. Schematic vertical section showing the distribution of density and the slope of the sea surface against a level surface. At $z = H$, $p_A = p_B$.

and B, respectively; they slope in the opposite direction to the sea surface. The average density of a vertical water column between ζ and the depth $z = H$ is $\bar{\rho}_A$ at station A, and $\bar{\rho}_B$ at station B. Hence, the pressure p_A at depth H is $p_A = g\bar{\rho}_A H + p_a$, and the pressure p_B at depth H is $p_B = g\bar{\rho}_B H + g\rho(\zeta)\Delta\zeta + p_a$. If $\bar{\rho}_A$, $\bar{\rho}_B$, and $\Delta\zeta$ are such that $p_A = p_B$ at depth H, it follows that $(\Delta p/\Delta n)_{z=H} = 0$ if:

$$\bar{\rho}_A H = \bar{\rho}_B H + \rho(\zeta)\Delta\zeta$$

This corresponds to eq.66 and shows that the depth H where compensation of the two terms in eq.66 occurs, is:

$$H = \frac{\rho(\zeta)\Delta\zeta}{\bar{\rho}_A - \bar{\rho}_B} \qquad (67)$$

This equation can be used to estimate the depth of the layer of no motion, provided $\Delta\zeta$ is known in addition to the vertical density distribution at two adjacent oceanographic stations (NEUMANN, 1952a). If accurate sea surface current observations between two stations A and B are available, and if these currents would follow the geostrophic assumption as expressed by eq.62, it would be relatively easy to find the depth of the "level of no motion". It is seen from eq.67 that for a given $\Delta\zeta$ the depth H increases as $\bar{\rho}_A - \bar{\rho}_B$ decreases. In strongly stratified water, the level of no motion is found at shallower depth than in a weaker stratification. In a very weakly stratified part of the ocean, H becomes very deep and approaches infinity if $\bar{\rho}_A - \bar{\rho}_B$ approaches zero. This can happen in nearly homogeneous water masses that are found in some higher latitudes, for example around the Antarctic Continent.

In practice, it is much more difficult to find this level, H, where eq.66 is fulfilled or nearly fulfilled. It will be shown later that friction plays an essential role in the balance of forces in wind-driven currents of the upper strata of the oceans. Also, these currents are often accelerated, and this, of course, destroys the geostrophic equilibrium.

Nevertheless, the geostrophic assumption has been, and still is, applied extensively in computing current speed and direction from the observed field of mass which allows the computation of the relative field of pressure by means of the hydrostatic equation. Direct application of this method in various forms has been given by DEFANT (1929a), WERENSKJÖLD (1937), and others, although the basic idea was proposed as early as 1903 by SANDSTRÖM and HELLAND-HANSEN.

Consider two oceanographic stations, A and B, for which the distribution of density with depth is known from observations of T and S with depth. The dynamic height difference D, in dynamic meters between two isobaric surfaces p_1 and p_2 is obtained by a numerical integration of eq.8. If the anomaly of the specific volume is introduced according to eq.8a, dynamic depth differences of isobaric surfaces between adjacent stations depend only on the terms that contain the anomaly of the specific volume. Thus:

$$D_A - D_B = \Delta D_A - \Delta D_B = \int_{p_1}^{p_2} \delta_A \, dp - \int_{p_1}^{p_2} \delta_B \, dp \quad (68)$$

The horizontal pressure gradient in eq.62 can be expressed in terms of the slope of isobaric surfaces according to eq.63. Introduction of the dynamic depth in units of a dynamic meter yields:

$$\frac{\partial p}{\partial n} = -10 \rho \frac{\partial D}{\partial n}$$

and:

$$2\omega \sin\phi \, c = -10 \frac{\partial D}{\partial n} = -10 \frac{\Delta D_B - \Delta D_A}{\Delta n} \quad (69)$$

where Δn is the horizontal distance when proceeding from station A to station B in the positive n-direction. Since ΔD_A and ΔD_B represent the (relative) dynamic depth anomaly between isobaric surfaces, the velocity $c = c_1 - c_2$ can also represent only the relative velocity between two isobaric surfaces p_1 and p_2. Thus, for practical uses, eq.69 can be written:

$$c_1 - c_2 = \frac{10}{2\omega \sin\phi \cdot \Delta n} (\Delta D_B - \Delta D_A) \quad (70)$$

The velocity difference, $c_1 - c_2$, is obtained in cm/sec if the dynamic depth difference is expressed in dynamic cm, and Δn in cm.

The application of eq.70 is explained by an example from a section across the Gulf Stream. In Fig.35 let station A be represented by "Atlantis" station 5298 and station B by "Atlantis" station 5299. Numerical evaluation of the integrals in eq.68 by stepwise summation of finite but small enough pressure (or depth) intervals yields ΔD_A and ΔD_B for isobaric surfaces p_2 with reference to p_1. If p_1 represents the constant pressure at the sea surface (sea pressure $p_1 = 0$), the anomalies of the dynamic depth of isobaric surfaces p_2 below the sea surface are given in Table I.

The first column of Table I gives the sea pressure p in dbar and the same values for the depth in m. This is not quite correct. However, the relative small difference between the depth in m and the pressure in dbar makes this approximation possible without any significant effect on the practical computation of

TABLE I

COMPUTATION OF RELATIVE GEOSTROPHIC CURRENTS NORMAL TO THE VERTICAL PLANE BETWEEN "ATLANTIS" STATIONS 5298 AND 5299. DISTANCE BETWEEN THE TWO STATIONS $\Delta n = 28.06$ KM; AVERAGE GEOGRAPHICAL LATITUDE $\phi = 36°18'$N

p (dbar) or depth (m)	ΔD_A (dyn.m)	ΔD_B (dyn.m)	$\Delta D_B - \Delta D_A$ (dyn.m)	Rel. speed (cm/sec)
0	0	0	0	0
25	0.093	0.102	0.009	− 2.9
50	0.173	0.190	0.017	− 7.0
75	0.243	0.264	0.021	− 9.5
100	0.298	0.332	0.034	− 14.0
150	0.391	0.450	0.059	− 24.3
200	0.466	0.546	0.080	− 33.0
300	0.581	0.720	0.139	− 57.4
400	0.675	0.891	0.216	− 89.1
500	0.753	1.042	0.289	−119.3
600	0.815	1.164	0.349	−144.0
800	0.919	1.362	0.443	−182.8
1,000	1.011	1.496	0.485	−200.1
1,200	1.097	1.594	0.497	−205.1
1,500	1.229	1.732	0.503	−207.6
2,000	1.449	1.947	0.498	−205.5

the anomalies of the dynamic depths of isobaric surfaces and the subsequent computation of relative currents. For example at a geometric depth of 1,000 m in a standard ocean, the pressure is 1,010 dbar.

The next three columns present the values ΔD_A and ΔD_B in dynamic meters and the difference $\Delta D_B - \Delta D_A$. Total values of the dynamic depth of a given isobaric surface are obtained by adding the standard dynamic depth interval between isobaric surfaces p_1 and p_2 as shown in eq.8a. According to the tables by BJERKNESS and SANDSTRÖM (1910) the dynamic depth of the 1,000 dbar surface in a standard ocean ($S = 35‰$, $T = 0°C$) is $D_{35,0,p} = 970.4032$ dyn.m. Thus, for example, at station A the total dynamic depth of the 1,000 dbar surface is $D = D_{35,0,p} + \Delta D_A = 970.403 + 1.011 = 971.414$ dyn.m.

The current speeds normal to the vertical plane between the two stations with reference to the sea surface are obtained from eq.70. Since $c_1 = 0$, and $\Delta D_B - \Delta D_A > 0$ at all levels, the relative currents have a negative sign, which means that they are directed such that station B is on the left hand side when one faces in the direction of the relative current.

In order to arrive at absolute currents, either the absolute pressure field or the absolute current velocity must be known with sufficient accuracy at least at one level in the sea. Such a level is called the *reference level*. Since the inclination of isobaric surfaces against level surfaces is very small, no method exists in oceanography at present to measure this inclination directly. Direct current measurements at sea are quite difficult, and in many cases it is questionable whether the measured mean currents are close enough to geostrophic equilibrium in order to relate the mean observed current velocity to the horizontal pressure gradient. Only in exceptional cases could measured current velocities be directly used to establish a reference level for transforming relative into absolute velocities (e.g., WÜST, 1924; DEFANT, 1940c; SWALLOW and WORTHINGTON, 1959). In the past, oceanographers have tried to find a zero level or a level of no motion in the ocean by indirect evidence. If such a level develops above the ocean bottom at a depth H according to eq.67, isobaric surfaces at that depth can be assumed to coincide with level surfaces and the geostrophic current becomes zero.

In weakly stratified or shallow waters, where the depth H is greater than the depth of the bottom, a zero level as a reference surface for transforming relative into absolute currents cannot easily be established without further assumptions. Several suggestions have been offered and attempts have been made to find evidence for a motionless or nearly motionless layer beneath the sea surface. In most of these cases, "motionless" is defined for the *horizontal* motion only. A layer of no *absolute* motion including the vertical component in the three-dimensional oceanic current field is even more difficult to establish. In spite of all efforts no reliable, objective method for determining the layer of no motion in the sea is at present available. The reader is referred to a critical study of the dynamic method in oceanography by FOMIN (1964) who also reviewed different methods for computing or estimating the level of no motion.

Among the existing methods, it appears that DEFANT's (1941) approach is most practical. Although his method may not yield unique results in regions of weak and variable currents between surface and bottom, it seems to be consistent and reliable enough for moderately or strongly stratified water bodies. Defant applied his method to the whole Atlantic Ocean and has shown that one arrives, indeed, at a reasonable depth distribution of a reference layer for transforming relative into absolute dynamic topographies.

Essentially, Defant starts with the hypothesis that it is more reasonable to assume that the strongest currents occur in the upper, most strongly stratified strata of the sea. Upon comparing the differences of the relative dynamic depth of given isobaric surfaces between adjacent oceanographic stations, it is found that in some deep-sea layers this difference was practically constant over a rather large depth interval. It seems unreasonable to assume that these deep layers are moving with uniform speed and that the surface layers are almost motionless. Thus, Defant suggested that these deep layers with constant or nearly constant horizontal pressure gradients are motionless in horizontal direction, rather than at a uniformly high speed as compared to the sea surface.

Defant's ideas will now be applied to the relative distribution of $\Delta D_B - \Delta D_A$ as a function of depth for the two "Atlantis" stations in Table I. Fig.38 shows the values of $\Delta D_B - \Delta D_A$ versus depth on the upper relative scale, and these values give a current to the south which increases its speed with depth and reaches more than 200 cm/sec between 1,000 and 2,000 m below the sea surface. This, of course, is not true, and the absolute

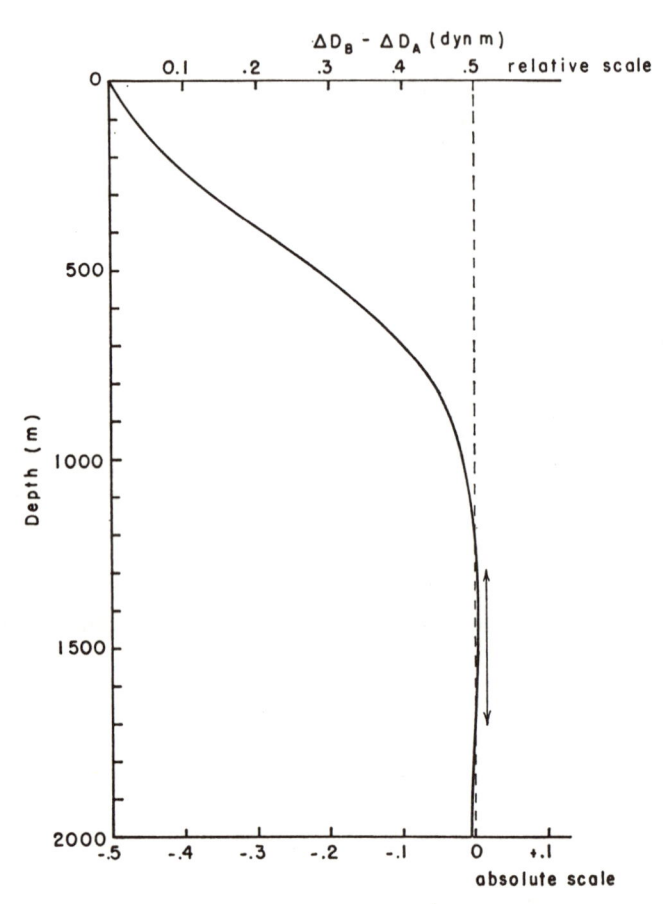

Fig.38. Dynamic depth differences, $\Delta D_A - \Delta D_B$, of isobaric surfaces for Atlantis station numbers 5298 and 5299 as a function of depth. The relative scale is shown at the top, the most probable absolute scale at the bottom of the graph.

zero point for the inclination of isobaric surfaces has to be fixed at a deeper level. It is obvious that the most reasonable assumption of selecting the absolute zero for $\Delta D_B - \Delta D_A$ is in the layer around a depth of 1,500 m. If the absolute zero for the differences of the anomalies of the dynamic depths of isobaric surfaces is placed around 1,500 m, the proper direction of the Gulf Stream will be obtained in the upper strata. In Fig.38 the most reasonable absolute scale for $\Delta D_B - \Delta D_A$ is given by the scale in the lower part of the graph.

Recent results of direct current observations in the Gulf Stream region (SWALLOW and WORTHINGTON, 1961) have shown that Defant's method is basically sound. In application to complete sections between the North American east coast and the Sargasso Sea region, DEFANT (1941) found that the level of no horizontal motion slopes downward from the continental side to the Sargasso Sea. On the average, the depth of the Gulf Stream increases from about 500 m on the continental side to about 2,000 m on the Sargasso Sea side. DEFANT's (1941) charts showing the geostrophic currents in the Gulf Stream region also indicate a countercurrent in the deep sea underneath the Gulf Stream flowing to the south. This deep-sea countercurrent has also been deduced by STOMMEL (1955), based on different arguments.

The particular depth distribution of the level of no motion for the Gulf Stream has other interesting dynamic consequences (NEUMANN, 1956; MARTINEAU, 1958). It can be shown that the sloping lower boundary of the Gulf Stream and the extreme baroclinicity of the Stream result in a nearly equivalent-barotropic flow. A similar vertical structure is indicated for the Kuroshio in the western Pacific Ocean along the Japanese Islands. This current reveals some of the same dynamical properties as the Gulf Stream. Both currents will be discussed in the following chapter.

If the depth of a reference layer or the depth of a zero layer is known, it is easy to transfer the relative currents shown in Table I to absolute geostrophic currents. According to Fig.38 the depth of 1,500 m is assumed to be motionless in the vertical structure of currents passing normally through the vertical plane between stations A and B. The relative current at 1,500 m depth is most likely equal to zero. Therefore, to arrive at absolute current speeds, the speed 207.6 cm/sec should be added to the relative current at this level and to all other levels because the relative pressure distribution has to be kept constant. Table II shows the result of this transformation of relative into absolute currents between "Atlantis" stations 5298 and 5299.

Speeds of more than 200 cm/sec have been observed in the upper layers of the Gulf Stream, and the absolute speeds of the current derived with the assumption of a geostrophic flow are quite realistic. In some parts of the Gulf Stream it is often observed that the highest speed of the stream is not at the sea surface but at a subsurface level. This feature will also be discussed in the following chapter, and a possible dynamic explanation will be offered.

In summary, it can be stated that in some oceanic regions like the Gulf Stream, the Kuroshio and other regions of strong currents (Agulhas Current, DIETRICH, 1935) a nearly geostrophic flow of dominant ocean currents is closely approximated, and the application of the classical dynamic method can lead to fairly good results if the zero layer or a reference layer for transforming relative into absolute current velocities can be established with sufficient accuracy. The reason why this simple method does not always give a satisfactory answer for all currents

TABLE II

TRANSFORMATION OF RELATIVE INTO ABSOLUTE CURRENT SPEEDS BETWEEN "ATLANTIS" STATIONS 5298 AND 5299; THE RELATIVE SPEEDS ARE ALSO SHOWN IN TABLE I

Depth (m)	Abs. speed (cm/sec)	Rel. speed (cm/sec)	Depth (m)	Abs. speed (cm/sec)	Rel. speed (cm/sec)
0	207.6	0	400	118.5	− 89.1
25	204.7	− 2.9	500	88.3	−119.3
50	200.6	− 7.0	600	63.6	−144.0
75	198.1	− 9.5	800	24.8	−182.8
100	193.6	−14.0	1,000	7.5	−200.1
150	183.3	−24.3	1,200	2.5	−205.1
200	174.6	−33.0	1,500	0	−207.6
300	150.2	−57.4	2,000	2.1	−205.5

is that oceanic water movements are often accelerated, and frictional forces and vertical motions must play a significant role. Particularly, the layers near the sea surface are often not in geostrophic balance. Here, both acceleration and friction can be dominant features of ocean currents. In waters where the currents reach the bottom, bottom friction also has to be included in the balance of forces that act upon a mass of water.

Accelerations caused by sudden changes in the driving forces, like the wind stress at the sea surface, are among the most important causes for a deviation from the geostrophic equilibrium. The problem of time dependent changes of ocean currents has not yet been solved satisfactorily, although some promising results have been obtained by recent workers in the field of theoretical dynamic oceanography (ICHIYE, 1951; VERONIS and MORGAN, 1955; see also FOFONOFF, 1962).

Practical techniques for transforming relative into absolute dynamic topographies

In the case of a reference surface of constant depth, the transformation of relative into absolute topographies of isobaric surfaces is easily performed. The first attempt to use a reference surface of variable depth was made by DIETRICH (1937b). The sloping reference surface in a dynamic section was approximated by a broken line changing its depth stepwise between adjacent pairs of oceanographic stations. Fig.39 represents a schematic vertical section containing stations A, B, C and D. The relative slope of isobaric surfaces $p_1(\text{rel})$, $p_2(\text{rel})$, . . ., $p_5(\text{rel})$ is shown by light lines with reference to the uppermost isobaric surface, $p_0(\text{rel})$, which is assumed to coincide with a level surface. The absolute slope of the sea surface, $p_0(\text{abs})$ will be different from $p_0(\text{rel})$ if it is found that at greater depths, say between p_3 dbar and p_5 dbar, the horizontal motion is zero. In this case, the absolute slope of isobaric surfaces in the "zero layer" must coincide with level surfaces, and it is necessary to adjust the relative pressure distribution to the absolute throughout the whole vertical column of water.

Assume that at stations A and B in Fig.39 the zero layer is found at depths p_A and p_B dbar, respectively. At station C it is at

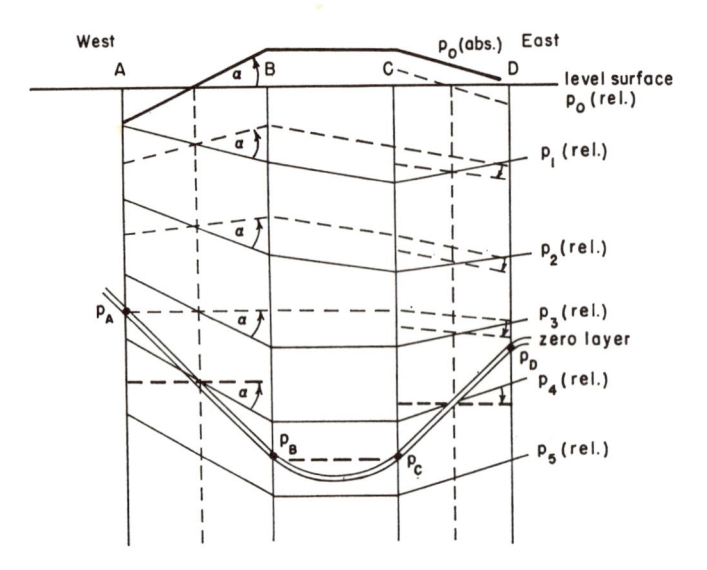

Fig.39. Transformation of relative into absolute slopes of isobaric surfaces.

$p_C = p_B$ dbar, and at station D at p_D dbar. The sloping reference layer is approximated by horizontal surfaces between adjacent stations as shown in the figure by heavy dashed lines such that:

$$p_{AB} = 1/2(p_A + p_B), \quad p_{BC} = p_C = p_B, \quad p_{CD} = 1/2(p_C + p_D)$$

are isobaric surfaces in the reference layer that coincide with level surfaces. To obtain the absolute slope of isobaric surfaces in the section it is necessary to adjust the slope of all isobars at the same rate between adjacent stations because the relative pressure distribution must be preserved. Between stations B and C no adjustment is needed, since the isobaric surface $p_{BC} = p_B = p_C$ coincides with a level surface. The absolute slope of isobaric surfaces in the section of Fig.39 is shown by light dashed lines with the exception of the sea surface, p_0(abs), which is shown by a heavy solid line. Thus, it is seen that the geostrophic currents perpendicular to the zonal section are to the north between stations A and B and to the south between stations C and D if the section is in the Northern Hemisphere. Between stations B and C, the surface current is zero, and in deeper layers it is directed to the south until the current becomes zero again at the depth of the reference layer.

If oceanographic stations are distributed over a larger oceanic region, say over a whole ocean, and the depth of the reference surface varies in both x and y-directions, a similar stepwise approximation of the variable depth of the reference surface can be applied. However, in this case the condition has to be satisfied that the same values for the absolute topographies of isobaric surfaces are obtained independent of different horizontal paths that are taken. DEFANT (1941) used a triangular network of oceanographic stations and connected three stations in the corners of a triangle in the same way as was explained in Fig.39 for a section. Thus, if the absolute slopes of isobaric surfaces for the three corner stations, A, B, C in a triangle have been adjusted to each other, the value at A obtained by connecting C to A should agree with the starting value at A. Usually a difference remains as a result of the stepwise approximation technique. This "triangulation error" must be distributed carefully with "a weight" to the three steps taken in the approximation. If this is done, a second triangle of oceanographic stations is connected and adjusted to the absolute values of the preceding triangle. In this way triangle to triangle was connected for the whole Atlantic Ocean. This useful but elaborate method can be considerably simplified as shown in Fig.40.

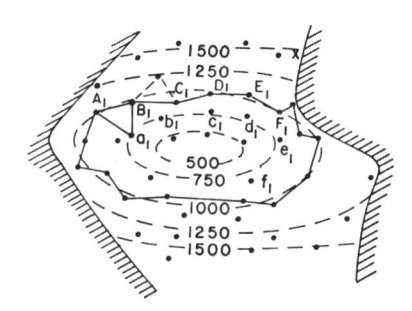

Fig.40. Transformation of relative into absolute dynamic topographies for the case of a reference surface of variable depth. The dashed lines are lines of equal depth of the reference surface.

With properly chosen intervals for the depth of the reference surface, oceanographic stations with the same or nearly the same reference depth can be connected. These isobaths of reference depth can be, but do not necessarily have to be, closed lines. In Fig.40 the dots indicate oceanographic stations, and the dashed lines represent lines of equal reference depth. If a station does not lie exactly on one of the drawn isobaths (as will be the case with most of the stations) the nearest isobath can be chosen for that station with sufficient accuracy if the depth intervals between isobaths representing the reference layer are small enough. Usually, in the ocean a depth interval of 250 m is adequate.

Starting at station A_1 in Fig.40 with an arbitrary value, say $A_1 = 0$ dyn cm for a given isobaric surface, stations B_1, C_1, D_1, E_1, and so on can be connected to each other without difficulty, because they all lie either on or very close to the 1,000 m depth reference surface. Closing the sequence of stations along the 1,000 m depth of the reference layer, one arrives at A_1, and the starting value at $A_1 = 0$ dyn cm should be obtained if there is no error in the computation. To find the connection with a station on the next reference level the most suitable triplet of stations can be used, for example the triplet A_1, B_1, a_1 in Fig.40. For this triangle, DEFANT's (1941) method of error distribution similar to the practice of precise leveling is employed to find the absolute connection of stations a_1, b_1, c_1, etc. around the 750 m isobath of the reference surface. Again, all stations lying on or near the reference depth can quickly be adjusted to each other. The same procedure is employed to one of the stations on the next reference level, and so on, until all stations in the area under consideration are related to each other with their absolute values of isobaric topographies. This method reduces the computational effort to a minimum (NEUMANN, 1942).

The practical applicability of the classical method of dynamic computations is restricted in many ways. Besides the question of whether or not the geostrophic assumption is adequately fulfilled to permit the computation of current velocity with sufficient confidence from a given pressure field, the accuracy of the pressure field itself is often subject to doubt. The pressure field in the sea is obtained from the observed field of density, which ultimately is based on observations of temperature and salinity at different depths. Unless these data are taken simultaneously at different places in the ocean, one cannot expect to obtain a truly representative pressure field for computing ocean currents. Most oceanographic stations in a part of the ocean do not fulfill the requirement of synopsis. Periodic and non-periodic disturbances affect the distribution of temperature and salinity, and the field of mass and pressure can be in error if not taken simultaneously over a larger area. The quest for synoptic oceanography becomes more and more urgent as physical oceanography progresses. SEIWELL (1937) has shown that significant differences in the dynamic height of isobaric surfaces at a fixed station can occur within a few days. DIETRICH (1937b) also found from repeated series at an anchor station that variations of 10 dyn cm can occur at intervals of only a few days. The effect of internal waves on the mass distribution and consequent pressure field computations was clearly demonstrated by DEFANT (1950). The uncertainty of the correct level of reference for transforming relative into absolute pressure fields can also lead to erroneous interpretations of the pressure field in terms of currents, even when the currents can be approximated by a simple geostrophic flow. Deviations from the geostrophic equilibrium will be considered in the following sections of this chapter.

INERTIA CURRENTS

A simple case of accelerated currents without friction is obtained from eq.12. For horizontal motion it follows that:

$$\frac{du}{dt} - 2\omega \sin\phi v = -\frac{1}{\rho}\frac{\partial p}{\partial x}$$

$$\frac{dv}{dt} + 2\omega \sin\phi u = -\frac{1}{\rho}\frac{\partial p}{\partial y}$$

(71)

The accelerations on the left side of eq.71 including the Coriolis accelerations are balanced by the horizontal pressure gradient. In a homogeneous ocean where ρ is constant, such pressure gradients can only be the result of a slope of the sea surface. Changes of the sea surface slope during relatively short time intervals can be caused by changes of the average meteorological wind field, for example with passages of meteorological fronts. Southwesterly winds in the Northern Hemisphere change often more or less suddenly to northwesterly winds with the passage of a "cold front".

Let us assume that for some reason the pressure gradient becomes zero, and $\partial p/\partial x = 0$, $\partial p/\partial y = 0$ in eq.71. In this case, without friction:

$$\frac{du}{dt} = 2\omega \sin\phi v$$

$$\frac{dv}{dt} = -2\omega \sin\phi u$$

(72)

represents the most simple example of an accelerated current on the rotating earth.

Currents, as described by eq.72 are called *inertia currents*. In most cases, horizontal pressure gradients and friction have to be added to eq.72. Inertia currents can be regarded as due to the lack of balance between the pressure gradient, friction and the resultant geostrophic acceleration.

If the first of eq.72 is multiplied by u and the second by v, subsequent addition of both equations yields:

$$u\frac{du}{dt} + v\frac{dv}{dt} = 0$$

This equation states that $1/2(dc^2/dt) = 0$ because:

$$\frac{dc^2}{dt} = \frac{d(u^2 + v^2)}{dt} = 2\left(u\frac{du}{dt} + v\frac{dv}{dt}\right)$$

Since $dc^2/dt = 0$, it follows that a water particle moves with constant speed. Therefore the acceleration must result from a change in the direction of the current vector. This is easily shown by multiplying the first of eq.72 by v and the second by u. Subtraction of the second equation from the first yields:

$$v \frac{du}{dt} - u \frac{dv}{dt} = 2\omega \sin\phi \, c^2$$

Since:

$$\frac{d(u/v)}{dt} = \frac{v \, du/dt - u \, dv/dt}{v^2}$$

it is seen that:

$$v^2 \frac{d(u/v)}{dt} = 2\omega \sin\phi \, c^2 \tag{73}$$

In a rectangular xy-coordinate system $u/v = \cot\alpha$ and $v^2 = c^2 \sin^2\alpha$ where α denotes the angle between the x-axis and the direction of the current. Hence, it follows from eq.73 that:

$$\frac{d(\cot\alpha)}{dt} = \frac{2\omega \sin\phi}{\sin^2\alpha}$$

or:

$$\frac{d\alpha}{dt} = -2\omega \sin\phi \tag{74}$$

This equation states that the moving water particle changes its direction at a constant rate if ϕ, the geographical latitude, can be considered constant. Therefore, inertia currents as described by eq.72 must move in a circle with constant speed. This circle is called the *circle of inertia*. In the Northern Hemisphere $d\alpha/dt$ (eq.74) is negative, and the motion around the inertia circle is clockwise. In the Southern Hemisphere $d\alpha/dt$ is positive and the motion is counterclockwise.

The equation for the inertia circle is easily derived from eq.72, which can be written as:

$$\frac{du}{dt} = f \frac{dy}{dt}; \quad \frac{dv}{dt} = -f \frac{dx}{dt}$$

where $f = 2\omega \sin\phi$. Upon integration, it follows that $u = fy + a$,

$v = -fx + b$ (a and b are constants of integration). Since $u^2 + v^2 = c^2$:

$$\left(x - \frac{b}{f}\right)^2 + \left(y + \frac{a}{f}\right)^2 = \left(\frac{c}{f}\right)^2$$

This is the equation of a circle of radius $r = c/f$ with its center at $x_0 = b/f, y_0 = -a/f$.

The solution for the velocity components is:

$$u(t) = u_0 \cos ft + v_0 \sin ft$$
$$v(t) = v_0 \cos ft - u_0 \sin ft$$

This set of equations satisfies eq.72, and it is apparent that $u^2 + v^2 = u_0^2 + v_0^2 =$ constant. The initial velocities and final velocities are related to each other by a simple rotation of the xy-axis.

The time needed to complete a full path around the circle of inertia is called the *inertia period*, $T_p = 2\pi/f$. If r is the radius of the inertia circle and c the speed of a water particle travelling around this circle, the inertia period is:

$$T_p = \frac{2\pi r}{c} \tag{75}$$

Since the Coriolis force always acts at a right angle to the motion, the only acceleration to balance the Coriolis acceleration in the absence of friction is the centrifugal acceleration c^2/r when the water moves in a circle of radius r. In order to maintain the inertia motion in a circle, both accelerations must be equal and must act in opposite directions. Thus, eq.72 requires that:

$$\frac{c^2}{r} = 2\omega \sin\phi \, c$$

and the radius of the inertia circle is:

$$r = \frac{c}{2\omega \sin\phi} \tag{76}$$

At the equator, r becomes infinite, and at the poles it is a minimum for a given speed c. If eq.76 is used to eliminate c in eq.75, it follows that:

$$T_p = \frac{\pi}{\omega \sin\phi} \qquad (77)$$

Thus, the inertia period depends only on the geographical latitude. $2\pi/\omega$ represents a sidereal day, which is about four minutes shorter than a solar day. The inertia period is also known as "half a pendulum day" since it is half the period of revolution of a Foucault's pendulum. Thus, at the poles, T_p is approx. 12 h, at latitude 30°, T_p is approx. 24 h, and at the equator T_p is infinite.

For latitudes around 45°, $2\omega \sin\phi$ equals approx. 10^{-4}, and for a current speed of 10 cm/sec, the radius of the inertia circle is approx. 1 km. The inertia period is about 17.4 h. For a current speed of 100 cm/sec the radius of the inertia circle would be 10 km, and it is seen that in middle latitudes the magnitude of r is 1–10 km with an inertia period of 17–18 h.

It was not until 1931 that the existence of inertia currents in the oceans was established by direct observations. HELLAND-HANSEN and EKMAN (1931) first succeeded in demonstrating the presence of inertia currents by direct measurements in middle latitudes of the North Atlantic Ocean. Although in these latitudes ($\phi = 30.2°$N) the inertia period is very close to the diurnal period of the tides, the current measurements after elimination of the tides revealed periodic motions with the required period.

The classical example of inertia currents with a translatory motion superimposed was obtained in the Baltic Sea by GUSTAFSON and KULLENBERG (1936). In the Baltic Sea, the tides are almost insignificant. If the observed currents are split into their components u and v, a time series representation of the observed periodic currents can be given, as shown in Fig.41. The phase difference between the two components is very close to a quarter of a period, and the ratio of the amplitudes is very near to one. If the first few cycles during the generation of the current are disregarded, the period of oscillation is about 14 h which agrees well with the theoretical inertia period of 14 h 8 min in the latitude of observation. Thus, the observations by Gustafson and Kullenberg, after elimination of the superimposed translatory motion, represent almost pure (damped) circular inertia currents. The decrease of the radius of the circle of inertia with

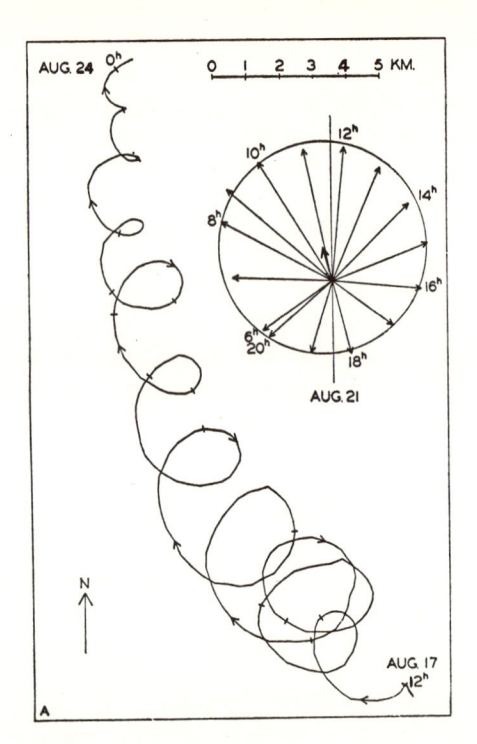

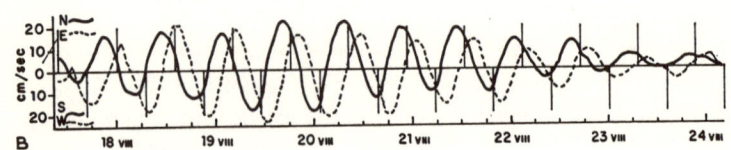

Fig.41A. Inertia currents, superimposed by a translatory motion observed in the Baltic Sea by GUSTAFSON and KULLENBERG (1933). The insert shows a central vector diagram for currents on August 21.
B. Velocity components of the observed currents according to EKMAN (1953). (After NEUMANN and PIERSON, 1966.)

time is due to dissipative forces (friction) which reduce the speed of the current.

More recently, an interesting example of almost pure inertia currents in the deep sea of the open Atlantic Ocean was reported by Pochapsky (1966). Results of one series of observations obtained by a pair of neutrally buoyant floats are shown in Fig.42A. The geographical position is about 28°N, 55°W, with the master float 67 settled at an average depth of 2,340 ± 20 m. Float 66 was 140 ± 20 m deeper. Float 66 moved in an ellipse having axes of 2.55 and 2.20 km, respectively. The average speed was 8.2 cm/sec. The center of the ellipse moved away from the master float at a speed of 0.23 cm/sec. The observed period of oscillation of 25.2 h is very close to the inertia period of 25 h, but it is also in the range of diurnal tidal periods. However, the observed average diameter of the nearly circular path in Fig.42A of about 2.4 km agrees with the theoretical diameter $2r = 2.4$ km, of an inertia circle in a latitude of 28°. Thus, Pochapsky's observations at a depth of about 2,500 m seem to represent almost pure inertia currents.

Other recent evidence for inertia currents in the deep sea was obtained by Nan'niti et al. (1964). These investigators also used neutrally-buoyant floats at 1,000 m and 2,000 m depths at a latitude of about 31.5°N and near 143°E, a distance of 300 km from Torishima. Significant oscillations with a period of about 23 h are exhibited in their observations. Plots of float tracks are shown in Fig.42B for depths of 1,000 m and 2,000 m.

CIRCULAR MOTION AND MEANDERING CURRENTS

Since geostrophic currents are non-accelerated frictionless water movements, they can only represent currents along straight horizontal streamlines. If the motion is circular or meandering with a radius of curvature r_g, pure geostrophic currents cannot exist even in the case of no friction, because there will be a component of acceleration relative to the earth of magnitude c^2/r_g to be considered. This is the centrifugal acceleration. Only if this component is negligible compared with $2\omega \sin\phi\, c$ can a geostrophic equilibrium be assumed. This requires $c^2/r_g \ll 2\omega \sin\phi\, c$, or $r_g \gg r$, according to eq.76, where r is the radius of the inertia circle. Thus, circular or meandering currents cannot be considered in geostrophic equilibrium if the streamline curvature approaches the radius of the inertia circle. Besides the Coriolis force, the centrifugal force must be considered. The centrifugal force is an *apparent* force that acts perpendicularly to the motion. Depending on the curvature it can act either in the direction of the Coriolis force or against it. The modification of the balance of forces in a geostrophic current when the centrifugal force is included is shown for curved isobars in Fig.43. A current of this type is called a *gradient current*. If the curvature of streamlines

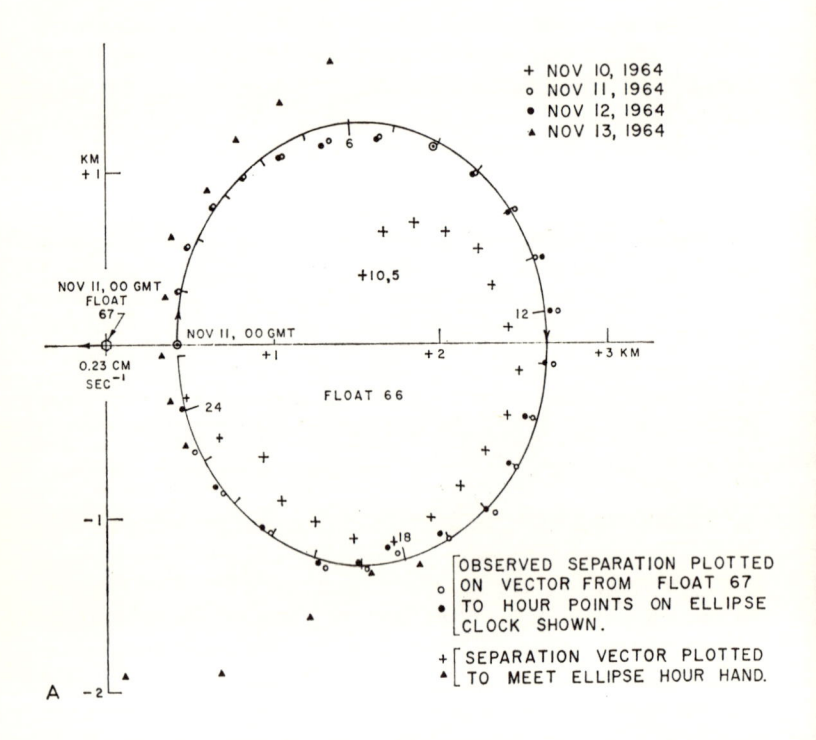

Fig.42A. Observed separations of floats 66 and 67 plotted relative to an assumed elliptical path. (After POCHAPSKY, 1966.)

B. Tracks of neutrally-buoyant floats at 31°37.5′N, 143°08.5′E. The upper graph is for a depth of 1,000 m, the lower graph for 2,000 m. The numbers along the tracks indicate the day (of July, 1964) and the time. (After NAN'NITI et al., 1964.)

CIRCULAR MOTION AND MEANDERING CURRENTS 157

(or isobars) in a gradient current is small when compared to the radius of the inertia circle, the gradient current is very close to a geostrophic current, and in many ocean currents the difference is negligible. However, in low latitudes, caution is indicated. When compared to Coriolis forces, centrifugal forces become more and more important as one approaches the equator, since the horizontal component of the Coriolis force vanishes at $\phi = 0$.

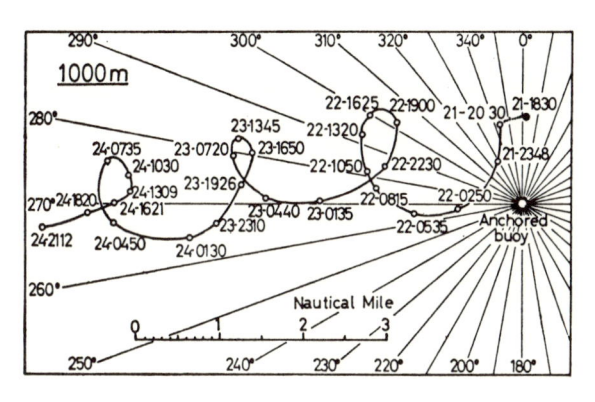

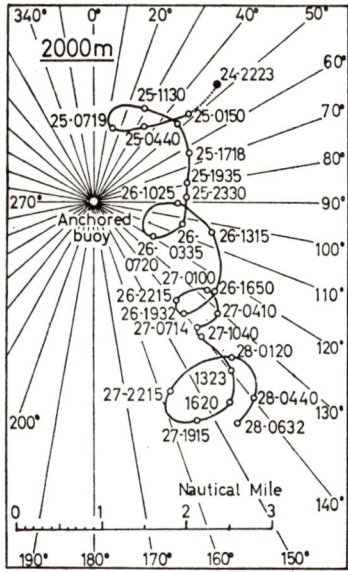

158   MAJOR TYPES OF OCEAN CURRENTS

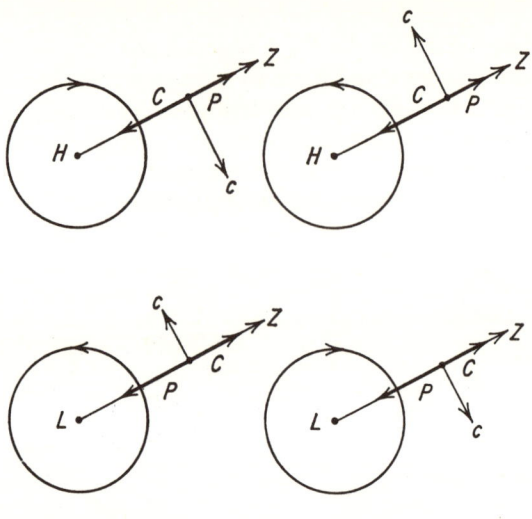

Northern Hemisphere   Southern Hemisphere

Fig.43. Balance of forces in a gradient current. P denotes the pressure gradient, C the Coriolis force, and Z the centrifugal force.

Gradient currents

For stationary horizontal motion ($\partial u/\partial t = 0$, $\partial v/\partial t = 0$) it follows from eq.71 that:

$$u\frac{\partial u}{\partial x} + v\frac{\partial u}{\partial y} - fv = -\frac{1}{\rho}\frac{\partial p}{\partial x}$$

$$u\frac{\partial v}{\partial x} + v\frac{\partial v}{\partial y} + fu = -\frac{1}{\rho}\frac{\partial p}{\partial y}$$

(78)

The balance of forces in curved or meandering currents is best examined by transforming eq.78 into polar coordinates. The position of any point $P(x, y)$ in the xy-plane can be determined in polar coordinates by the distance r (radius vector) from the pole or origin, O, to the point $P(r, \theta)$ together with the angle θ measured from the direction OX–OP (Fig.44). θ is positive if measured counterclockwise, negative if measured clockwise. Relations between rectangular and polar coordinates are:

$$x = r\cos\theta \qquad y = r\sin\theta$$
$$r = (x^2 + y^2)^{\frac{1}{2}} \qquad \theta = \tan^{-1} y/x$$

The velocity components, dx/dt and dy/dt, are:

$$u = c_r \cos\theta - c_\theta \sin\theta$$
$$v = c_r \sin\theta + c_\theta \cos\theta \qquad (79)$$

where $c_r = dr/dt$ is the radial and $c_\theta = r\, d\theta/dt$ the tangential velocity in a circular motion. Since both x and y are functions of r and θ, the x and y derivatives of a function $F(x,y) \equiv F(r,\theta)$ are:

$$\frac{\partial F}{\partial x} = \frac{\partial F}{\partial r}\frac{\partial r}{\partial x} + \frac{\partial F}{\partial \theta}\frac{\partial \theta}{\partial x} = \cos\theta \frac{\partial F}{\partial r} - \frac{\sin\theta}{r}\frac{\partial F}{\partial \theta}$$

$$(80)$$

$$\frac{\partial F}{\partial y} = \frac{\partial F}{\partial r}\frac{\partial r}{\partial y} + \frac{\partial F}{\partial \theta}\frac{\partial \theta}{\partial y} = \sin\theta \frac{\partial F}{\partial r} + \frac{\cos\theta}{r}\frac{\partial F}{\partial \theta}$$

If the first of eq.80 is multiplied by u and the second by v it follows with consideration of eq.79 that the sum:

$$u\frac{\partial F}{\partial x} + v\frac{\partial F}{\partial y} = c_r \frac{\partial F}{\partial r} + \frac{c_\theta}{r}\frac{\partial F}{\partial \theta} \qquad (81)$$

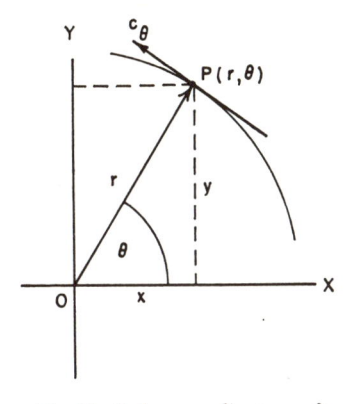

Fig.44. Polar coordinates r, θ.

This result can be applied to eq.78 when the velocity components u and v, respectively, are used instead of F. Thus, eq.78 transforms into eq.82:

$$c_r \frac{\partial c_r}{\partial r} + c_\theta \frac{\partial c_r}{r\partial \theta} - \frac{c_\theta^2}{r} = fc_\theta - \frac{1}{\rho}\frac{\partial p}{\partial r}$$

$$c_r \frac{\partial c_\theta}{\partial r} + c_\theta \frac{\partial c_\theta}{r\partial \theta} + \frac{c_\theta c_r}{r} = -fc_r - \frac{1}{\rho}\frac{\partial p}{r\partial \theta}$$
(82)

For the special case of circular concentric isobars with their centers at $r = 0$, $\partial p/\partial \theta = 0$, and with a stationary motion that is parallel to the isobars, also $c_r = 0$ and $\partial c_\theta/\partial \theta = 0$. These conditions satisfy the second of eq.82 identically. The first equation can be written as:

$$fc_\theta + \frac{c_\theta^2}{r} = \frac{1}{\rho}\frac{\partial p}{\partial r}$$
(83)

This equation governs circular, frictionless motion parallel to the isobars. It expresses a balance of pressure gradient, Coriolis and centrifugal forces per unit mass. Since the pressure gradient can be positive or negative, there are four cases of possible motion to be distinguished as shown in Fig.43. Depending on the sign of the pressure gradient, the motion can be cyclonic or anticyclonic. In the Northern Hemisphere a cyclonic motion is counterclockwise, in the Southern Hemisphere it is clockwise. The anticyclonic motion around a high pressure center in the Northern Hemisphere is clockwise and counterclockwise in the Southern Hemisphere. EKMAN (1923) introduced the terms *cum sole* and *contra solem* to describe the motions around high pressure centers and low pressure centers, respectively.

The balance of forces for the four cases shown in Fig.43 indicates that a gradient current in cyclonically curved motion requires a greater pressure gradient than a gradient current of equal velocity in anticyclonically curved motion. Or, if the absolute value of the pressure gradient is the same for both cases of rotation, the gradient current speed is smaller around a cyclonic pressure field than it is around an anticyclonic pressure field.

The solution to eq.83 is:

$$c_\theta = -\frac{fr}{2} \pm \sqrt{\left(\frac{f^2 r^2}{4} + \frac{r}{\rho}\frac{\partial p}{\partial r}\right)} \qquad (84)$$

A positive value of c_θ means contra solem motion and a negative value cum sole motion. If the positive root of eq.84 is considered, the sense of rotation agrees with a motion around a low pressure center (contra solem) where c_θ is positive. This is necessary, since cyclonic rotation should approach zero for a vanishing pressure gradient ($\partial p/\partial r = 0$). For a motion around a high pressure center, $\partial p/\partial r$ is negative. With this, the square root in eq.84 is smaller than $fr/2$ and c_θ is negative (anticyclonic).

Next, consider a negative root of eq.84. This means a cum sole motion (anticyclonic), because the positive root has been shown to be necessary for the opposite sense of rotation. As $\partial p/\partial r$ approaches zero, c_θ approaches $-fr$. This agrees with eq.76 as a solution of eq.72 which represent inertia currents. An anticyclonic motion in both hemispheres is possible for the case $\partial p/\partial r = 0$.

With $\partial p/\partial r < 0$, it is seen that the root in eq.84 becomes zero if:

$$\frac{f^2 r}{4} = -\frac{1}{\rho}\left|\frac{\partial p}{\partial r}\right|$$

and in order to maintain a real value of c_θ for cum sole motion, it is necessary that:

$$\left|\frac{\partial p}{\partial r}\right| < r\rho\omega^2 \sin^2\phi \qquad (85)$$

If the pressure gradient and, therefore, the current speed becomes too high, the Coriolis force cannot balance the pressure gradient force and the centrifugal force which increases as the square of the current speed. This holds especially where r is small, that is, near the center of the anticyclonic gyres.

Such a restriction does not exist for low pressure areas, and with cyclonic motion, where both pressure gradient and wind (current) can, and very often do, increase toward the center. This has been clearly demonstrated for atmospheric circular motions. (See, e.g., HAURWITZ, 1941.)

Theoretically, in a current around a low pressure area, $\partial p/\partial r$ and, therefore, the current speed can increase indefinitely toward the center, since the root in eq.84 always remains positive. This explains why in cyclonic atmospheric vortices, like hurricanes, the pressure falls rapidly toward the center. There is no restriction imposed on the increase of the pressure gradient toward the center of the cyclonic gyre. With increasing speed of the circular motion and decreasing r the centrifugal force increases and its effect on the dynamic balance becomes more and more significant compared to the Coriolis force. Also frictional forces become more and more important as the speed of the current and its lateral gradient increases. Under such conditions, the assumptions made for the derivation of a gradient current no longer hold, and the balance of forces is *essentially* determined by pressure gradient, centrifugal and frictional forces. In particular, this applies to small scale vortices like "sand devils" and tornados in the atmosphere, and to the whirlpools in the Strait of Messina[1] where r is very small. If the horizontal component of the Coriolis force becomes insignificant, small-scale vortices can even rotate temporarily in anticyclonic direction around a low pressure center, although this case is not stable because everywhere on the rotating planet Earth, with the exception of the equator ($\phi = 0$), the horizontal component of the Coriolis force is present.

Cyclostrophic motion

A criterion for the relative significance of the Coriolis force on horizontal motions is the *Rossby number*:

$$R_o = \frac{\text{characteristic velocity}}{\Omega R_e}$$

where Ω is the angular velocity of the earth's rotation and R_e is

---

[1] MAZZARELLI (1938) has given an interesting report about the eddies, current convergences and rip currents in the Strait of Messina. Scylla and Charybdis, and the tidal currents in the Strait of Messina have been dealt with in detail by DEFANT (1940a).

the radius of the earth. A dimensionless number corresponding to the Rossby number can also be written:

$$R_o = \frac{\text{current velocity}}{\Omega L} \qquad (86)$$

where L is a characteristic length parameter of the current, for example its width. If $L = r$, where r is the radius of the inertia circle, the Rossby number can be compared with $c/2r\omega \sin\phi$ which follows from eq.76. The Rossby number is the ratio of inertial force to Coriolis force. Very large Rossby numbers mean that the Coriolis effect may be neglected. This occurs if the rotation rate is slow (small Ω) or if the dimensions of the system characterized by eq.86, L, are small.

In small scale systems where c_θ^2/r becomes much greater than the term fc_θ, the Coriolis acceleration may be neglected, and a cyclostrophic motion can be approximated. For this type of motion the centrifugal force is balanced by the pressure gradient, such that:

$$\frac{c_\theta^2}{r} = \frac{1}{\rho}\frac{\partial p}{\partial r} \qquad (87)$$

Depending on the initial rotation, pure cyclostrophic motion can be cyclonic or anticyclonic but the pressure gradient must always be positive, that is, the pressure must be lowest in the center of the vortex. In equatorial regions, where exactly at the equator $2\omega \sin\phi = 0$, cyclostrophic (or approximately cyclostrophic) motion can be expected to occur more frequently for larger vortices than in higher latitudes. The small scale structure of ocean currents and the associated pressure field are not yet adequately known to illustrate this type of motion by direct observations.

It is seen that in middle latitudes where $f \approx 10^{-4}$, and $c_\theta = 100$ cm/sec, the absolute values of both terms on the left side of eq.83 become equal if $r = 10$ km. However, in lower latitudes around $4°$ or $5°$ where $f \approx 10^{-5}$, r is as large as 100 km for the same speed around curved streamlines when the centrifugal force equals the Coriolis force.

CURRENTS INCLUDING FRICTION AND DIFFUSION

In a general way frictional forces were introduced in Chapter III. It is difficult to include accurately their effect in the dynamic balance of forces because the magnitude of friction can vary considerably in ocean currents, depending on various factors, for example, current shear, size of currents, density stratification of oceanic water masses, and other parameters. However, if frictional forces are to be included, their effect on ocean currents can be estimated, and it can be shown that the current structure changes significantly. Without specifying the character of friction, first consider frictional forces in a non-accelerated current. From eq.12 it follows that in the case of horizontal currents the balance is given by eq.88:

$$2\omega \sin\phi \, v + F_x = \frac{1}{\rho} \frac{\partial p}{\partial x}$$

$$-2\omega \sin\phi \, u + F_y = \frac{1}{\rho} \frac{\partial p}{\partial y}$$

(88)

If compared with horizontal geostrophic currents (eq.61), it is seen that the components of the frictional force disturb the geostrophic equilibrium such that the current can no longer be parallel to the isobars.

A simple assumption about friction is the Guldberg–Mohn assumption (Chapter III). If it is assumed that $F_x = -ku$ and $F_y = -kv$ are the components of the frictional force per unit mass in the x- and y-directions, respectively, eq.88 can be written as:

$$\frac{1}{\rho} \frac{\partial p}{\partial x} = fv - ku$$

$$\frac{1}{\rho} \frac{\partial p}{\partial y} = -fu - kv$$

(89)

The Guldberg-Mohn friction assumes that the frictional force acts in the opposite direction to the velocity. With a more general assumption about friction, as in the form of equations 23 or 24, it can be shown that the vector of the frictional force is not necessarily opposite to the current vector. Let the horizontal components of the frictional force be F_x and F_y, respectively.

If u_g and v_g are the velocity components of a geostrophic flow parallel to the isobars according to eq.61, and defined by the pressure gradient, then:

$$fv_g = \frac{1}{\rho}\frac{\partial p}{\partial x}; \quad fu_g = -\frac{1}{\rho}\frac{\partial p}{\partial y} \qquad (90)$$

Substitution of eq.90 into eq.88 yields:

$$\begin{aligned} fm_y &= = F_x \\ fm_x &= -F_y \end{aligned} \qquad (91)$$

where $m_y = v_g - v$ and $m_x = u_g - u$ are the components of the difference vector, $\mathbf{m} = \mathbf{c}_g - \mathbf{c}$, between the geostrophic current vector $\mathbf{c}_g$ and the actual current vector $\mathbf{c}$ which is influenced by friction. If the first equation in eq.91 is multiplied by m_x and the second by m_y, it follows, after substracting the second from the first equation, that:

$$F_x m_x + F_y m_y = 0 \qquad (92)$$

This equation represents a scalar product of the horizontal vectors $\mathbf{F}$ and $\mathbf{m}$. Since this product is zero, the vector $\mathbf{F}$, the

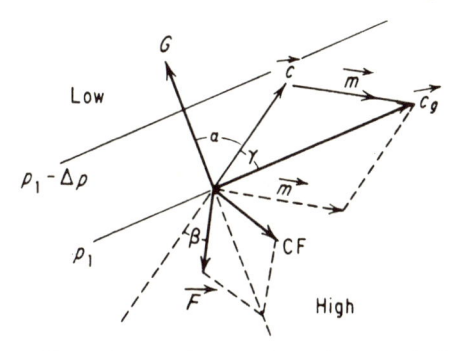

Fig.45. Balance of forces in a straight horizontal current including friction.

frictional force, must be perpendicular to the vector m. Fig.45 illustrates the balance of forces per unit of mass. It is seen that, in general, the vector, F, cannot be opposite the current vector c but forms an angle, β, with the current direction. This angle is zero only in the exceptional case where m is perpendicular to c. For atmospheric motions, SVERDRUP (1916a) provided empirical evidence for the deflection angle β.

The approximate value of k in eq.89 for ocean currents was estimated by DEFANT (1929a) between 10^{-6} and 10^{-7}/sec. NEUMANN (1954) found an average of $k \approx 3 \cdot 10^{-6}$ for the upper one thousand meters of the oceanic circulation. The order of magnitude $k \approx 10^{-6}$ corresponds to a deflection angle $\gamma = 90° - \alpha$ (see Fig.45) of about $2°$ in middle latitudes. A similar order of magnitude of k seems to apply to atmospheric motions.

Of course, the value of k can vary from place to place in the oceans as well as in the atmosphere. An interesting example of smoothly decaying inertia currents was given by the observations of GUSTAFSON and KULLENBERG (1936) which are reproduced in Fig.41. The decrease of the velocity between August 20 (c_0) and August 24 (c) is, approximately $c_0/c = 1.5/0.15 = 10$, over a time interval of four days. As will be shown in the following section, this rate of velocity decrease suggests $k = 0.7 \cdot 10^{-5}$/sec.

Decay of a current under the effect of friction

If currents are accelerated, and the simple Guldberg–Mohn assumption about friction is applied (DEFANT, 1929a), the equations for horizontal motion are:

$$\frac{du}{dt} = fv - ku - \alpha \frac{\partial p}{\partial x}$$

$$\frac{dv}{dt} = -fu - kv - \alpha \frac{\partial p}{\partial y}$$

(93)

After multiplication of the first equation by u and of the second by v, addition yields:

$$u\frac{\mathrm{d}u}{\mathrm{d}t}+v\frac{\mathrm{d}v}{\mathrm{d}t}=\frac{1}{2}\frac{\mathrm{d}c^2}{\mathrm{d}t}=-\alpha\left(\frac{\partial p}{\partial x}u+\frac{\partial p}{\partial y}v\right)-kc^2 \quad (94)$$

where $c^2 = u^2 + v^2$. Now, let the horizontal pressure gradient, which has caused the motion, suddenly vanish for some reason. The motion will continue, following the equation:

$$\frac{1}{c}\frac{\mathrm{d}c}{\mathrm{d}t}=-k \quad (95)$$

If at the time $t = 0$ the current velocity is c_0, integration of eq.95 yields:

$$c = c_0{}^{-kt} \quad (96)$$

For the case where $k = 0$, eq.95 states that with $(\mathrm{d}c/\mathrm{d}t)/c = 0$ pure frictionless inertia currents would be the result. The motion as described by eq.95, or its integrated form, eq.96, represents decaying (damped) inertia currents as a result of friction. It is seen that the velocity c decreases exponentially with time, and the rate of decrease depends on the value of k. As c decreases, the radius of the inertia circle also decreases as shown in eq.76.

For the case of damped inertia currents in the Baltic Sea (Fig.41) $c_0/c = 10$ for a time interval of four days. Thus, from $kt = -\ln c/c_0$, it follows that k equals $0.7 \cdot 10^{-5}$. This value of k is about ten times larger than the values estimated by DEFANT (1929a) and about two times larger than the value obtained by NEUMANN (1954) for an average of the upper 1,000 m of the ocean. The greater value of k for the Baltic Sea seems reasonable because of the shallow water. Other cases of decaying inertia currents were evaluated by GARNER (1961a) for oceanic as well as atmospheric currents. The results are that for the upper strata of the oceans and the lower tropospheric layers of the atmosphere where friction plays a dominant role, the order of magnitude $k \approx 10^{-6}$ through 10^{-5} is the right order of magnitude to estimate friction when the simple approach of GULDBERG and MOHN (1876) is used.

The law of parallel fields and its disturbance by friction and diffusion

In the case of frictionless ocean currents and without mixing a non-accelerated current must always be parallel to the isobars and to the isopycnals. This follows from the law that governs geostrophic flow. From $\partial p/\partial x - f\rho v = 0$ and $\partial p/\partial y + f\rho u = 0$, it follows after elimination of the pressure by cross differentiation of the two equations, taking f = constant, and subtraction of one equation from the other that $u\partial\rho/\partial x + v\partial\rho/\partial y = 0$ if $\partial u/\partial x + \partial v/\partial y = 0$. Introduction of the stream function ψ, (eq.37) for horizontal flow ($w = 0$) yields $(\partial\rho/\partial x)(\partial\psi/\partial y) - (\partial\rho/\partial y)(\partial\psi/\partial x) = 0$. This equation states that the lines of ρ = constant (isopycnals) must be parallel to the lines ψ = constant (streamlines). In a stratified ocean, it also follows that the isobars and isopycnals at one level must be parallel to each other and to those at other levels if the density stratification is continuous.

The law of parallel fields (or solenoids) was first mentioned and explained by Helland-Hansen and Ekman (EKMAN, 1923). Ekman states[1]: "Several years ago when examining oceanographic charts of the Norwegian Sea, my friend Björn Helland-Hansen and I noticed to our surprise that the spatial distribution of temperature and salinity had a remarkable tendency to follow the topographical features of the ocean bottom."

In a stratified ocean, stationarity requires that streamlines of geostrophic flow and isopycnals be parallel. This is evident because, otherwise, surfaces of equal density would be displaced and this would not be compatible with a stationary density distribution. EKMAN (1923) has clearly shown the limitations of the law of parallel fields by stating where and when it does not apply: It does not apply when friction or diffusion dominate the structure of currents and the associated mass (density) field. It also excludes gains or losses of heat or salt amount in sea water (sources or sinks of the physical–chemical properties of sea water) and discontinuities in the structure of the field of mass or currents.

The study of large scale current conditions in the ocean has shown that over large areas the isopycnals at different levels are

---

[1] Translation from the original text on p.66 of EKMAN (1923).

indeed nearly parallel to each other and parallel to the average large scale circulation. In many cases, however, the law of parallel fields seems to be disturbed or even completely destroyed.

This is not surprising, particularly not in the surface layers of the sea, where currents are often neither stationary nor frictionless. Also the field of mass as given by the fields of temperature and salinity at the sea surface cannot always be treated as stationary. Even in the case of a stationary distribution of both currents and mass, the law of parallel fields cannot apply if friction or diffusion are significant factors to be considered in the mutual adjustment of the two fields. This, of course, means a radical departure from the idea of a geostrophic current, or gradient current. Both of these currents are stationary; but friction and/or diffusion can change the relationship between currents and the field of mass significantly.

The effects of friction and diffusion on the relationship between the field of current and the field of mass can easily be demonstrated, according to DEFANT (1931), for stationary horizontal currents. Fig.46 shows the horizontal distribution of streamlines and isopycnals for the case where these lines intersect.

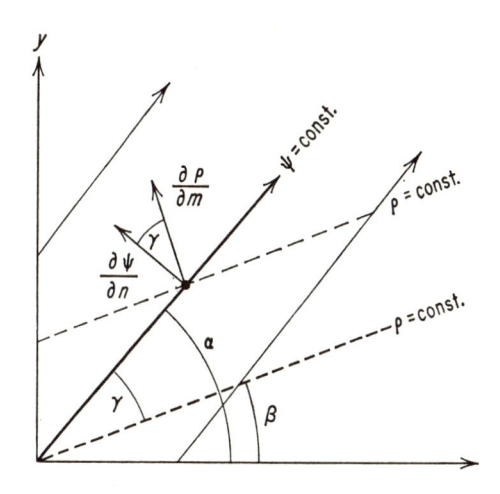

Fig.46. Streamlines, ψ, and isopycnal lines, ρ, in a horizontal current field. Both lines intersect, and form the angle of intersection, γ.

Both fields are considered to be stationary. The system of horizontal streamlines follows the equation $\psi(x, y) =$ constant and the system of isopycnals follows the equation $\rho(x, y) =$ constant. With the neglect of vertical motion $\partial u/\partial x + \partial v/\partial y = 0$, and the velocity components can be expressed by the stream function ψ (eq.37).

In Fig.46, the angle α represents the inclination of streamlines against the x-axis, and β is the corresponding inclination angle for the isopycnals, Since:

$$\tan\alpha = \left(\frac{dy}{dx}\right)_\psi = \frac{v}{u} = -\frac{\partial\psi/\partial x}{\partial\psi/\partial y}$$

$$\tan\beta = \left(\frac{dy}{dx}\right)_\rho = -\frac{\partial\rho/\partial x}{\partial\rho/\partial y}$$
(97)

it follows that the "crossing angle" between streamlines and isopycnals is $\alpha - \beta = \gamma$, and:

$$\tan\gamma = \tan(\alpha - \beta) = \frac{\tan\alpha - \tan\beta}{1 + \tan\alpha \tan\beta} \tag{98}$$

Substitution of eq.97 into eq.98 yields:

$$\tan\gamma = -\frac{\dfrac{\partial\psi}{\partial x}\dfrac{\partial\rho}{\partial y} - \dfrac{\partial\psi}{\partial y}\dfrac{\partial\rho}{\partial x}}{\dfrac{\partial\psi}{\partial y}\dfrac{\partial\rho}{\partial y} + \dfrac{\partial\psi}{\partial x}\dfrac{\partial\rho}{\partial x}} = \frac{u\dfrac{\partial\rho}{\partial x} + v\dfrac{\partial\rho}{\partial y}}{u\dfrac{\partial\rho}{\partial y} - v\dfrac{\partial\rho}{\partial x}} \tag{99}$$

For geostrophic flow, the numerator in eq.99 is zero and, therefore, $\gamma = 0$.

Non-accelerated horizontal motion, including friction, is given by eq.100:

$$\frac{\partial p}{\partial x} - f\rho v = R_x$$

$$\frac{\partial p}{\partial y} + f\rho u = R_y$$
(100)

where:

$$R_x = \frac{\partial}{\partial y}\left(A_h \frac{\partial u}{\partial y}\right) + \frac{\partial}{\partial z}\left(A_z \frac{\partial u}{\partial z}\right) = \rho F_x$$

$$R_y = \frac{\partial}{\partial x}\left(A_h \frac{\partial v}{\partial x}\right) + \frac{\partial}{\partial z}\left(A_z \frac{\partial v}{\partial z}\right) = \rho F_y$$

(101)

If the first equation in eq.100 is differentiated with respect to y and the second with respect to x, keeping $f = 2\omega \sin\phi$ constant, subtraction of the first from the second equation yields:

$$v\frac{\partial \rho}{\partial y} + u\frac{\partial \rho}{\partial x} = \frac{1}{f}\left(\frac{\partial}{\partial x} R_y - \frac{\partial}{\partial y} R_x\right) \quad (102)$$

Eq.102 can be substituted into eq.99, and it follows that:

$$\tan\gamma = \frac{\frac{1}{f}\left(\frac{\partial}{\partial x} R_y - \frac{\partial}{\partial y} R_x\right)}{u\frac{\partial \rho}{\partial y} - v\frac{\partial \rho}{\partial x}} \quad (103)$$

Eq.103 shows that the angle γ depends on friction, such that in the case of no friction ($R_x = R_y = 0$), $\tan\gamma = 0$ and the parallelism between streamlines and isopycnals is preserved. The denominator is not equal to zero since it represents the scalar product $(\partial\psi/\partial y)(\partial\rho/\partial y) + (\partial\psi/\partial x)(\partial\rho/\partial x)$ which has a maximum value for $\gamma = 0$. When friction is present, $\tan\gamma \neq 0$, and the value of γ depends on the changes in the horizontal space of the frictional forces. However, under such conditions, vertical motion should be considered. Eq.103 can serve as a rough estimate of γ in regions outside of strong convergences and divergences.

If friction resulting from lateral shearing stresses is neglected and the vertical eddy viscosity coefficient A_z in eq.101 is constant, it follows that:

$$\frac{\partial \psi}{\partial x}\frac{\partial \rho}{\partial y} - \frac{\partial \psi}{\partial y}\frac{\partial \rho}{\partial x} = \frac{A_z}{f}\frac{\partial^2}{\partial z^2}\left(\frac{\partial^2 \psi}{\partial x^2} + \frac{\partial^2 \psi}{\partial y^2}\right) \quad (104)$$

172 MAJOR TYPES OF OCEAN CURRENTS

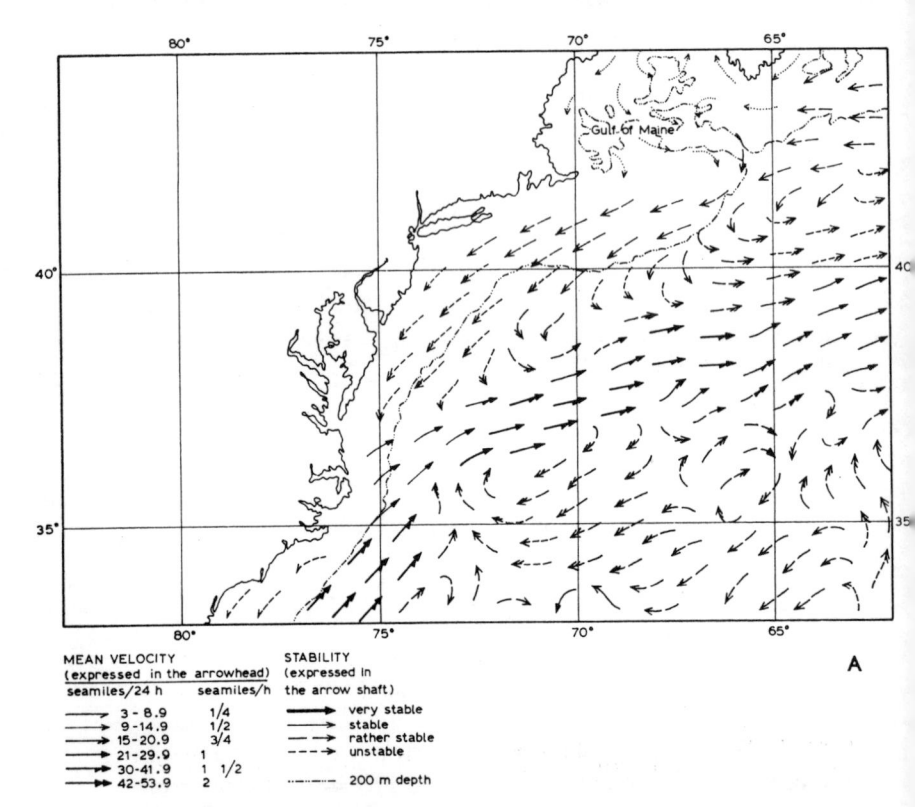

Fig.47. Mean distribution of currents and density (σ_t) at the sea surface off the North American east coast (April, May, June). (After NEUMANN and SCHUMACHER, 1944.)
A. Mean velocity and stability. B. Density distribution of the currents.

For potential flow, the term in parentheses is always zero, and in this case the parallelism of streamlines and isopycnals is preserved.

Fig.47 shows an example of the mean distribution of density and currents at the sea surface off the North American east coast (NEUMANN and SCHUMACHER, 1944). The mean surface current vectors for the month of May and the mean isopycnals for the

LINES OF EQUAL DENSITY

B

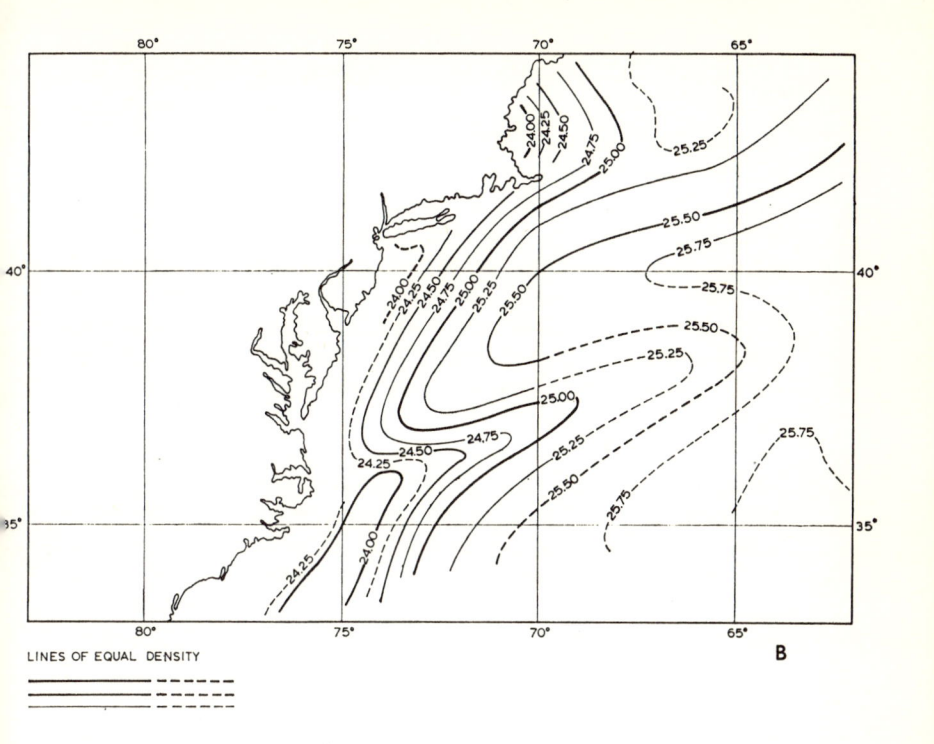

months April–June were constructed independently, and later both fields were combined.[1] It is seen that the current vectors (or streamlines which can be drawn from the vector field) and the isopycnals intersect at different angles. This is not surprising because mixing and friction are most significant in the sea surface layers. With the assumption that the two fields are stationary (or quasi-stationary), such charts can be used to estimate the magnitude of terms like $\partial R_y/\partial x$ and $\partial R_x/\partial y$, approximately, in different parts of the current system with the exception of regions where the vertical velocity component becomes significant or where accelerations have to be considered.

---

[1] Similar charts for February (January–March), August (July–September) and November (October–December) were also presented by NEUMANN and SCHUMACHER (1944).

Tongue-like distribution of water properties as a result of currents (advection) and diffusion

A fact that seriously restricts the application of the "core method" and complicates the interpretation of results obtained by this method is that ocean currents are never completely frictionless. In addition, mixing of water masses has to be considered in the dynamic analysis of currents. This is the reason why the "core method" cannot be applied to the surface strata of the oceans where eddy viscosity and eddy diffusion are significant factors. For deeper strata, WÜST (1936b) successfully applied this method to the analysis of the deep sea stratification and the spreading (flow) of characteristic water masses. However, even in the deep sea layers, caution is indicated if no other evidence is available to support the assumption that the "core" or "tongue" of a certain sea water property (e.g., the salinity) is indeed the result of a current.

CASTENS (1931), THORADE (1931), and DEFANT (1929b, 1936c) among others considered the effects of both advection (currents) and mixing of water masses on the distribution of conservative properties in ocean water. SVERDRUP (1940b) has shown that even in a motionless ocean basic features of the actual vertical and meridional temperature structure can be explained on the basis of pure heat conduction in vertical and meridional direction if realistic assumptions are made for the temperature distribution along the boundaries of the oceans.

Most interesting results were first presented by DEFANT (1929b) and THORADE (1931). Assume that the distribution of a property, for example, the salinity, S, is constant with time. Consider a vertical section (xz plane) showing a tongue-like pattern as presented in Fig.48. Let the x-axis point in the direction of a current with the velocity u ($v = 0$, $w = 0$) and allow eddy diffusion in the vertical direction with a constant coefficient K_z such that $K_x = 0$ and $K_y = 0$. With these assumptions and for stationary conditions, eq.60 reduces to eq.105:

$$u \frac{\partial S}{\partial x} = K_z \frac{\partial^2 S}{\partial z^2} \qquad (105)$$

if $K_z = A_z/\rho$ = constant and if the property s is replaced by the property S, the salinity of the sea water.

Eq.105 is the most simple equation that can be used to relate advection and mixing by eddy diffusion, and to study their combined effects on the distribution of S. For a solution to this

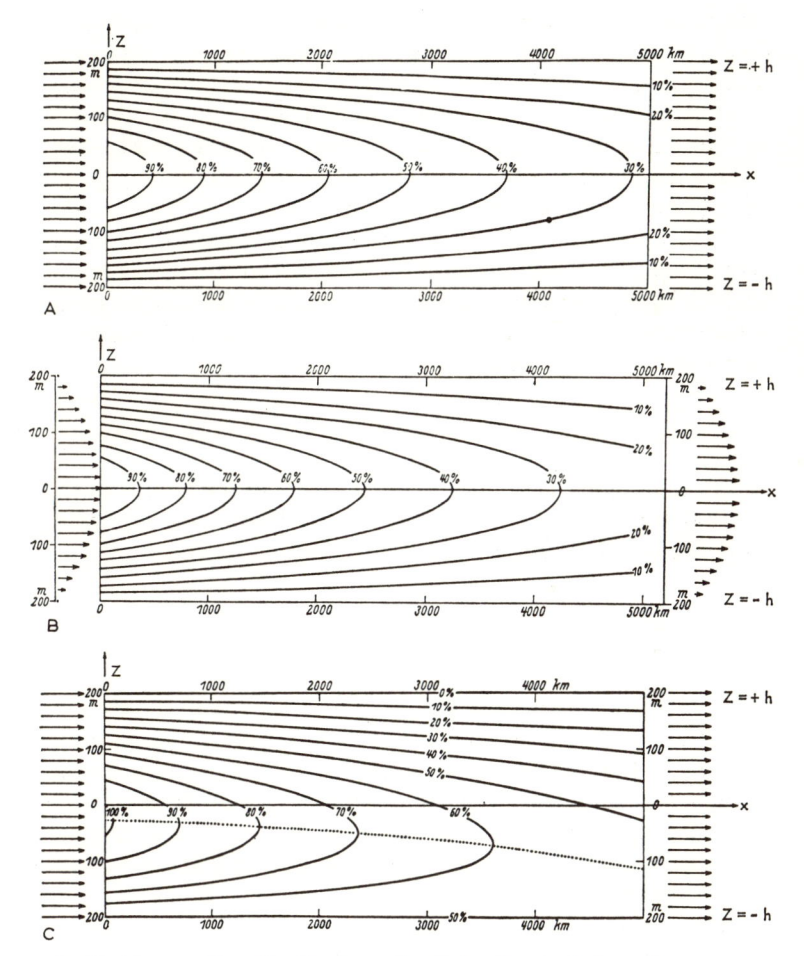

Fig.48. Tongue-like distribution of salinity in vertical sections corresponding to different boundary conditions and currents. (After THORADE, 1931, from NEUMANN and PIERSON, 1966.)

equation, statements about boundary conditions have to be made. The solution represents the distribution of S as a function of x and z, and this pattern is what the oceanographer observes when measuring the salinity at different depths in a vertical section along a line in the x-direction. From such sections, the observer is often inclined to draw conclusions about major directions of ocean current (see "Core method", Chapter I). Without any further evidence about the character of currents in this region, this method can, however, lead to erroneous results.

First, consider a horizontal current flowing in the positive x-direction with a constant velocity $u = u_o$ as shown in Fig.48A. Along the upper and lower boundaries of the section, at $z = +h$ and $z = -h$, respectively, the salinity is $S = S_o$ = constant with x. Such a boundary condition may be possible in nature as a result of certain physical processes that maintain a constant salinity along horizontal surfaces even within a uniform current. Further, assume that at $x = 0$ the vertical salinity distribution is given by the equation:

$$S = S_o \cos \frac{\pi}{2} \frac{z}{h}$$

Thus, the salinity has a maximum at $z = 0$ in the middle of the section between $z = +h$ and $z = -h$. With these boundary conditions, the solution to eq.105 is:

$$S = S_o \exp\left[-\frac{\pi^2 K_z}{4 u_o h^2} x\right] \cos \frac{\pi}{2} \frac{z}{h} \qquad (106)$$

The relative stationary distribution of the salinity in the section, according to eq.106, is shown in Fig.48A with the assumption that $K_z = 4$ cm^2/sec, $u_o = 10$ cm/sec, $h = 2 \cdot 10^4$ cm). A definite tongue-like arrangement of isohalines is evident even though the current is uniform thoughout the whole section. With increasing distance x, the salinity concentration decreases and the vertical and horizontal gradients of S diminish as a result of eddy diffusion. Transport of the property S into the section takes place at $x = 0$ and transport out of the section takes place through the other boundaries. The tongue-like pattern in the distribution of S may lead to the erroneous assumption that the current velocity has a maximum along the center line of the tongue. In an investigation of the circulation of the Atlantic Intermediate Water, KIRWAN (1963) has shown that the maximum current and the salinity minimum in general do not coincide.

Nearly the same qualitative salinity distribution in the section can be obtained by assuming a current, as shown in Fig.48B, which has indeed a maximum in the axis of the tongue. Thus from the mere appearance of tongue-like distributions no definite conclusions can be drawn as to the currents.

Using the same equation (eq.105) but different boundary conditions, THORADE (1931) has shown that different types of tongues can be derived. Even a tilt of the axis of maximum property concentration may appear in a *uniform* current if the distribution of property transport into the section at $x = 0$ is asymmetrical with respect to $z = 0$. This is shown in Fig.48C where the distribution of S indicates a downward curving tongue which, of course, has no relationship to a current with an axis of maximum speed that "tilts" towards $z = -h$. Similar results can be derived for the flow in horizontal sections with lateral mixing if eq.105 is applied to the xy-plane.

With care and judgment the interpretation of the distribution of water properties in oceanic sections can yield valuable results for the analysis of water movements even if simple concepts such as constant eddy diffusion coefficients are involved. More recently, GARNER (1962) used observations of salinity and a quantity related to the amount of ^{14}C in the water for a meridional section near 180° longitude in the South Pacific Ocean to derive the major water movements. His analysis was based on the diffusion equation including advection which can be applied separately to both of the observed water properties in the section. For the distribution of ^{14}C it was also necessary to introduce a term that accounted for the radioactive decay of this property. With the observed distribution of salinity and ^{14}C, and the assumption that the magnitudes of the vertical and lateral eddy diffusion coefficients are 10 (cm^{-1} g/sec) and 10^6 (cm^{-1} g/sec), respectively, Garner was able to evaluate two partial differential equations and to solve these equations for the current components in meridional (v) and vertical (w) directions along the section. The result was a quite realistic velocity field showing an area of subsiding water in the sub-Antarctic regions and a layer of northward spreading water at subsurface levels indicating the flow of Antarctic Intermediate Water towards the equator.

WIND-DRIVEN CURRENTS IN A HOMOGENEOUS OCEAN

If a wind blows over a water surface, pure wind-driven currents are produced by transfer of momentum from wind to water (wind stress) at the air–sea interface and by friction between moving layers within the water.

The mechanism of momentum transfer between the moving air and the sea surface is not yet fully understood. Part of the momentum transfer provides the energy used in generating waves, and there is also a net forward motion of water associated with the propagation of waves at the sea surface.

As a result of our lack of understanding of the physical mechanism of momentum transfer at the sea surface, the fundamental relationship between the wind stress, τ, at the air–sea interface, and the wind speed, w, is not known with the desired degree of accuracy. For a given wind speed measured at a given height over the sea surface, the wind stress depends also on other parameters like the vertical stability of the air immediately above the sea surface and sea state conditions, that is, the state of development of waves when a wind of a given speed starts to act. This introduces the time, or the duration of wind action, as an important factor in the wind stress–wind relationship. Also, the horizontal distance over which the wind blows, or the size of the ocean area covered by the wind (the "fetch"), is an important parameter which needs to be taken into account, because both duration and fetch of wind action determine the state of wave motion generated by the wind. Only in the case where duration and fetch are large enough, and where the waves are fully arisen for a given wind speed, does the dependency of τ on duration and fetch of wind action vanish. Sea state conditions are important in the wind stress–wind relationship because the waves at the sea surface provide a certain hydrodynamical "roughness" of the interface and interact with the wind field above the waves (STEWART, 1961a,b; SCHMITZ, 1962a,b). Such problems have to be investigated in much greater detail before definite relationships between the wind stress and the wind over wavy sea surfaces can be established.

It is customary to express the wind stress τ at a given wind speed w of the air over water in the form of eq.107:

$$\tau = \rho' C(w) w^2 \qquad (107)$$

where ρ' is the density of the air and the dimensionless factor $C(w)$ is called "drag" or "resistance" coefficient. Its value depends on the parameters mentioned above and on the height over the sea surface where the wind is measured. At present, many investigators prefer to use a constant value of $C(w)$. Others suggest an increase of $C(w)$ with increasing wind speed and still others suggest a decrease of $C(w)$ with increasing wind speed. A summary of the present state of our knowledge of this problem was recently given by ROLL (1965) and NEUMANN and PIERSON (1966). DELEONIBUS (1966) has shown that $C(w)$ can behave either way, depending on the Richardson number of the air flow over the waves. According to DeLeonibus, $C(w)$ decreases with increasing wind speed for one range of Richardson numbers and increases with increasing wind speed for another range.

The first one who suggested an empirical relationship between wind stress and wind was EKMAN (1905), using a constant drag coefficient. From observations of the slope of the sea surface during a storm in the Baltic Sea in 1872 he arrived at:

$$\tau = 3.2 \cdot 10^{-6} \, w^2 \, (\text{cm}^{-1} \, \text{g/sec}^2 \text{ or dynes/cm}^2)$$

if the wind speed w is measured in cm/sec. With an average air density of $\rho' = 1.25 \cdot 10^{-3}$, Ekman's result is:

$$\tau = \rho' \cdot 2.6 \cdot 10^{-3} \, w^2$$

When compared with eq.107 the value of $C(w) = 2.6 \cdot 10^{-3}$ is probably applicable to an anemometer height of about 15 m above the water. This value of $C(w)$ is approximately in the middle of the estimates that followed from the results of later investigations. Ekman's estimate seems to apply with fair accuracy for wind speeds around 20 m/sec measured at an anemometer height of about 15 m. However, the question of how $C(w)$ varies with wind speed has not yet been answered. The uncertainties of the correct value of $C(w)$ and its dependence on the wind speed are particularly great at low wind speeds.

Pure drift currents in deep and shallow water

During the Norwegian North Polar Expedition 1893–96 with the famous research vessel "Fram", NANSEN (1902) observed that the drift of ice with respect to the wind did not follow the wind direction but deviated to the right by 20° to 40° when looking in the direction of the wind. He suggested an explanation that considers both friction between the wind and the sea surface, and friction within the water, together with the effect of Coriolis forces acting upon the moving water layers. On Nansen's suggestion, EKMAN (1902) investigated the problem mathematically and laid the foundation for one of the most important theoretical developments in dynamic oceanography. His first paper was followed by a remarkable and thorough study of the problem of wind-generated ocean currents (EKMAN, 1905) which was the forerunner of basic contributions to the theory of ocean currents. Nansen's observations and ideas were in sharp contrast to ZÖPPRITZ's theory (1878) of wind-driven ocean currents; however, Ekman's analysis proved that Nansen's hypothetical explanation was essentially correct.

A simple non-mathematical explanation of some basic results obtained by Ekman can be given by mere physical reasoning. Consider the water as composed of infinitesimally thin horizontal layers and assume that each layer has a constant velocity. However, the velocity changes from layer to layer. It is reasonable to assume that the velocity decreases with depth if the wind stress as a driving force acts at the surface while friction within the water acts as an agent of energy dissipation. On the rotating earth and with consideration of Coriolis forces acting always perpendicular to the motion at each depth, it can be shown that the horizontal current must also change its direction while its speed decreases with increasing depth.

Fig.49 represents three current vectors at different levels projected on a horizontal plane. These vectors are drawn from point 0 to points a_1, a_2, a_3. The current vector 0–a_3 is just above the current vector 0–a_2, and the current vector 0–a_1 is just below the vector $0-a_2$. An observer drifting with the current $0-a_2$ would notice immediately above a resultant water movement in the direction a_2-a_3 and immediately below a resultant current in the direction a_2-a_1. It is reasonable to assume that the frictional force (or drag) between two layers acts in the direction of the velocity difference between the layers. This is indicated in Fig.49 by arrows pointing in the directions a_2-b_3 and a_2-b_1, respectively. If no forces other than friction (besides the gravitational force) act on the moving water, the only force to balance the resultant frictional force a_2-a_4 at the level of the current vector $0-a_2$ is the Coriolis force.

In the Northern Hemisphere the horizontal component of the Coriolis force acts to the right when facing in the direction of the motion. For the current vector $0-a_2$ the direction and magnitude of the Coriolis force (CF) is indicated by the dashed arrow a_2-a_5. With stationary and non-accelerated motion at each level, it follows that the resultant of frictional forces must balance the horizontal component of the Coriolis force. Thus, the resultant of frictional forces at the level $0-a_2$ must point in the direction a_2-a_4 as shown in Fig.49. This direction is necessarily opposite to the Coriolis force. Thus, it is seen that the

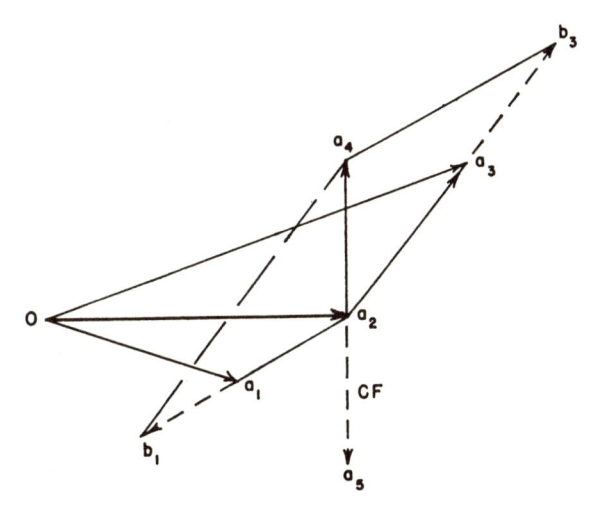

Fig.49. Graphical explanation of the Ekman spiral (see text).

requirement for non-accelerated flow can only be fulfilled when the current in the Northern Hemisphere turns to the right while decreasing with increasing depth.

The deflection of the current from the wind direction and the increase of this angle with depth is one of the most outstanding features obtained by Ekman for wind-driven currents and it was not surprising that his results were exposed to considerable criticism immediately after publication of his theory (e.g., see EKMAN, 1908, 1909).

Pure horizontal drift currents in a homogeneous ocean are stationary as long as frictional forces balance the horizontal components of the Coriolis force. Horizontal pressure gradients within the water do not occur. The only driving force is the wind stress at the sea surface and frictional coupling between layers within the water. If friction resulting from vertical shear is considered, and if the assumption is made that the vertical eddy viscosity coefficient A_z in eq.24 is constant with depth, it follows from eq.12 that for pure non-accelerated wind-driven currents:

$$2\omega \sin\phi \, u = \frac{A}{\rho} \frac{\partial^2 v}{\partial z^2}$$

$$-2\omega \sin\phi \, v = \frac{A}{\rho} \frac{\partial^2 u}{\partial z^2}$$

(108)

The subscript z at A_z has been omitted in eq.108 with the understanding that $A = A_z$. The solution of this set of simultaneous differential equations requires statements about two boundary conditions. The boundary condition for the sea surface (upper boundary condition) is that a wind stress acts in the direction of the wind. Since the resulting currents are independent of the orientation of the rectangular (Cartesian) coordinate system, it is of advantage to orient one coordinate axis, say the y-axis, in the wind direction. The x-axis is perpendicular to y pointing to the right of the y-axis while the z-axis points downward in a positive direction.

With the wind stress acting in the direction of the positive y-axis, $\tau_x = 0$, and $\tau_y = -A(\mathrm{d}v/\mathrm{d}z)_{z=0}$. This constitutes the upper boundary condition, while $u = 0$, $v = 0$ at $z = \infty$ (for infinitely deep water) constitutes the lower boundary condition. The solution to eq.108 with consideration of these boundary conditions is:

$$u = V_0 \, e^{-(\pi/D)z} \cos\left(45° - \frac{\pi}{D}z\right) \tag{109}$$

$$v = V_0 \, e^{-(\pi/D)z} \sin\left(45° - \frac{\pi}{D}z\right)$$

where:

$$D = \pi \sqrt{\frac{A}{\rho \omega \sin\phi}} \tag{110}$$

$$V_0 = \frac{\tau}{Aa} \tag{111}$$

$$a = \sqrt{\frac{2\rho\omega \sin\phi}{A}} \tag{112}$$

V_0 represents the speed of the surface current.

Eq.109 describes the character of pure wind-driven currents in a homogeneous ocean. The subscript y for τ_y in eq.111 has been omitted if it is understood that the wind direction is oriented in the positive y-direction or in the direction of the v-component of the drift current.

The structure of Ekman's pure wind drift currents (eq.109) shows that for the sea surface ($z = 0$), $u = V_0 \cos 45°$ and $v = V_0 \sin 45°$. Hence, the surface current, $V_0^2 = u^2 + v^2$, points in a direction 45° cum sole to the wind direction. In the Northern Hemisphere the surface water moves in a direction that is 45° to the right, and in the Southern Hemisphere 45° to the left when one is looking in the direction toward which the wind is blowing.

At a depth $z = D$, which is given by eq.110, it is seen that:

$$\begin{aligned} u_D &= V_0 \, e^{-\pi} \cos(45° - \pi) \\ v_D &= V_0 \, e^{-\pi} \sin(45° - \pi) \end{aligned} \tag{113}$$

This shows that at $z = D$ the current vector has decreased to the value $e^{-\pi}$ times the surface speed (ca. 1/23 V_o), and that the current direction is exactly opposite to the surface current direction. The depth D is called the *depth of frictional influence*.

From eq.110 it is seen that D increases proportionally to the square root of the vertical eddy viscosity coefficient. Thus, it can be expected that with increasing wind speed and increasing wave motion which produces stronger vertical mixing in the surface layers (wind-stirring) the vertical extent of the pure drift current increases.

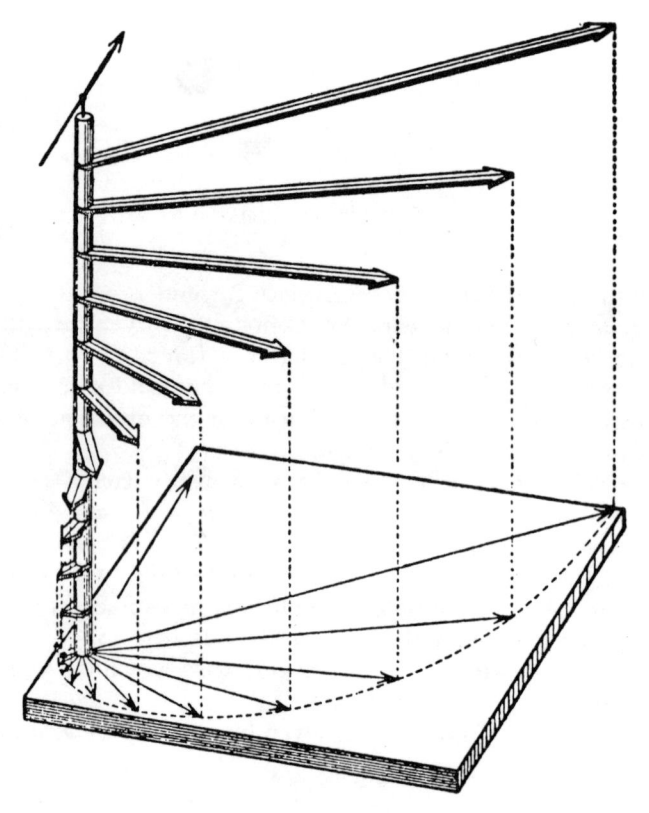

Fig.50. Vertical structure of a pure drift current. (After Ekman, 1905.)

D also depends on the geographical latitude, ϕ, and decreases with the square root of $\sin\phi$. At the geographical equator where $\phi = 0$, D becomes infinite, and the quantity a in eq.112 becomes zero. Thus, V_o in eq.111 should become infinite theoretically, while the current direction changes from a right-hand deflection in the Northern Hemisphere to a left-hand deflection in the Southern Hemisphere. At first glance, these results may appear strange. However, it must be kept in mind that one of the most important assumptions in deriving eq.109 was that the ocean is infinitely deep and that $u = 0$, $v = 0$ at $z = \infty$. Since D approaches infinity at the equator, even a very deep ocean is not deep enough to satisfy the lower boundary condition as stated for the derivation of eq.109. At the equator, the case of finite depth of a pure drift current should be considered which shows that in fact no singularity exists where the latitude becomes zero.

The vertical structure of a pure drift current, according to Ekman, is shown in Fig.50. The arrows represent velocity vectors at depths of equal intervals. When projected on a horizontal plane, the end points of the current vectors form a logarithmic spiral, which is called the *Ekman spiral*.

Although the current approaches zero for infinitely deep water, by far the major horizontal water transport is found in the layer above the depth of frictional influence. The order of magnitude of this depth can be estimated if reasonable assumptions are made about the magnitude of eddy viscosity coefficients in the surface strata of the oceans. With an average value of $A = 100$ (cm^{-1} g/sec) and $\rho = 1$, $D = 36.7$ m/$\sqrt{\sin\phi}$. At a latitude $\phi = 45°$, the depth of frictional influence has increased to $D = 52$ m, and at $\phi = 5°$ to $D = 126$ m. Closer to the equator D increases rapidly to infinity at $\phi = 0°$.

The velocity of a pure drift current at the sea surface is given by eq.111. If this equation is combined with eq.110 and eq.112, it follows that:

$$V_o = \frac{\tau}{\sqrt{(2A\rho\omega \sin\phi)}} = \frac{\pi\tau}{(D\rho\omega \sin\phi)\sqrt{2}} \quad (114)$$

Theoretically, the surface velocity of a pure drift current in an infinitely deep, homogeneous ocean should increase directly proportionally to the wind stress. Direct numerical evaluation of eq.114 is difficult because of inadequate knowledge of the wind stress–wind relationship and of the value of eddy viscosity coefficients in turbulent currents. THORADE (1914a,b), DURST (1924), and others have studied the dependence of the drift current velocity at the sea surface on the wind speed empirically by using carefully selected current observations. Although no definite conclusions can be drawn as to the quantitive relationship expressed by eq.114, THORADE (1914a,b) was able to confirm the dependence of V_0 on the geographical latitude.

The deflection to the right of the current with increasing depth (Northern Hemisphere) has been demonstrated, at least qualitatively. More recently, interesting examples of this kind have been given by OKUBO (1964a) and ICHIYE (1965a) who observed the spreading pattern of dye (Rhodamine B) introduced at the surface and subsurface levels in the ocean. Fig.51 is a reproduction of such a spreading dye pattern some time after the dye was injected into the water. The dye pattern illustrates the turning of the current to the right with increasing depth.

Of special importance is the total vertically integrated horizontal water volume or mass transport in pure wind-driven currents. The total mass transport is obtained by integrating eq.109 over the depth between $z = 0$ and $z = \infty$, after multiplication by the constant water density ρ. If S_x and S_y are the x- and y-components of the total mass transport vector, $S^2 = S_x^2 + S_y^2$, across a section of 1 cm width:

$$S_x = \rho \int_0^\infty u \, dz = \rho \frac{V_0 D}{\pi \sqrt{2}} = \frac{\tau_y}{2\omega \sin\phi}$$

(115)

$$S_y = \rho \int_0^\infty v \, dz = 0$$

It should be kept in mind that eq.109 was obtained for the case where the y-axis points in the direction of the wind stress. Therefore, the subscript y has been added to τ in the first of eq.115. Hence, the remarkable result follows that the total transport in a pure drift current is directed 90° cum sole to the wind direction. In the Northern Hemisphere the total transport is to the right, in the Southern Hemisphere to the left when one

Fig.51. Aerial photograph showing the curvature of a dye pattern. The curvature is clockwise from the "head" (lighter part of the patch) to the "tail" as the patch increases its depth. Photograph taken about 10–15 min. after dumping of dye. Location: About 3 nautical miles southeast of "Scotland Lightship". Water depth 30 m. Wind south-southwest, 10 knots. The ship seen in the photograph is about 65 ft. long. (Courtesy of Dr. T. ICHIYE, 1965a.)

faces in the direction toward which the wind is blowing. The total transport is proportional to the wind stress and independent of the value of the eddy viscosity coefficient in the water.

EKMAN (1905) also studied the question of how long it takes for the pure drift current to develop stationarity after a wind of a certain speed has started to blow over an infinitely deep ocean. It was found that at the onset of the wind the current started immediately in wind direction. The further development of the drift current depends on the geographical latitude (inertia effects on the rotating earth). Therefore, the time for the development of a drift current is expressed in pendulum hours. At the poles, the pendulum day is equal to a sidereal day of about 23 h 56 min; in other latitudes a pendulum day is equal to a sidereal day divided by the sine of the geographical latitude. Half of a pendulum day equals the time a Foucault pendulum needs to change its plane of oscillation by 180°.

Fig.52 shows, according to Ekman, the development of the sea surface current when the time is expressed in "pendulum hours". At time zero the wind starts to blow suddenly in the positive y-direction. After 3 pendulum hours from the start of the wind, the surface current would be as shown by an arrow drawn from the origin of the coordinate system to the point 3 on the spiraling curve; after 6 pendulum hours the current would be to the point 6 on the curve, and so forth. This result shows that in the early stages of development, the drift current "overshoots" its stationary velocity and spirals around it during the following pendulum hours before it reaches stationarity. However, the *average* velocity taken over the first pendulum day will be very near to the stationary velocity, although oscillations around the mean motion will continue for days. Such oscillations may appear as damped inertia oscillations superimposed on the mean current. For deeper layers, Ekman was able to show that it takes longer to reach the stationary state. The oscillations around the stationary drift current in a homogeneous ocean show characteristic inertia currents with equal phase at different depths. In shallow water the stationary state of motion will be established more quickly than in deep water. Other most

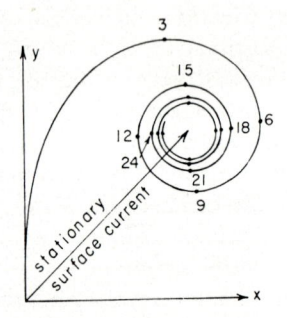

Fig.52. Hodograph showing the development of a pure drift current at the sea surface. The wind is blowing in the direction of the positive y-axis. The time after the wind started to blow with constant speed is given in pendulum hours. (After EKMAN, 1905.)

interesting cases of non-stationary wind-driven currents were theoretically investigated by HIDAKA (1933). These studies include not only the case where the wind gradually increases its speed, but also the development of slope currents with time.

The results obtained for infinitely deep water are significantly modified if the water depth is finite and becomes comparable to the depth of frictional influence as defined by eq.110. For the *case of finite depth, d,* EKMAN (1905) introduces the lower boundary condition $u = 0$, $v = 0$ at the bottom ($z = d$) and retains all other assumptions that have led to eq.109. The mathematical solution to this problem shows that the structure of a pure drift current in water of limited depth is essentially controlled by the ratio d/D. The solution appears more complicated and the exponential function in eq.109 is replaced by hyperbolic functions. Its analytical form will not be given in this text, and the reader is referred to EKMAN's (1905) classical publication or to the review of Ekman's work by NEUMANN and PIERSON (1966).

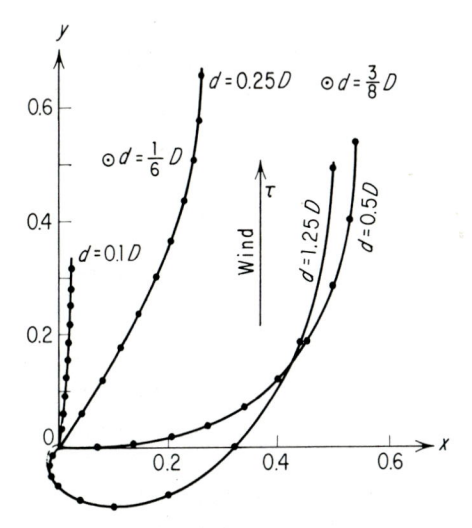

Fig.53. Vertical structure of a drift current in an ocean of finite depth projected on a horizontal plane. The depth, d, is expressed in terms of the thickness, D, of the upper layer of frictional influence. The wind is blowing in the direction of the positive *y*-axis. (After EKMAN, 1905.)

Fig.53 represents the vertical velocity distribution in pure drift currents for different ratios d/D when projected on a horizontal plane. The velocity vectors at different depths are not shown. They can easily be added by drawing arrows from the origin of the coordinate system to points indicated on the curves. These points refer to different depths and proceed from the end point of each curve (sea surface) downward to equidistant levels $0.1d$, $0.2d$, $0.3d$, etc., to the depth d which is at the origin ($u = 0$, $v = 0$). Each curve is labelled for different ratios d/D, and it is seen that the curve for $d = 1.25D$ looks very much like the Ekman spiral in infinitely deep water. Minor deviations are found near the origin of the coordinate system. The small dashed curve near $x = 0$, $y = 0$ at the curve for $d = 1.25D$ refers to $d = 2.5D$. In infinitely deep water, the logarithmic spiral continues to wind around the origin, whereas in water of finite depth the velocity must vanish at $z = d$.

The deflection angle α between the direction of the surface current in shallow water and the wind direction can be computed from the formula:

$$\tan \alpha = \left(\frac{u}{v}\right)_{z=0} = \frac{\sinh 2ad - \sin 2ad}{\sinh 2ad + \sin 2ad} \qquad (116)$$

where $ad = \pi(d/D)$. If d/D is small, α is small, and the surface current is nearly in the wind direction. As d/D increases, α is alternately a little smaller and larger but finally approaches the value $45°$ when $d/D \to \infty$. With increasing depth the value $\alpha = 45°$ is first reached for $d/D = 0.5$. Thus it is seen that for water depths, d, that are approximately equal to D or greater, the deviation from the case of an infinitely deep ocean is very small.

Another interesting result is that with very small ratios of d/D the current at different levels is almost in the same vertical plane while the speed decreases toward the bottom. For $d/D = 0.1$ the current at all levels between the surface and the bottom is nearly in the wind direction; however, the speed has reduced considerably as a result of friction. At the surface it is only about half of the speed in deep water.

EKMAN (1905, 1928a) and other investigators (notably JEFFREYS, 1923; FJELDSTAD, 1929; ROSSBY and MONTGOMERY, 1935; PRANDTL, 1942) have also dealt with the problem of other frictional laws in the theory of pure drift currents, especially with the effect of a variable friction with depth. Significant modifications of Ekman's original results have been found especially when the water depth is shallow (SVERDRUP, 1929). However, in deep water, Ekman's theory appears to give a satisfactory approximation of actual conditions. Our present knowledge of vertical changes of the eddy viscosity coefficient in the sea and its dependence on wind speed, vertical density stratification and other factors is inadequate. General improvements of Ekman's original results hinge on a better understanding of the character of turbulence in the oceans. Among the more recent papers on Ekman's theory are summaries and further developments by ICHIYE (1949, 1950), SARKISIAN (1954, 1957), FELSENBAUM (1956), WELANDER (1957), OZMIDOV (1959), and SAINT-GUILY (1959, 1962).

If the water density increases with depth, certain modifications of Ekman's original theory have to be expected, especially when a density discontinuity is found above the depth of frictional influence. In the case of a normal continuous increase of density with depth, deviations from the homogeneous ocean are small (DEFANT, 1927). However, a density discontinuity layer may act as a "Sperrschicht" for the vertical exchange of momentum. Below the density discontinuity the vertical eddy viscosity coefficient can be strongly reduced compared to the layer above the discontinuity. This means that the upper layer will slide over the denser lower layer, and the wind effect is essentially confined to the top layer. A secondary "internal drift current" may, however, develop below the transition layer. SVERDRUP (1928, 1929) has shown that a density discontinuity has indeed a significant effect on the vertical structure of a drift current.

In the real oceans where the water density increases with depth, where the wind field over the water is not constant but changes direction and speed, and where lateral boundaries block the horizontal water transport produced by pure drift currents, other complications arise. The actual current structure will be modified significantly by the appearance of slope currents and relative currents in stratified water. Such changes will be considered in the following sections.

Slope currents

It is evident that the classical type of *pure* drift currents according to Ekman cannot develop in the oceans. Horizontal water transports cannot continue indefinitely. Changes in the wind field and the interference of coasts must produce either a piling up of water in the regions of converging transports or a depression of the water level in regions of diverging transports. This leads to a slope of the sea surface compared to a level surface and to the appearance of horizontal pressure gradients in the sea which are not considered in the theory of pure drift currents.

The structure of slope currents was also investigated for the first time by EKMAN (1905). Assume that a slope of the sea surface in a homogeneous ocean has developed, for example, by the piling up of water against a coast or in a convergence between winds of opposite direction. For simplicity, let the inclination of the sea surface against a level surface, β, be everywhere in the same direction and let the motion be non-accelerated. This motion would then be geostrophic, except for the influence of friction. EKMAN (1905) introduced frictional forces resulting from vertical shearing stresses. In this case, the hydrodynamic equations are:

$$2\omega \sin\phi \, u = -\frac{1}{\rho}\frac{\partial p}{\partial y} + \frac{A}{\rho}\frac{\partial^2 v}{\partial z^2}$$

$$-2\omega \sin\phi \, v = \frac{A}{\rho}\frac{\partial^2 u}{\partial z^2}$$

(117)

if, again for convenience sake, the y-axis of the rectangular coordinate system is placed in the direction of the horizontal pressure gradient (sea surface slope). In a homogeneous ocean, the slope of isobaric surfaces, or the horizontal pressure gradient, is constant with depth and $\partial p/\partial y = -g\rho \tan\beta \approx -g\rho\beta$ for small angles β. Since u and v are functions of z only, eq.117 can be written:

$$\frac{d^2v}{dz^2} = a^2 u - \frac{\rho g \beta}{A}$$

$$\frac{d^2u}{dz^2} = -a^2 v$$

(118a)

where a has the same meaning as in eq.112.

A solution to eq.118a was given by Ekman for the following boundary conditions:

(*a*) Upper boundary condition at $z = 0$: a sea surface slope exists but there is no wind at the sea surface. This requires $A(\partial u/\partial z) = 0$, $A(\partial v/\partial z) = 0$.

(*b*) Lower boundary condition at $z = d$, where d is the depth of the sea: $u = 0$, $v = 0$.

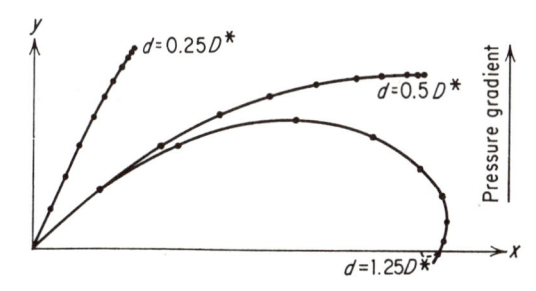

Fig.54. Vertical structure of a slope current according to EKMAN (1905). The water depth d is expressed in terms of the thickness, D^*, of the bottom layer of frictional influence. (After NEUMANN and PIERSON, 1966.)

The solution to this problem will not be given in an analytical form in this text. The reader is referred to EKMAN (1905). Fig.54 demonstrates some important facts of the vertical structure of Ekman's slope currents. Again, the current vectors at different levels are computed for equidistant depths and projected on a horizontal plane. As a result of friction due to vertical shearing stresses, and the lower boundary condition, a bottom layer of frictional influence develops in analogy to the upper layer of frictional influence (eq.110) in a pure drift current. The thickness of the bottom layer of frictional influence is D^*, and its functional form is the same as the depth D in eq.110:

$$D^* = \pi \sqrt{\frac{A}{\rho \omega \sin \phi}} \qquad (118)$$

Since the eddy viscosity coefficient, A, may differ in the upper and lower layers of frictional influence the numerical values of D^* and D may not be the same.

Depending on the ratio of the water depth, d, and the depth D^*, the vertical structure of the slope current can be very different. This is shown in Fig.54 where curves are drawn for ratios $d/D^* = 0.25, 0.5$ and 1.25. The dots on each curve indicate the end points of the velocity vector to be drawn from the origin. It is seen that the deeper the water compared to D^*, the more is the slope current deflected from the direction of the pressure gradient, while its velocity increases from the bottom upward. The dashed portion near the x-axis at the curve for $d = 1.25D^*$ indicates the difference between this curve and the curve for $d = 2.5D^*$. Thus, if the water depth is approximately equal to the depth D^*, or greater, the vertical structure of the slope current is essentially the same as in the case where d is many times greater than D^*.

The bottom layer in a slope current of depth $d > D^*$ is occupied by the bottom current in which the velocity decreases nearly exponentially toward the bottom. In infinitely deep water, the end points of the bottom current vectors, when projected on a horizontal plane, form a logarithmic spiral. Sufficiently far above the bottom, the slope current flows essentially in a direction parallel to the isobars, and the current is very nearly a geostrophic current (Ekman's "mid-water" current). This current cannot contribute to the equalization of sea surface slopes; only the bottom current has a component in the direction of the pressure gradient which produces an "anisobaric" water transport. If this water transport is blocked, for example, by the interference of a coast parallel to the isobars or by an opposite transport in the bottom layer of frictional influence, water will be "piled up" in the region of bottom water convergence, and the result is an equalization of the horizontal pressure gradient or a "filling up" of the low pressure area. Consequently, the current must gradually weaken and finally disappear. Stationary conditions are possible only if the anisobaric water transport near the bottom is completely balanced by the same but opposite anisobaric transport in other layers.

In Ekman's model, this balance is provided by the drift current in the surface strata of the ocean which, actually, produces the slope of the sea surface. The superposition of a pure drift current and a slope current in homogeneous water leads to Ekman's model of an elementary current system.

The elementary current system according to Ekman

Consider the case of a non-uniform wind distribution over the open ocean and assume a zonal wind field as shown in Fig.55. Such conditions could be realized in the Northern Hemisphere around 30° latitude between the prevailing west winds of middle latitudes and the Trade winds. These winds produce pure drift currents with a total horizontal mass transport in a direction 90° cum sole from the wind direction. The southward transport in the region dominated by the west wind and the northward transport in the Trade wind region converge in mid-latitudes around 30°. These transports are indicated in the figure by arrow S. If there is no compensation for these transports by opposite water displacements, the water must pile up in the convergence zone. The resulting slope of the sea surface produces horizontal pressure gradients in meridional direction which, in turn, cause slope currents to develop. If the water is deep enough, these slope currents will consist of quasi-geostrophic currents at mid-depths and bottom currents.

The anisobaric transport of the slope current, M, occurs essentially in the bottom layer of frictional influence, D^*, and the anisobaric transport in the pure drift current occurs essentially in the upper layer of frictional influence, D. Stationary conditions are possible, if both transports, S and M, respectively, balance each other. Similar considerations apply near lateral boundaries of the oceans (coasts) where convergences or divergences of wind-driven surface water can produce sea surface slopes.

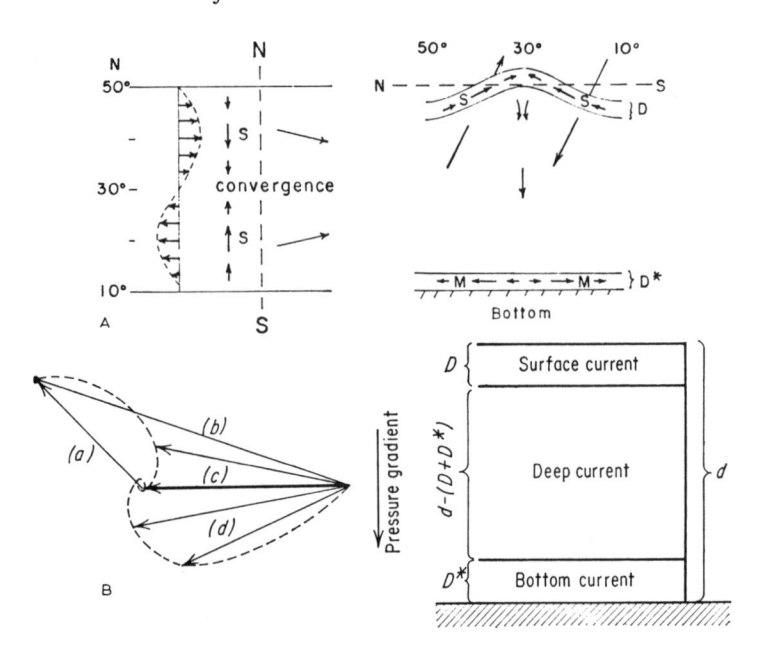

Fig.55A. The left part shows a zonal wind field between 10° and 50°N. In the region of the westerlies, the water transport, S, produced by pure drift currents in the upper layer of frictional influence is to the south, in the region of the easterlies to the north. The right part presents a schematic meridional section along the line N–S through the homogeneous water. The anisobaric transport, S, in the upper layer of frictional influence is balanced by the anisobaric transport, M, in the bottom layer of frictional influence. The slope current (deep current) between the depths D and D^* is essentially geostrophic.
B. Vertical structure of Ekman's elementary current system between surface and bottom. The hodograph on the left shows the vertical current structure projected on a horizontal plane.

The resulting current system in deep water consists of three basic components, the bottom current, the geostrophic or nearly geostrophic deep current, and the surface layer current. This system is called the *elementary current system*. The bottom current occupies essentially the layer between the bottom and the depth D^* above the bottom. The deep current is found between the upper layer of frictional influence, D, and the bottom layer of frictional influence, D^*. The surface layer current is a superposition of the pure drift current and the deep current and occupies, essentially, the upper layer of frictional influence. If the current vectors at different depth are projected on a horizontal plane, the vertical structure 'of Ekman's elementary current system is shown in Fig.55B which applies to the Northern Hemisphere. It is seen that the surface current vector, as the result of superposition of the pure drift current and the deep current is deflected from the wind direction by an angle smaller than 45°.

CURRENTS IN A NON-HOMOGENEOUS OCEAN

The density of the water in the oceans generally increases with depth. Only in exceptional cases, especially near the sea surface, can the density of the water slightly decrease with depth before the normal increase starts, while static stability is still maintained in layers with a negative vertical density gradient (NEUMANN, 1948b).

The rate of change of density with depth can be very different in different oceanic regions. Fig.56 represents some examples for low, middle, and high latitudes showing the vertical distribution of σ_t with depth. In general, the increase of density with depth is much smaller in polar or subpolar regions than in tropical or subtropical regions. In some equatorial waters, the transition between the upper layer of low density and the deeper layer of higher density is so sudden that the term "density discontinuity" has been applied to the depth of the strongest gradient $\partial \rho / \partial z$. In such cases, it is possible to approximate the vertical oceanic stratification by a "two-layer ocean". However, this approximation is not satisfactory when oceanic areas of greater extent—for example a whole hemisphere or a larger part of it—are under consideration.

The spatial distribution of density in the oceans is largely controlled by external factors such as the "conditioning" of water masses at the sea surface, that is, heating, cooling, precipitation, evaporation, and other air–sea interaction effects, and the spreading of such water masses. In addition, the field of

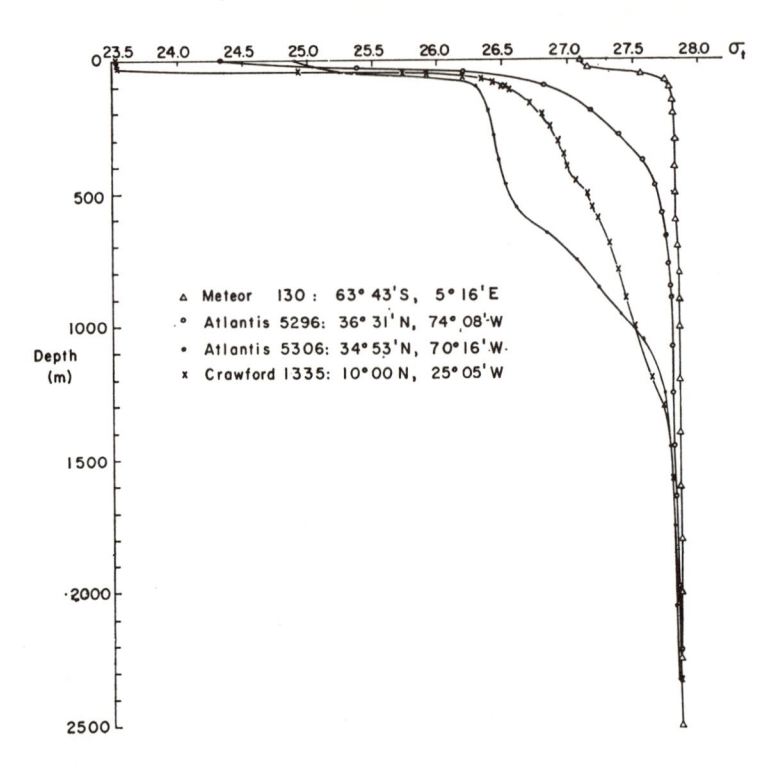

Fig.56. Examples of the vertical density distribution (σ_t) in low, middle and high latitudes.

density is also significantly affected by the structure of ocean currents and mutual relationships between the field of mass and the field of currents.

Mass stratification in geostrophic and gradient currents

For non-accelerated, frictionless ocean currents in a stratified ocean, an important relationship exists between the mass distribution and the relative velocity distribution in vertical direction. Orient a rectangular x, y, z coordinate system such that the y-axis points in the direction of a geostrophic current, the x-axis to the right of the positive y-axis, and the z-axis downward. If $f = 2\omega \sin\phi$ and $u = 0$, the balance of forces in lateral and vertical directions is, respectively:

$$\rho f v = \frac{\partial p}{\partial x}; \quad g\rho = \frac{\partial p}{\partial z} \tag{119}$$

Differentiation of the first equation with respect to z yields:

$$\rho f \frac{\partial v}{\partial z} + f v \frac{\partial \rho}{\partial z} = \frac{\partial^2 p}{\partial x \, \partial z} \tag{120}$$

Differentiation of the second of eq.119 with respect to x yields:

$$g \frac{\partial \rho}{\partial x} = \frac{\partial^2 p}{\partial x \, \partial z} \tag{121}$$

Hence, it follows from eq.120 and 121 that:

$$\frac{\partial v}{\partial z} = \frac{g}{\rho f} \frac{\partial \rho}{\partial x} - \frac{v}{\rho} \frac{\partial \rho}{\partial z} \tag{122}$$

Eq.122 shows that in a geostrophic current the vertical velocity shear is related to the changes of density in lateral and vertical directions. In most practical cases, the last term in eq.122 is small when compared with the first term on the right hand side, and can be neglected. For example, with a current speed of 100 cm/sec and $\partial \rho / \partial z \approx 10^3 (\partial \rho / \partial x)$, the magnitude of the last term is $10^5 (\partial \rho / \partial x)$, whereas the factor $g/\rho f$ is of the magnitude 10^7 in middle latitudes. Thus, the first term on the right side of eq.122 is about 100 times larger than the second term. When approaching the equator, the first term becomes even larger, and $\partial v / \partial z$ should, theoretically, increase to infinity right at the equator if $\partial \rho / \partial x$ would not approach zero. It is known that the Equatorial Undercurrent in the oceans reveals extremely strong vertical velocity gradients, especially in the upper strata above its maximum speed, and that $\partial \rho / \partial x$ indeed approaches zero at the equator. Although there are other important factors (e.g., nonlinear field accelerations and friction) involved in the dynamic explanation of this current, it is evident that the simple relationship shown in eq.122 plays a role even close to the equator. It is also of interest to note that oceanographic observations have shown that a near-geostrophic balance of forces in non-accelerated currents can be found very close to the geographical equator (MONTGOMERY and STROUP, 1962).

Eq.122, or its simplified (approximate) form:

$$\frac{\partial v}{\partial z} \approx \frac{g}{\rho f} \frac{\partial \rho}{\partial x} \qquad (122a)$$

corresponds to the "thermal wind equation" in dynamic meteorology. It expresses the fact that in the Northern Hemisphere with a geostrophic balance of the flow the denser water is found to the left of the current if the current speed decreases with depth. If the current speed increases with depth, the denser water is found to the right when one faces in current direction. In the Southern Hemisphere, the directions are reversed, and the denser water is found to the right of the current if the

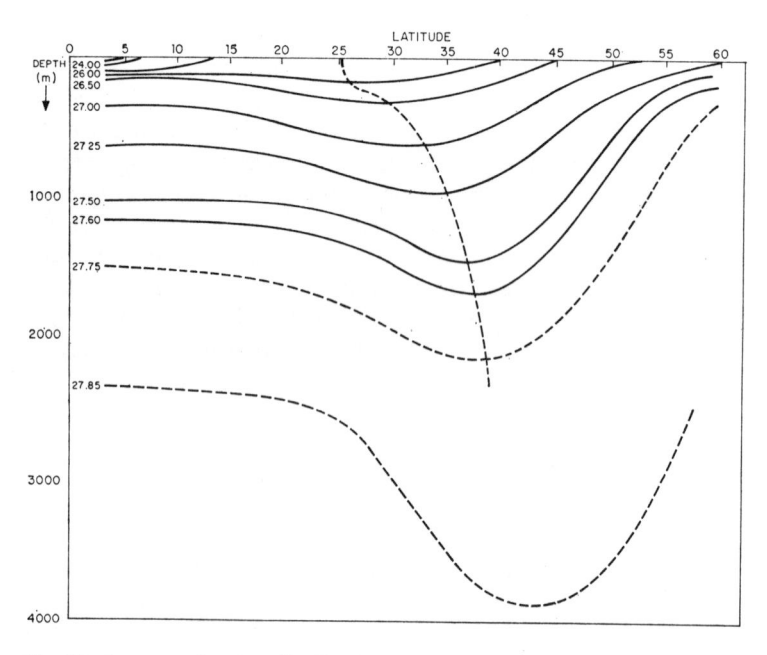

Fig.57. Average density distribution (σ_t) in a meridional section of the South Atlantic Ocean.

velocity decreases with depth, and to the left of the current if the velocity increases with depth.

The accumulation of warm (less dense) water in the centers of the great anticyclonic gyres of ocean currents in mid-latitudes is essentially the result of an adjustment of the field of mass to the field of the general oceanic circulation. This is demonstrated in Fig.57 which represents an average meridional density section for the South Atlantic Ocean. The inclination of isopycnals toward deeper layers between the westward flowing currents in subtropical latitudes and the eastward flowing currents in the region of middle and higher latitudes is indicated. Note that the axis of the accumulation of less dense water in latitudes around 35° tilts toward the poleward side of the ocean with increasing depth. A similar displacement is found in the North Atlantic Ocean (DEFANT, 1941).

The magnitude of the inclination of isopycnal surfaces in the oceans is about one hundred times greater than the inclination of isobaric surfaces.

In regions where a nearly homogeneous surface layer is separated from a nearly homogeneous deep layer by a density discontinuity, the density stratification can be approximated by a two-layer ocean. It has been shown by WITTE (1878) that the density discontinuity between the upper lighter water and the deeper heavier water must be inclined against a level surface if the two layers move relative to each other. This relationship was later rediscovered by MARGULES (1906).

Although density discontinuities of this kind (where the density gradient approaches infinity) occur neither in the oceans nor in the atmosphere, the approximation of a region of extremely strong density gradients by such a "discontinuity" is often quite useful. By analogy to the atmosphere, such regions are called *oceanic frontal zones*, and their intersection with any horizontal surface is called a *front*.

An idealized (infinitely "sharp") front is shown in the vertical section of Fig.58A which applies to the Northern Hemisphere. The coordinate system is oriented in such a way that the y-axis points in the direction of the (geostrophic) current, and the z-axis points positively downward. The dynamic boundary condition along the density discontinuity, s, between the upper and

202   MAJOR TYPES OF OCEAN CURRENTS

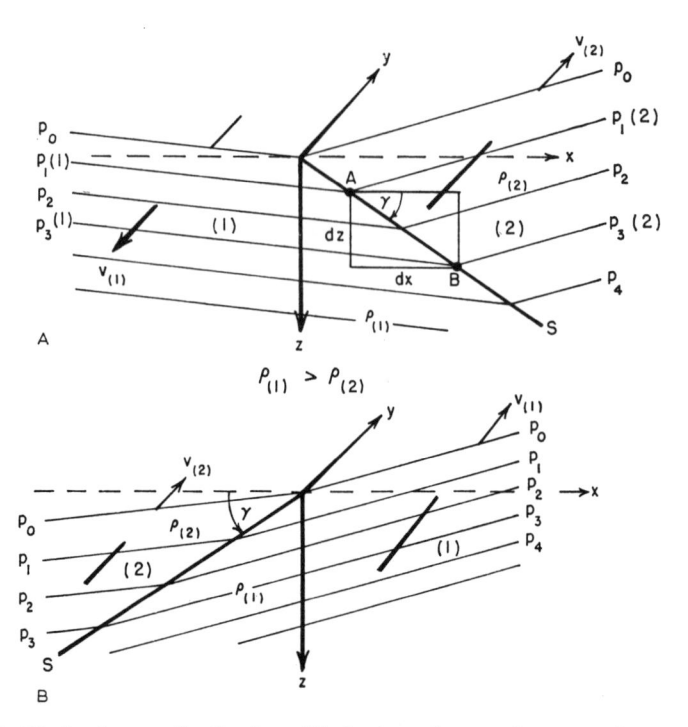

Fig.58. Stationary distribution of isobaric surfaces and currents in a two-layer ocean. The density discontinuity, *s*, separates two water bodies with densities $\rho_{(1)}$ and $\rho_{(2)}$, respectively, where $\rho_{(1)} > \rho_{(2)}$ (see text).

the lower layers with density $\rho_{(1)}$ and $\rho_{(2)}$, respectively, requires continuity of pressure when crossing the discontinuity, although the pressure gradient be discontinuous. Thus, at point *A* of the discontinuity surface, the pressure $p_1(1) = p_1(2)$ and at point *B* $p_3(2) = p_3(1)$. Since:

$$p_1(1) + \left(\frac{\partial p}{\partial z}\right)_{(1)} dz + \left(\frac{\partial p}{\partial x}\right)_{(1)} dx - \left(\frac{\partial p}{\partial z}\right)_{(2)} dz -$$
$$- \left(\frac{\partial p}{\partial x}\right)_{(2)} dx = p_1(2)$$

the dynamic boundary condition states that:

$$\left[\left(\frac{\partial p}{\partial z}\right)_{(1)} - \left(\frac{\partial p}{\partial z}\right)_{(2)}\right] dz + \left[\left(\frac{\partial p}{\partial x}\right)_{(1)} - \left(\frac{\partial p}{\partial x}\right)_{(2)}\right] dx = 0$$

or:

$$\frac{dz}{dx} = \tan\gamma = -\frac{(\partial p/\partial x)_{(1)} - (\partial p/\partial x)_{(2)}}{(\partial p/\partial z)_{(1)} - (\partial p/\partial z)_{(2)}} \quad (123)$$

Eq.123 expresses a relationship between horizontal and vertical pressure gradients in the two water bodies and the slope, γ, of the density discontinuity which has to be fulfilled in order to satisfy the dynamic boundary condition. With the z-axis pointing downward, the slope of the density discontinuity (frontal slope) in Fig.58A is positive. Since $(\partial p/\partial z)_{(1)} = g\rho_{(1)}$ and $(\partial p/\partial z)_{(2)} = g\rho_{(2)}$, the denominator of eq.123 is positive, because $\rho_1 > \rho_2$. Thus it follows that the numerator must be negative. For the case of currents shown in Fig.58A where the upper layer flows in the positive y-direction (velocity $v_{(2)}$) and the lower layer in the negative y-direction (velocity $v_{(1)}$), the numerator is always negative, independent of the absolute velocity $v_{(1)}$. With opposing currents in the two layers, the velocity of the lower layer can even be greater than the velocity of the upper layer. However, if the geostrophic velocity $v_{(1)}$ in the lower layer is in the same direction as the geostrophic velocity $v_{(2)}$ in the upper layer, a positive stationary slope of the density discontinuity surface compatible with the dynamic boundary condition can be maintained only if $(\partial p/\partial x)_{(1)} < (\partial p/\partial x)_{(2)}$. This means that the velocity $v_{(1)}$ must be smaller than the velocity $v_{(2)}$. If the velocity $v_{(1)}$ is greater than the velocity $v_{(2)}$ and in the same direction, $(\partial p/\partial x)_{(1)} > (\partial p/\partial x)_{(2)}$, and $\tan\gamma$ is negative. This means that the slope of the density discontinuity has reversed. Such a case of a stationary mass stratification in a two-layer ocean is shown in Fig.58B. If the two water bodies are in relative motion to each other and stationary, the heavier water always spreads in a wedge-like form below the lighter water; however, the direction of the slope of the density discontinuity depends on the vertical shear of the current velocity between the two layers. This, of course, agrees with the results obtained from eq.122 for a continuous density stratification.

Introduction of the geostrophic assumption according to eq. 62, and of the hydrostatic equation, eq.123 yields:

$$\tan \gamma = -\frac{f}{g}\left(\frac{\rho_1 v_1 - \rho_2 v_2}{\rho_1 - \rho_2}\right) \qquad (124)$$

The subscript (*1*) applies to the water body below, and the subscript (*2*) to the water above the discontinuity. In eq.124 and below, the parentheses around the subscript have been omitted.

Eq.124 is known in meteorology and oceanography as the "Margules equation". Following theoretical studies "Über atmosphärische Bewegungen" by VON HELMHOLTZ (1888), MARGULES (1906) applied eq.124 in its present form to atmospheric frontal zones. However, before Von Helmholtz, WITTE (1878) derived exactly the same equation for oceanic frontal zones. He applied

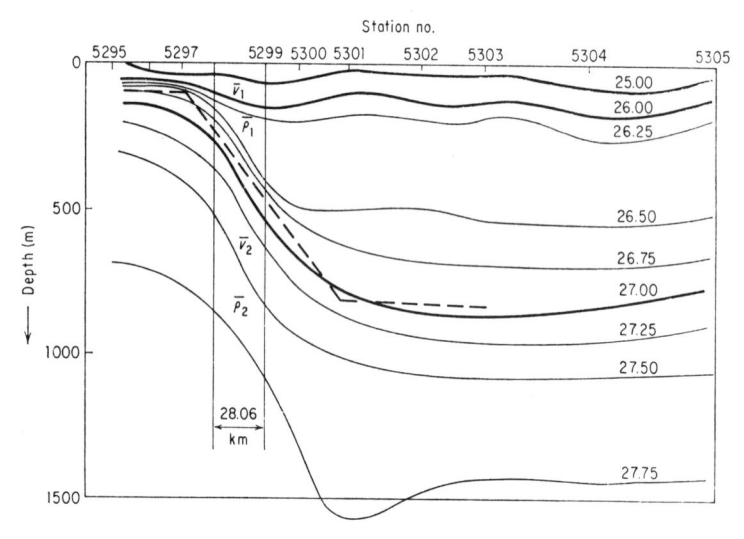

Fig.59. Density section (σ_t) across the Gulf Stream (Chesapeake Bay–Bermuda) with indication of an oceanic front. The slope of this front is approximated by the broken line between "Atlantis" stations 5297 and 5301. (After NEUMANN and PIERSON, 1966.)

his equation to observations obtained by the "Challenger"-Expedition (1872–1876) in the Gulf Stream region. WITTE (1879) concluded: "The results of the "Challenger" expedition have confirmed my ideas in a splendid way". Witte's result is probably one of the first essential contributions to the dynamics of ocean currents, and it appears proper to call eq.124 the "Witte equation" or at least "Witte–Margules equation" (NEUMANN and PIERSON, 1966).

The density section across the Gulf Stream shown in Fig.59 indicates an oceanographic "front" by the crowding of isopycnals. The front is approximated by a broken line between stations 5297 and 5301. It separates a lighter upper water layer of average density $\bar{\rho}_2 = 1.0254$ from a heavier lower water layer of average density $\bar{\rho}_1 = 1.0274$. The average velocity of the upper layer (see p.144) is $\bar{v}_2 = 171.5$ cm/sec, and the average velocity of the lower layer is $\bar{v}_1 = 44.0$ cm/sec. Application of eq.124 to an approximate two-layer ocean with these average velocities and densities yields the slope $\Delta z/\Delta x = \tan \gamma = 6.57 \cdot 10^{-3}$. This compares favorably with the observed slope of $\Delta z = 200$ m along the distance $\Delta x = 28.1$ km between stations 5298 and 5299. The observed slope is $200/28,100 = 7.1 \cdot 10^{-3}$.

The vertical density stratification of the upper strata in tropical and in some subtropical oceanic regions can well be approximated by a two-layer ocean (see Fig.56). Thus, it is to be expected that the topography, or slope, of the tropical density discontinuity surface reflects to some degree the dynamic structure of the stationary current system in these regions (SVERDRUP, 1932, 1934a; DEFANT, 1936b). With the assumption that the currents flow essentially in zonal direction and that their speed decreases with depth, meridional sections across the equator as shown in Fig.60 should be observed. Currents flowing in the upper (lighter) layer toward the west are indicated by the letter W and currents flowing toward the east by E. The sea surface slope corresponding to these currents is also indicated in the figure. The heavier water of the layer below the density discontinuity (shaded area) is assumed to be at rest.

The slope of the density discontinuity in the examples of Fig.60 agrees qualitatively with eq.124 and shows that the

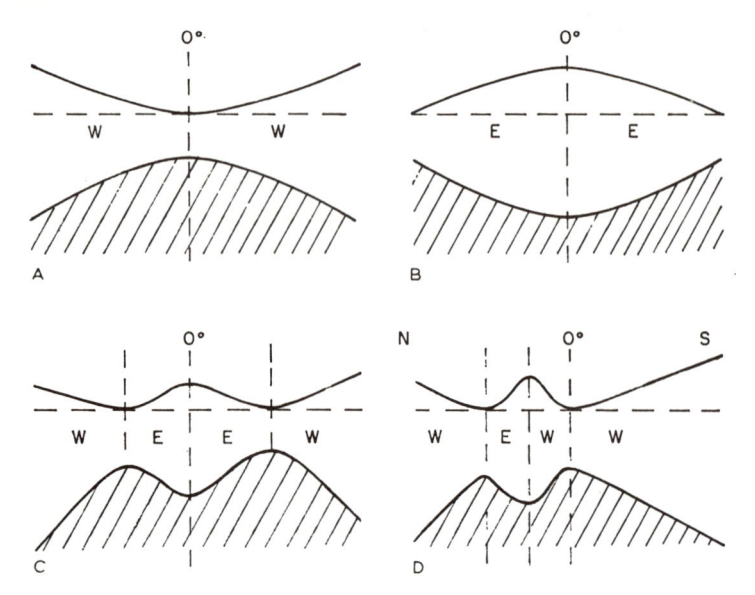

Fig.60. Stationary distribution of the sea surface slope and of the slope of a density discontinuity in meridional sections crossing the equator. The heavier water body below the density discontinuity is shaded, and at rest. (According to SVERDRUP, 1932, 1934a, and DEFANT, 1936b.)

density discontinuity slopes in the opposite direction from the sea surface. (The sea surface slope is greatly exaggerated with respect to the slope of the density discontinuity.) At the equator, tan γ must be zero, and, therefore, the discontinuity surface always crosses the equator as a horizontal surface.

Of particular interest and practical importance is case (D) in Fig.60 where a narrow eastward current in the Northern Hemisphere is embedded in the westward flowing currents of both hemispheres. This case represents in a schematic way some major features of the equatorial current system in the central part of the Atlantic Ocean and of the Pacific Ocean. The westward flowing currents represent the North and South Equatorial Currents, and the eastward flowing current is the Equatorial Countercurrent.

Observations in the tropical and subtropical parts of the oceans have verified this structure. The meridional profiles in

Fig.61 according to DEFANT (1961), illustrate these conditions for the central part of the tropical Atlantic Ocean. Equatorial current systems will be discussed in greater detail in Chapter V.

Gradient currents and stationary vortices have been dealt with in a similar way by EXNER (1917), BJERKNES (1921), and DEFANT (1929b). With curved streamlines, centrifugal forces have to be considered in addition to the Coriolis force. The slope of a density discontinuity surface in a two-layer ocean with circular motion is obtained from:

$$\tan \gamma = -\frac{f}{g}\left(\frac{\rho_1 c_1 - \rho_2 c_2}{\rho_1 - \rho_2}\right) - \frac{1}{rg}\left(\frac{\rho_1 c_1{}^2 - \rho_2 c_2{}^2}{\rho_1 - \rho_2}\right)$$

Since the densities ρ_1 and ρ_2 differ only slightly, the equation can be written, with sufficient accuracy:

$$\tan \gamma \approx -\frac{f}{g}\frac{\rho(c_1 - c_2)}{\rho_1 - \rho_2}\left(1 + \frac{c_1 + c_2}{fr}\right) \tag{125}$$

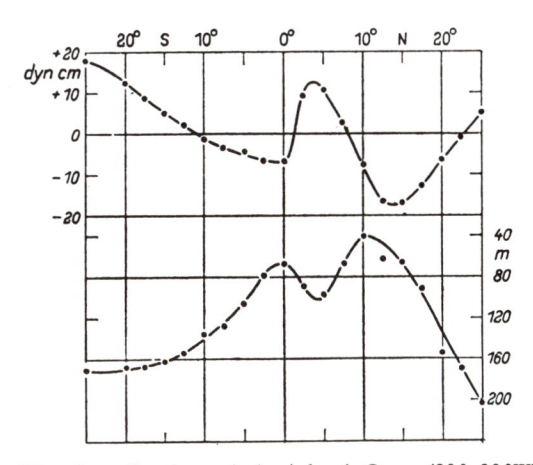

Fig.61. Meridional profiles through the Atlantic Ocean (20°–30°W) showing the slope of the sea surface (upper curve) and of the thermocline (discontinuity) (lower curve). Vertical scales are shown on the left for the deviation of the sea surface from a level surface, and on the right for the depth of the thermocline. (After DEFANT, 1961.)

where $\rho \approx \rho_1 \approx \rho_2$, and r is the radius of curvature of the streamline along which c_1 and c_2, respectively, are measured.

The effect of the centrifugal force is shown in eq.125 in a correction term (second term in parentheses). With the exception of low latitudes, this correction term is usually small; however, with small radii, r, it can become very significant. Assuming that $\rho_1 = 1.027$, $\rho_2 = 1.025$, $c_1 = 10$ cm/sec, $c_2 = 50$ cm/sec and $r = 100$ km, it follows from eq.125 that in mid-latitudes with $f = 10^{-4}$ the slope $\gamma \approx 2.1 \cdot 10^{-3} (1 + 0.06)$. The slope compared to the case of a straight (geostrophic) flow is only about 6% larger. However, for $r = 10$ km it would be 60%! When approaching the equator, caution is indicated with greater radii of curvature. At a geographical latitude of 5°, this example shows that even with $r = 100$ km, $\gamma \approx 2.67 \cdot 10^{-3} (1 + 0.46)$.

In a continuously stratified ocean, the relationship between the vertical velocity shear, $\partial c/\partial z$, and the mass stratification in gradient currents, or vortices, is obtained from:

$$\frac{\partial c}{\partial z} = -\frac{g}{f - 2c/r}(\tan\gamma - \tan\beta)\frac{1}{\rho}\frac{\partial \rho}{\partial z} \qquad (126)$$

Here, γ is the slope of isopycnals and β is the slope of the isobars. Usually, $\tan\beta$ can be neglected when compared with $\tan\gamma$. Since $\tan\gamma \, \partial\rho/\partial z = -\partial\rho/\partial x$, eq.126 can be written, approximately, as eq.127:

$$\frac{\partial c}{\partial z} = \left(\frac{g}{f - 2c/r}\right)\frac{1}{\rho}\frac{\partial \rho}{\partial x} \qquad (127)$$

For $r \to \infty$, eq.127 reduces to eq.122a.

The structure of symmetrical vortices, showing the mass and pressure distribution for the case of *decreasing* velocity with depth is shown in Fig.62. It is seen that in the case of a cyclonic rotation, denser (usually colder) water is found in the center of the vortex. With anticyclonic rotation, less dense (usually warmer) water of the upper layers is forced into greater depths in the center of the vortex. This is the reason for the fact that in the center of the great anticyclonic gyres in the oceans (20°–30° latitude) the water at depths between about 100 m and 1,000 m or more is warmer than in the outer regions of the gyre centers.

If the speed of ocean currents increases with depth, the in-

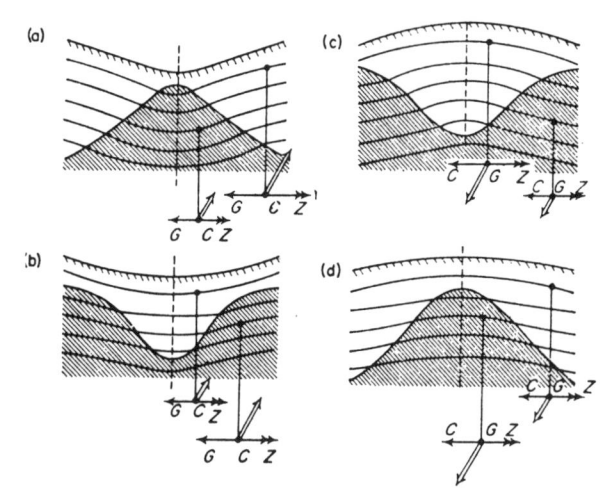

Fig.62. Structure of symmetrical vortices in a two-layer ocean showing isobars in a vertical section across the vortices and the density discontinuity between the two layers. The heavier water of the lower layer is shaded. In (a) and (c) the upper layer rotates faster than the lower layer. In (b) and (d) the lower layer rotates faster than the upper layer. In the force diagrams, G represents the horizontal pressure gradient, C the Coriolis force and Z the centrifugal force. (After DEFANT, 1961.)

clination angle of isopycnals around the center of the vortex is reversed. Less dense (or warmer) water is forced into deeper layers if the circulation is cyclonic, and denser (or colder) water is forced to the upper layers if the circulation is anticyclonic.

Although it is an exceptional case in the oceans where the velocity of currents increases continuously with depth, some regions of the upper strata may show such an increase before the general decrease of speed of stationary currents with increasing depth begins. An interesting example of a smaller-scale vortex of this kind was found north of the Azores during the International Gulf Stream Survey (1938), approximately over the peak of the Altair-cone.

The Altair-cone is, apparently, an inactive submarine volcano rising from the Mid-Atlantic Ridge to about 900 m depth below the sea surface. The R.V. "Altair" had anchored on the peak

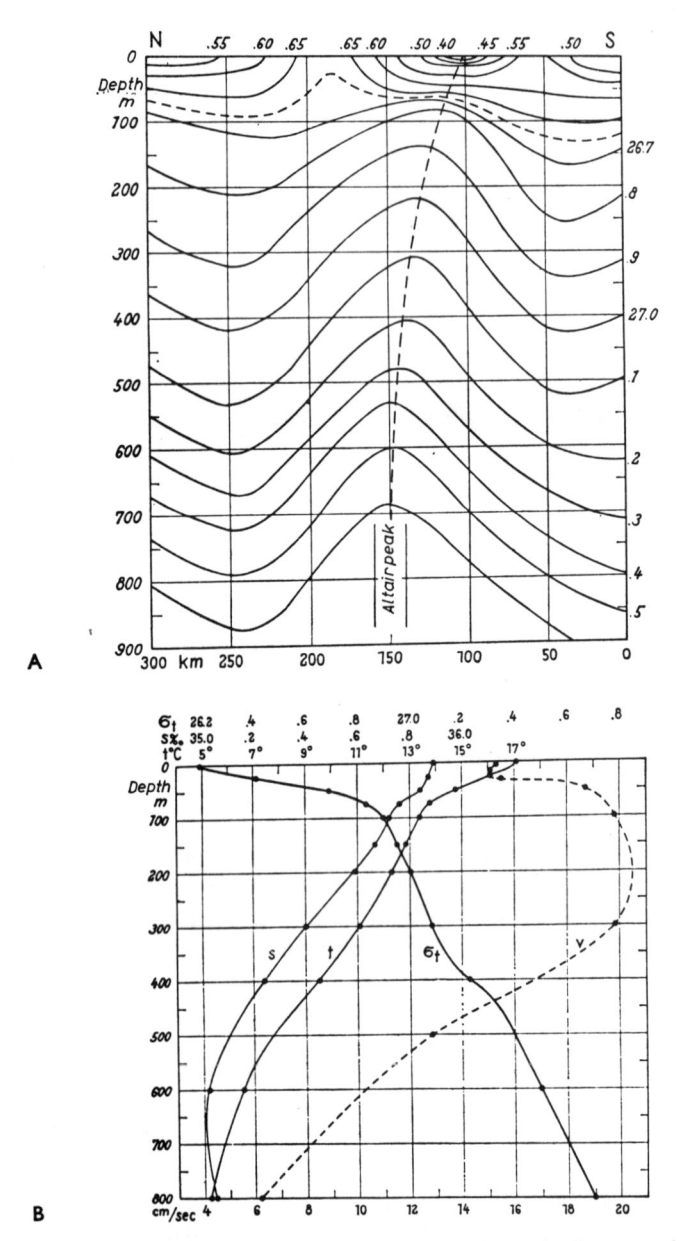

Fig.63A. Meridional density section across the cyclonic vortex above the "Altair" cone. The observed average vertical distribution of current speed (June 16–20, 1938) is shown in B together with the vertical distribution of temperature, salinity, and density (σ_t). (After DEFANT, 1940b.)

of this volcano and measured currents and other oceanographic parameters at regular intervals while the Norwegian research vessel "Armauer Hansen" made systematic oceanographic sections around, and crossing close to the anchoring position of the "Altair". These data provided an excellent opportunity for investigating among other questions, the relationship between currents and mass distribution. Evaluation of these data by DEFANT (1940b) resulted in a greater comprehension of the structure of smaller-scale vortices in the oceans.

Fig.63A shows an average meridional density section across the cyclonic vortex above the Altair-cone, based on the observations during the anchoring period (about 4 days). Current measurements and the observed mass distribution show clearly that the vortex is divided into two parts. Around the center of the cyclonic vortex down to about 75–100 m depth, the slope of isopycnal surfaces is *upward* from the center of rotation. Below that depth it is *downward* from the center of rotation. It is also seen that the axis of the vortex inclines to the south.

Such a density distribution requires increasing (stationary) current speed with increasing depth in the upper layer and decreasing current speed with increasing depth in the lower layer. This has been confirmed by direct current observations which are shown in the graph Fig.63B by the dashed curve, v.

The observed average isopycnal slopes are in good agreement with eq.125 (DEFANT, 1940c), although the cyclonic vortex performed pulsations which were recorded by continuous current measurements and also detected in the mass stratification. The period of the pulsations corresponded to the inertia period (about 17.1 h in a latitude of 44°33'). Periodic current variations of as much as half the velocity of the basic current occurred. Therefore, the changes in time of the distribution of isopycnals in this cyclonic gyre must have been quite significant.

Oscillations of a circular two-layered vortex have also been dealt with theoretically by DEFANT (1940c). The free periods of oscillation (*Eigen*-periods) which characteristically depend on the angular velocity of the Earth rotation play an important role in such a system. Defant has shown that the Eigen-period of 16.76 h of the Altair-cone vortex closely approaches the period of an inertia oscillation. A storm that had passed over the region just before the R.V. "Altair" occupied her anchor station seems to have caused the observed pulsations of the cyclonic vortex.

Ekman's relative currents and the elementary current system in a non-homogeneous ocean

EKMAN (1905, 1906) studied theoretically the structure of relative currents including friction resulting from vertical shearing stresses. First, he considered the simple case where the density increased uniformly with depth. At a certain depth $z = d$, he assumed that the isobaric surfaces are horizontal, and that at this depth the absolute current velocity vanishes. For simplicity, isopycnal surfaces were considered as planes parallel to each other and situated with respect to the rectangular x, y, z coordinate system in such a way that in the relative pressure field:

$$\partial p/\partial x = 0; \quad -\partial p/\partial y = b(d - z)$$

where b is a constant. Also, the vertical eddy viscosity coefficient, A_z, was assumed to be constant.

For stationary conditions, the equations of motion are:

$$\frac{d^2 u}{dz^2} + \frac{\rho f}{A_z} v = 0 \qquad (128)$$

$$\frac{d^2 v}{dz^2} - \frac{\rho f}{A_z} u + \frac{b}{A_z}(d - z) = 0$$

with the boundary conditions:

$$\frac{du}{dz} = \frac{dv}{dz} = 0 \text{ for } z = 0$$

$$u = v = 0 \text{ for } z = d$$

The solution to this problem is represented in Fig.64, showing the vertical structure of the relative current for different ratios d/D^* when projected on a horizontal plane. Current vectors can be thought of as being drawn from the origin of the xy-system to points on the curves. Vectors drawn from the origin to the end points of each curve represent surface currents. It is seen

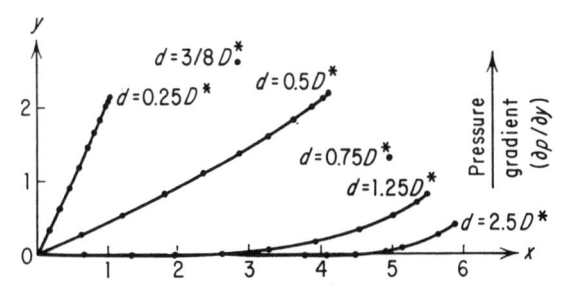

Fig.64. Vertical structure of relative currents in water of different depth. (After EKMAN, 1905. From NEUMANN and PIERSON, 1966.)

that in deep water, where d is greater than the depth D^* (eq. 118), the stationary relative current is nearly perpendicular to the pressure gradient $\partial p/\partial y$. For $d/D^* \to \infty$, the relative current flows at all levels in the isobaric direction. With small ratios d/D^*, the relative current becomes more and more deflected from the isobaric direction as the depth d decreases, and if $d \ll D^*$, the current is almost in the direction of the pressure gradient.

A remarkable result obtained from this simple model is that the currents at different depths are nearly in a vertical plane. This shows that in the first place friction affects the direction of this plane with respect to the horizontal pressure gradient, while there is not much turning of the current direction with depth. The rate of decrease of velocity in the vertical between the surface and the depth d is smallest in the surface layers and increases in deeper layers where it becomes almost linear.

EKMAN (1906) also investigated other cases of vertical density stratification: for example, the case where a homogeneous lighter top layer spreads out over a denser, but also homogeneous deeper layer in a wedge-like form. It is assumed that the boundary between the two layers is an inclined plane along a straight coast which extends in the x-direction. The depth, d, of the upper layer is constant in a direction parallel to the coast; whereas perpendicular to the coast d decreases with increasing offshore distance. Such a stratification is often found in near-shore waters where less saline coastal water overlies denser but more homogeneous deep water.

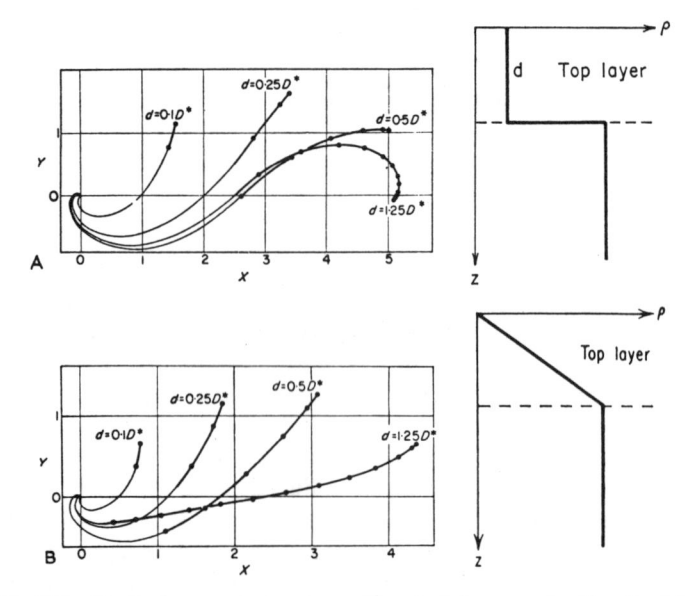

Fig.65A. Vertical current structure off a straight coast (x-direction) in a two-layer ocean projected on a horizontal plane. The depth of the homogeneous upper layer is d. The graph to the right shows the vertical density distribution.
B. Same as in A for a stratified upper layer as shown in the graph to the right. (After EKMAN, 1906.)

For the simple case of a two-layer model with constant density in the top layer and constant density in the deeper layer, Ekman computed vertical current velocity distributions as shown in Fig.65A. Again, the vertical current structure depends on the ratio of the thickness, d, of the upper layer and the quantity D^* as defined by eq.118. D^* is, essentially, an expression of the effect of friction resulting from vertical shearing stresses. The curves in Fig.65A are computed with the assumption that in the regime of the current the ratio d/D^* is constant. The heavy part of the curves refers to points in the upper layer and the light part to the lower layer. Velocity vectors can be thought of as being drawn from the origin of the coordinate system to points on the curves.

The currents are strongest at the surface. However, while

decreasing with depth, the motion of the top layer exerts vertical shearing stresses and friction on the underlying, heavier, deep water. This causes an "internal drift current" underneath the density discontinuity. The effect of the top layer on the deep water depends on the ratio d/D^*. For great depth, $d > 0.5D^*$, the currents in the upper layer have a seaward component (positive y-direction) while the currents in the lower layer are all directed toward the coast. With small ratios d/D^*, the upper strata of the lower layer are dragged away from the coast and the depth of reversal of direction of the component perpendicular to the coast is somewhere beneath the boundary between the two layers of different density.

A better approximation to a real density stratification is the assumption that the density in the top layer increases linearly with depth between the surface and depth d, while the lower layer is homogeneous. EKMAN (1906) assumed a vertical density stratification as shown in the right part of Fig.65B and computed the current structure under similar conditions as in the preceding model. The curves in Fig.65B show that significant changes have occurred compared to the case of a homogeneous top layer. In general, the offshore transport in the deeper layer is smaller. If d/D^* is large, say $d > 1.25D^*$, the deep water is only slightly affected by the motion of the top layer which is almost parallel to the coast. The water below the depth d remains almost motionless. For small ratios of d/D^*, part of the upper strata of the lower, homogeneous layer is dragged offshore by the currents in the upper layer; however, for d equal to and greater than $0.3D^*$, all the offshore transport is carried by currents in the top layer.

Although simple mass distributions and associated relative current systems as studied by Ekman rarely occur in nature, the results obtained about sixty years ago illustrate some basic facts and are still of great interest in dynamic oceanography. From his basic studies, Ekman arrived at the important conclusion that the sea surface under the influence of external disturbances will adjust its slope in such a manner that the horizontal pressure gradient produced by density differences gradually decreases with depth and can vanish entirely at some depth beneath the strongly stratified upper layer.

Ekman's investigations have also served to demonstrate the importance of vertical density stratification on wind-driven currents. If the effect of a pure wind-driven current in addition to relative and slope currents is considered, three constituents of an elementary current system in a non-homogeneous ocean can be distinguished: the slope current, the relative current, and the pure drift current.

The slope current can reach the ocean bottom when the slope of the sea surface is large compared to the relative pressure field. If the inhomogeneity of the density field is strong enough to compensate for the slope field of pressure, the current can vanish at some depth above the bottom. In this case, the bottom current is missing.

The pure drift current will not differ significantly from the drift current in homogeneous water if the homogeneous top layer is deep enough compared to the depth of the upper layer of frictional influence, or if the density in the top layer increases gradually with depth and does not include density discontinuities.

Wind-driven currents and slope currents

The total pressure field in a stratified ocean is the sum of the relative field of pressure due to differences in the distribution of mass and the slope field of pressure due to the slope of the sea surface. The total horizontal pressure gradient is obtained from eq.65. It was shown in eq.66 that the velocity of a non-accelerated current in a stratified ocean must decrease with depth if the slope of isopycnic surfaces is opposite to the sea surface slope. In this case, the slope field of pressure and the relative field of pressure compensate each other at a certain depth such that a level (or layer) of no motion develops above the ocean bottom if the ocean is deep enough to make complete compensation possible (see eq.67).

Consider a vertical section in Fig.66 perpendicular to an infinitely long coast and assume that water in the upper layer of frictional influence is steadily transported toward the coast from B to A by a pure wind drift current. This produces a slope of the sea surface from A to B (line *a* in Fig.66) and a slope current parallel to the coast. The direction of this current is such that

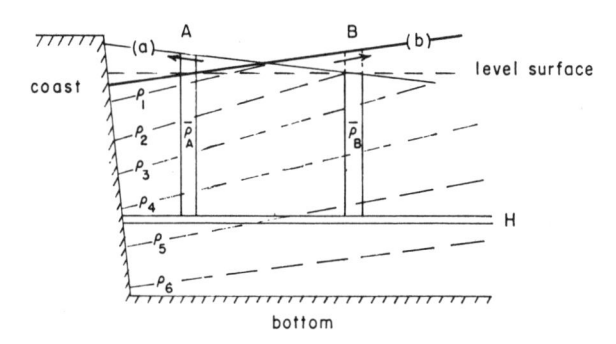

Fig.66. Vertical section perpendicular to an infinitely long coast showing the field of density and two different slopes of the sea surface (see text).

station A is on the right when one faces in the current direction. The field of mass, shown by the isopycnals $\rho_1, \rho_2, \ldots, \rho_6$ will slowly adjust to the current field, and with stationary conditions, the slope of the isopycnals must be opposite to the slope of the sea surface as required by eq.122. According to eq.67, compensation of the slope field of pressure by the relative field of pressure produced by the distribution of mass may occur at a depth H above the bottom, and the bottom layer below H can be motionless.

Next assume that, suddenly, the wind changes direction by 180°. The transport of water in the surface layer toward the coast vanishes, and almost immediately an offshore drift current begins to transport surface water towards the open sea. At some distance from the coast, surface water may converge and pile up, while the sea surface near the coast is lowered. This reversal of surface drift in the upper layer of frictional influence responds quickly to the wind shift, whereas the deep water layer, including the deep current, is much more sluggish in the adjustment of its field of mass. Thus, during this transient state, where the sea surface slopes upward from station A to station B, as shown by the heavy line *b* in Fig.66, the slope pressure field and the relative pressure field reinforce each other. The horizontal gradient of the total field of pressure as given by eq.65 increases with depth and can reach its maximum near the bottom. The result is that bottom currents with speeds exceeding those at the surface can occur even in the deep sea during the transient state of wind-driven currents in a stratified ocean. Relatively strong bottom currents in the deep sea have been inferred from sedimentary structures, particularly ripple marks, in bottom photographs and sediment cores (HEEZEN and HOLLISTER, 1964). More evidence of strong bottom currents in the deep ocean will be given in Chapter V.

A strict theoretical analysis of wind-driven currents in a baroclinic ocean involves not only mathematical difficulties but also difficulties in the proper physical formulation of the problem. The observed density stratification of the ocean is partly the result of an adjustment of the field of mass to the field of currents, and partly directly related to thermohaline processes that locally alter the field of density and produce a system of currents that is generally referred to as thermohaline circulations. This latter constituent of the general oceanic circulation is by far not so well understood as the wind-driven currents, because it is even difficult to construct a satisfactory model of the distribution of net heating and cooling and salinity changes due to evaporation and precipitation at the sea surface (STOMMEL, 1957b). Some interesting theoretical results on the subject of the three-dimensional ocean circulation by means of computer techniques have recently been obtained by BRYAN et al. (1967).

Some conclusions about the mechanism of formation of wind-driven currents in a non-homogeneous ocean can be drawn from a theoretical solution obtained by LINEIKIN (1955a), although some features of his model are not quite realistic.

Lineikin considered an infinitely long and infinitely deep channel with vertical shore boundaries. A rectangular coordinate system is oriented in such a way that the y-axis points in the direction of the channel, and the x-axis points across the channel such that one boundary is at $x = 0$ and the other at $x = L$. Thus L is the width of the channel. Besides friction resulting from vertical shearing stresses, Lineikin also included friction resulting from lateral shearing stresses, with the assumption that the eddy viscosity coefficients, A_z and A_h, respectively, are constant. The variation of the Coriolis parameter with latitude is neglected. This fact restricts the general application of Lineikin's results.

At the initial moment, $t = 0$, the sea is assumed to be motionless ($u = v = w = 0$), and the density as a function of depth is given by $\rho = \rho_0 (1 + bz)$. Thus, there is a linear increase of density with depth, and $\rho_0 b$ represents the vertical gradient. On the average, b is of the order 10^{-8} (cm^{-1}), although the vertical density gradient varies considerably in the oceans.

The boundary conditions are such that at the sea surface, $z = \zeta$, the wind exerts a stress, τ, with the components $\tau_x = -A_z \, \partial u/\partial z$, $\tau_y = -A_z \, \partial v/\partial z$. At $z \to \infty$, u, v, and w approach zero and isobaric surfaces, isopycnal surfaces and level surfaces coincide.

Along the shores at $x = 0$ and at $x = L$, $u = v = 0$, but $w \neq 0$, and $\partial \rho/\partial x = 0$.

The approximate solution to this problem, for the steady state, agrees in many points with Ekman's classical results. The pure drift current in the surface strata, however, depends on the width of the channel and, of course, on the geographical latitude. For example, the deflection angle between the wind and the surface current in mid-latitudes is only about 15° for

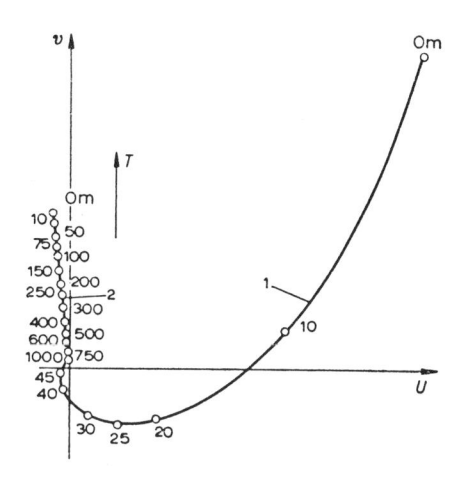

Fig.67. Hodograph of (*1*) the pure drift current and (*2*) the gradient current components of a steady wind-driven current, according to LINEIKIN (1955a). Numbers at circles indicate the depth in meters. The arrow T shows wind direction. (After FOMIN, 1964, fig.3.)

a narrow channel of 1 km width. For a width of 100 km it is about 42°, and for a width of 1,000 km very close to 45°, approaching this value for an infinitely wide channel.

The constituents of the total current and its vertical structure can also be compared with Ekman's elementary current system in a stratified ocean. Besides the pure drift component, which is essential only in the upper strata, the deep water is kept in motion by secondary wind effects, that is, by slope currents and relative currents, resulting from a readjustment of the density structure in the non-homogeneous sea. This constituent of the total current field was called by Lineikin the "convective-gradient current".

Fig.67 represents the vertical structure of the current system, according to LINEIKIN (1955a) when current vectors at different depths are projected on a horizontal plane. The figure shows the pure drift current and "gradient" components[1] separately for different depths. In computing this example, it was assumed that the wind blew in the direction of the channel (positive y-axis) with a speed of 10 m/sec, that the geographical latitude was $\phi = 45°$ (Northern Hemisphere), $A_h = 10^8$ (cm^{-1} g/sec), $A_z = 10^2$ (cm^{-1} g/sec) and $L = 10^7$ (cm).

For a steady state, the equation of continuity requires that the transport of water by the pure drift current in the upper strata (which is from left to right in Fig.67) must be compensated for by an opposite water transport in the deeper layers. This is shown in the figure by the vertical distribution of the "gradient" current between the surface and 1,000 m depth. This current transports water with a component in the minus x-direction. The "gradient" current decreases rapidly with depth in the surface layers and less rapidly in deeper layers. For the example

---

[1] The "gradient" component of the total current in Lineikin's model, or the "gradient" current as it may be called here, should not be confused with the gradient current as defined in section "Gradient currents" (pp. 158–162) of this chapter. Following Lineikin, the term "convective-gradient" current, or briefly, "gradient" current applied to this model serves to emphasize the fact that it depends on both the slope of the sea surface and the inhomogeneous density structure. It is induced by the absolute horizontal pressure gradient.

shown in Fig.67 its velocity is almost zero at 1,000 m depth. According to Lineikin's results, the decrease with depth follows an exponential function such that absolute motionless water is found at $z \to \infty$, in accordance with the lower boundary condition.

Since this current structure does not imply a "bottom current" in the sense of Ekman's analysis for a homogeneous ocean, frictional forces in the deep water, resulting from lateral shearing stresses, are most essential. It is also seen that in the deep layer of at least 1,000 m thickness, the "gradient" current does not need much to be deflected from the isobaric direction in order to provide the necessary balance of water transport in cross-channel direction. In mid-latitudes, a deflection angle of 2° or 3° of the "gradient" current direction from the direction of the isobars (or from the direction of the geostrophic current) is usually sufficient (NEUMANN, 1952a, 1955). Thus, the "gradient" component of the total wind-driven current can be considered with a fair degree of accuracy as a geostrophic current. However, the effect of friction and the slight deflection of this current are decisive factors in the mutual adjustment of the fields of density and currents, and in providing stationary conditions. In a homogeneous ocean (see Fig.55), according to Ekman, the balance of anisobaric mass transports is made possible by the appearance of a bottom current in the bottom layer of frictional influence.

In the preceding model, the "convective-gradient" current decreases exponentially with depth. The depth at which the current speed has decreased to the value $e^{-\pi}$ times the "gradient" component of current velocity at the sea surface was called the "depth of baroclinicity" (LINEIKIN, 1955b).

Attempts to apply the results of Lineikin's model to large-scale oceanic motions, however, failed. For horizontal scales of motion of 1,000 km, the "depth of baroclinicity" is much too great (about 30 km in a latitude of 40°). More realistic assumptions about the vertical density structure, a more realistic shaped ocean, and the introduction of a variable Coriolis parameter with latitude might help to improve the results. Indeed, STOMMEL and VERONIS (1957) have shown that the variation of the Coriolis parameter can limit the currents to a much more realistic depth. In a subsequent paper on the dynamics of the baroclinic layer in the ocean, LINEIKIN (1957) also included a variable Coriolis parameter.

Thermohaline circulations

Besides the wind stress at the sea surface, another important cause for ocean currents is the difference in temperature and salinity maintained by differential heating, evaporation and precipitation at the sea surface. These factors affect the sea water density. Vertical convection, overturning circulations and mixing provide the mechanism for distributing such density differences from the sea surface into deeper strata and for creating a certain "climate" at subsurface levels in the oceans. The result is horizontal density differences, and consequently, horizontal pressure differences and currents. Such currents are called *thermohaline currents*.

From the beginning of dynamic oceanography, thermohaline circulations and their role in the general oceanic circulation have been studied with more or less success. For some time, the question was raised and vigorously debated whether the wind-driven or the thermohaline circulations were of greater importance in explaining the general oceanic current field (Croll versus Carpenter during the 1870's; see KRÜMMEL, 1911, p.48). Today we believe that both are important and that they are not independent of each other. This makes the analysis of ocean currents more complicated than the analysis of atmospheric motions. The motions of the atmosphere can, essentially, be explained as a thermally driven "engine". Also, the interference of coastlines as lateral boundaries makes the dynamic study of ocean currents more complicated.

EKMAN (1926a) and GOLDSBROUGH (1933) investigated the circulation produced by an uneven distribution of precipitation and evaporation over the sea surface. In general, their conclusions can be summarized by saying that in the open ocean the velocity of currents produced by density differences of this kind is rather insignificant. SHULEIKIN (1945), STOMMEL (1950), TAKANO (1955b), and others have studied currents resulting from differential heating in various parts of the sea. The results also indicate relatively small velocities for such thermal circulations.

SANDSTRÖM (1908) demonstrated in his simple, but nevertheless informative, experiments that conditions in the ocean are not very favorable for the development of strong thermal circulations. Theoretically, his conclusion was supported by BJERKNES (1916). A "thermal engine" can work efficiently only when heating and expansion take place at higher pressure, and cooling and contraction at lower pressure. This is, in general, the case in the atmosphere. However, in contrast to the atmosphere, the thermal "engine" in the oceans is both heated and cooled at the sea surface, that is, at lowest sea pressure.

One of the earliest models of a thermal circulation in the oceans is the model by LENZ (1847). He assumed heating at the equator and cooling at the poles. The colder, polar water will sink to the bottom when sufficiently dense, and warmer water from equatorial regions will move poleward. Thus, his scheme of thermal oceanic circulations produced two big gyres in a vertical meridional section with sinking water in polar regions and rising water in equatorial regions, while the upper strata of the oceans moved poleward and the lower strata toward the equator. We know that this scheme is not verified in the oceans, although sinking of cold water in polar regions occurs and plays an important role in the explanation of the stratification and circulation of the oceans. However, in some parts of the polar or subpolar regions of the oceans, cooler water from deeper layers can also rise to the upper strata. Such conditions will be discussed in Chapter V, together with the deep sea stratification and circulation of the oceans.

The fact that the simple scheme of Lenz is untenable was shown by SANDSTRÖM (1909). His experiments have demonstrated that neither a "heat source" nor a "cold source", or both, can produce effective deep-reaching thermal circulations when both are at the same level, that is, under equal pressure. Although JEFFREYS (1925) has rightfully questioned the generality of Sandström's conclusions, the fact remains that thermally produced circulations are much more effective when the heat source is deep in the oceans and the cold source is in the upper strata.

In addition to differential heating and cooling, evaporation and precipitation over the oceans change the surface salinity and, therefore, the sea water density. Since increasing salinity produces denser water, the resulting "haline" circulation combines with and is superimposed on the "thermal" circulation. The interesting fact is that "thermal" and "haline" circulations counteract each other in high latitudes. Only in lower latitudes between about 25° of latitude and the equator do they act in the same sense to produce a thermohaline circulation by "overturning" of water masses (DEFANT, 1929a). Since all of these factors act from the sea surface, it seems that thermohaline circulations are limited to some regions in the oceans with significant effects only in the upper strata.

Both the wind-driven circulation and the thermohaline circulation in the oceans cannot be independent of each other. We are far from understanding the combined action of the two. This understanding involves a better knowledge of turbulent exchange processes in the sea for momentum, salinity and heat content. A model that satisfactorily considers all three has not yet been devised.

In spite of the fact that a thermal "engine" in the ocean cannot be very efficient, the thermohaline circulation in the ocean, especially when combined with the wind-driven circulation system, may prove to be quite significant. This was suggested and demonstrated by STOMMEL (1957b). ROBINSON and STOMMEL (1959), WELANDER (1959), STOMMEL and WEBSTER (1962), ROBINSON and WELANDER (1963), and BLANDFORD (1965b) carried on with the investigation of thermohaline circulations, particularly in connection with the theory of the ocean thermocline.

More recent work by BRYAN et al. (1967) is of special interest. The thermohaline circulation has been modeled in both the absence of any wind-driven component and with consideration of a wind stress applied to the surface. The thermohaline component was found to be surprisingly strong, showing a strong poleward current along the *western* side of the Northern Hemisphere model ocean. When combined with a wind-driven circulation the total circulation looks quite realistic.

Some recent results of Bryan's work will be discussed in the following chapter. Among the more recent papers, a theoretical analysis on thermally maintained circulation in a closed ocean basin by NIILER et al. (1965) is of interest. Their results seem to confirm the analysis of deep circulations in the North Atlantic Basin by WORTHINGTON (1965).

CHAPTER V

THE GENERAL CIRCULATION OF THE OCEANS

HORIZONTAL CIRCULATION OF WIND-DRIVEN OCEAN CURRENTS

A complete mathematical analysis of the general three-dimensional circulation in a baroclinic ocean offers considerable difficulties. This problem requires the consideration of all possible driving forces, the introduction of adequate dissipative forces between the surface and bottom, and lateral and vertical mixing processes. In some oceanic regions, the variable topography of the ocean bottom also has to be included.

Modern, electronic computing techniques can aid significantly in numerical solutions of coupled, nonlinear, partial differential equations that evolve from such a problem. However, one of the important quantities that needs further investigation is the effect of non-homogeneous, non-isotropic friction. Another difficulty in solving the problem is the fact that even the initial and boundary conditions cannot always be realistically stated. For example, when starting with an ocean at rest, the question is: what kind of initial density distribution should be included? If one introduces a density distribution similar to the average observed density field in the oceans, this initial condition already imposes a certain current field, because the density field is partly the result of the general circulation in the three-dimensional ocean space.

In the past, attempts have been made to solve the problem of the general oceanic circulation step-wise, by proceeding from simple models to more complicated ones. Proper combination, or superposition, of results obtained from such simple models may, finally, lead to a better physical understanding of the dynamics of ocean currents.

The foundation for the theoretical solution to the problem of the general oceanic circulation was laid by Walfrid Ekman.

Some of his most important results were mentioned in the preceding chapters. His further studies (e.g., EKMAN, 1923, 1928a,b, 1932a) were directed toward the calculation of the total current system in the oceans for a given wind field. As early as 1902, EKMAN introduced for the first time frictional forces resulting from vertical shearing stresses. This was a remarkable step forward compared to the Guldberg–Mohn assumption about friction which was used in early dynamic meteorology. The Ekman spiral could never have been derived with the Guldberg–Mohn assumption about friction.

In 1923, the important effect of a variable Coriolis parameter was discussed by Ekman. At the same time, EKMAN (1923) deduced some important laws concerning the curl of the deep current for the case of steady motion. Of special importance are EKMAN's (1923) theoretical investigations of the effects of non-uniform wind systems over the sea, of variable ocean depths, and of a variable Coriolis parameter on the horizontal circulation of wind-driven currents.

Ekman's results for a homogeneous ocean

For the case of stationary ocean currents in a homogeneous ocean, EKMAN (1923) derived a remarkable differential equation for horizontal wind-driven ocean currents. It describes the effects of a variable wind stress, a variable Coriolis parameter with latitude, a variable friction (resulting from vertical shearing stresses), and a variable ocean depth.

Our knowledge of friction in the oceans is still very limited. In Ekman's theory, only frictional forces resulting from vertical shearing stresses were considered, and their effects were implicitly shown in the quantities D and D^* (eq.110 and 118), respectively. With a constant vertical eddy viscosity coefficient, $D = D^*$, and the quantities:

$$B = \frac{kD}{2\pi}$$
$$b = kd - B$$
(129)

in Ekman's theory contain the essential effects of friction. In eq.129, $k = \rho g/f$, and d represents the depth of the sea. (In Ekman's equations, the z-coordinate points downward.)

The velocity, G, of Ekman's "deep current" in a homogeneous ocean is obtained from the sea surface slope, β, according to:

$$G = \frac{g}{f}\beta \qquad (130)$$

where $f = 2\omega \sin\phi$. This deep current is, essentially, a geostrophic current if the ocean depth, d, is greater than the sum $D + D^*$. EKMAN (1923) has shown that this deep current (although geostrophic) is a divergent current. Its divergence is:

$$\text{div} G = \frac{k\beta_x}{\rho R_e \tan\phi} \qquad (131)$$

where $\beta_x = \partial \zeta/\partial x$ is the x-component of the slope ζ of the sea surface, and R_e is the average radius of the earth. The divergence shown in eq.131 increases with increasing latitude as a result of the variation of f with latitude. Since:

$$\frac{\partial f^{-1}}{\partial y} = -\frac{1}{f R_e \tan\phi}$$

and: $\qquad (132)$

$$\frac{1}{f}\frac{\partial f}{\partial y} = \frac{\cot\phi}{R_e}$$

it is seen that the divergence in eq.131 is the result of a variable Coriolis parameter with latitude. In modern oceanography, the approximate variation of f with latitude, $\partial f/\partial y$ (= constant), is often called the "beta effect", and a plane system of coordinates in which the Coriolis parameter is considered constant except where differentiated in the meridional direction is called a "beta plane". Thus EKMAN (1923) introduced the important variation of f with latitude for the first time.

The z-component of the curl of the deep current, according to Ekman, is:

$$\text{curl} G = -\frac{k}{\rho}\text{div}\beta + \frac{k\beta_y}{\rho R_e \tan\phi} \qquad (133)$$

In 1923, EKMAN derived a remarkable differential equation for the horizontal transport of wind driven currents in a homogeneous ocean:

$$\text{curl} G = \frac{k}{\rho B f} \text{curl} \tau + \frac{k^2}{\rho B} \left(\beta_y \frac{\partial d}{\partial x} - \beta_x \frac{\partial d}{\partial y} \right)$$

$$+ \frac{k}{\rho D} \left[(\beta_x - \beta_y) \frac{\partial D}{\partial x} + (\beta_x + \beta_y) \frac{\partial D}{\partial y} \right]$$

$$+ \frac{k}{\rho R_e \tan \phi} \left(\beta_x \tan \alpha + \frac{\tau_x}{Bf} \right) \quad (134)$$

where $\tan \alpha = b/B$ is defined by eq.129.

The first term on the right hand side of eq.134 contains the effect of a variable wind stress, the second term contains the effect of a variable ocean depth, the third term contains the effect of a variable friction, and the last term contains the effect of the variation of the Coriolis parameter with latitude.

All effects may be of equal importance. The most cumbersome term, however, appears to be the term that describes the effect of a variable friction, although Ekman included in this term only frictional forces resulting from vertical shearing stresses. If lateral shearing forces are also considered, the basic differential equation would become even more complicated than is already expressed in eq.134.

Our knowledge about regional changes of friction in ocean currents is still very limited. Since we do not know exactly how to introduce variable exchange coefficients for eddy momentum transfer, these coefficients are often considered constants. With the assumption that in the horizontal wind-driven circulation of a whole ocean D is constant, terms in eq.134 containing the derivatives of D disappear. For this simplified case, Ekman's differential equation can be written:

$$\text{curl} G = \frac{k}{\rho B} \left[\frac{1}{f} \text{curl} \tau + \rho \left(G_x \frac{\partial d}{\partial x} + G_y \frac{\partial d}{\partial y} \right) - \frac{1}{R_e \tan \phi} \left(\frac{\rho G_y b}{k} - \frac{\tau_x}{f} \right) \right] \quad (135)$$

In eq.135, G is the geostrophic deep current with the components:

$$G_x = \frac{k}{\rho}\beta_y$$
$$G_y = -\frac{k}{\rho}\beta_x \tag{136}$$

Eq.135 represents a "vorticity tendency" equation. On the left hand side stands the curl of the deep current, and the right hand side contains in the first term the vertical component of the curl of the wind stress at the sea surface which plays such a dominant role in modern theories (e.g., SVERDRUP, 1947; MUNK, 1950). The second term in parentheses on the right hand side represents the scalar product $(G \cdot \nabla d)$ which considers a variable depth, d, of the ocean, and the last term represents the planetary curl effect, or the effect of the variation of the Coriolis parameter with latitude. The importance of the variation of the Coriolis parameter with latitude was later most clearly demonstrated by STOMMEL (1948a) for a simple horizontal circulation in a regular ocean of constant depth.

Ekman's eq.135 can be symbolically expressed in the form:

$$W = W_\tau + W_d + W_\phi \tag{137}$$

The "quasi curl" (EKMAN, 1923) on the right side of eq.137 contains the "anemogenic curl effect" (W_τ), the "topographic (quasi) curl effect" (W_d), and the "planetary curl effect" (W_ϕ).

The effect of bottom topography

The effect of the term W_d on the deep current can be demonstrated by examining it separately from the other terms. If:

$$\text{curl}\,G = \frac{k}{B}\left(G_x \frac{\partial d}{\partial x} + G_y \frac{\partial d}{\partial y}\right) \tag{138}$$

it is seen that $\text{curl}\,G = 0$ if G is perpendicular to ∇d (∇ is the gradient operator, $\partial/\partial x + \partial/\partial y$ for horizontal coordinates).

Eq.138 says that the current G must follow the depth contour lines (isobaths) of the ocean bottom in the case where curl$G = 0$. If s denotes the direction of the current G, and if this current crosses the isobaths, then:

$$\text{curl}G = \frac{2\pi}{D} G \frac{\partial d}{\partial s} \tag{139}$$

This equation states that a positive curl$G$ (contra solem) results where the current flows over increasing depth, and a negative curl$G$ (cum sole) results where the depth decreases in the direction of the current.

The effect of bottom topography on currents was demonstrated by EKMAN (1923) by studying special cases, although simultaneous effects of a varying Coriolis parameter with latitude were neglected in some (but not all) of his models. In a revised theory, EKMAN (1932a) included the effect of nonlinear field accelerations.

The term W_d is, obviously, most important where currents reach the ocean bottom. Especially along continental margins (continental shelves, slopes and rises), over and around seamounts, mid ocean ridges and rises, and other obstacles in the path of currents, the term W_d must play a significant part. Currents like the Gulf Stream and other *boundary currents* "hugging" the continental margin are affected by the bottom topography in addition to the effect of a variable Coriolis parameter with latitude.

The Gulf Stream south of Cape Hatteras keeps close to the coast, and after leaving Cape Hatteras in a more northeasterly direction, the Stream still seems to "feel" the bottom at least in some parts along its track, particularly when crossing a submarine ridge that extends southward from the Grand Banks. Over this ridge, the Gulf Stream is deflected to the south when flowing over decreasing depth, and back to the north when flowing east off the ridge over increasing depth. An explanation for the observed streamline deflection of the Gulf Stream in this region was first offered by EKMAN (1923). As a result of this deflection, colder water from the northern side of the Gulf Stream extends into southern latitudes, forming a "tongue" of cold water along about 50° longitude. Many detailed oceano-

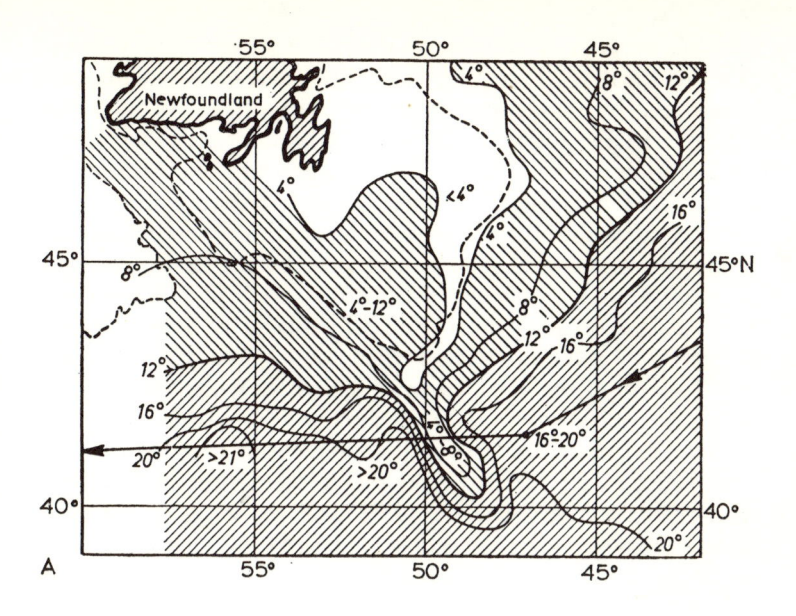

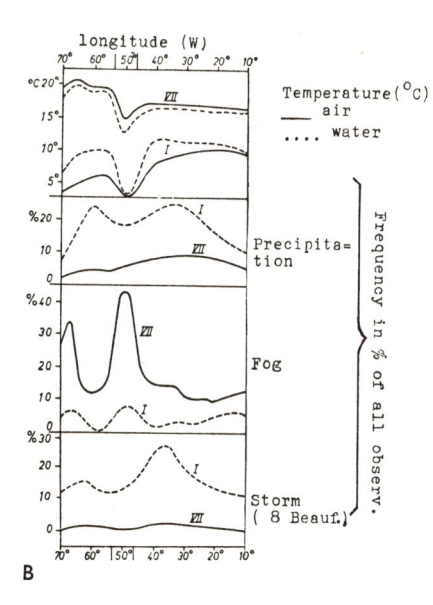

HORIZONTAL CIRCULATION OF WIND-DRIVEN CURRENTS

Fig.68A. Sea surface temperature distribution south of the Grand Banks May 31–June 16, 1938 (International Ice Patrol Service). The ships' tracks (April–August) are shown by the line with an arrow. (After NEUMANN, 1953.)

B. Some meteorological anomalies along the most frequently travelled ships' routes between the English Channel and New York (After KUHLBRODT, 1941.)

graphic surveys in this region by U.S. Coast Guard vessels (International Ice Patrol Service) and other observations have confirmed this temperature anomaly south of the Grand Banks. One example is shown in Fig.68. The isotherms in the vicinity of longitude 50°W bend sharply to the south and back to the north. This contrast is indicated over a relatively short zonal distance of about 60 nautical miles where the water temperature along the ship's track (shown in Fig.68 by the heavy line with an arrow) varies between about 20° and 5°C. The regional temperature anomaly near 50°W longitude is caused by the deflection of the Gulf Stream when passing over the ridge near 50°W south of the Grand Banks. This cold water anomaly has significant effects on the local climate and weather over this region. Relationships between oceanic and atmospheric anomalies south of the Grand Banks region were discussed by NEUMANN (1953). The lower part of Fig.68 represents some mean meteorological anomalies as observed on the frequently travelled ships' routes between the English Channel and New York (KUHLBRODT, 1941).

The original Ekman type of streamline deflection (EKMAN, 1923) for a current passing over a ridge in the bottom topography is shown in Fig.69A for the Northern Hemisphere. It disregards the variation of the Coriolis parameter with latitude; that is, the term W_ϕ in eq.137. SVERDRUP (1941) pointed out that in the case of a non-homogeneous ocean, certain modifications of Ekman's type of streamline deflection have to be expected as a result of a rearrangement of the field of mass to the field of currents. Essentially, Sverdrup's model applies a geostrophic equilibrium between the fields of mass and currents. When a frictionless current, according to Sverdrup, approaches a submarine ridge, a current deflection as shown in Fig.69B should occur. This type of streamline deflection appears more realistic than the original Ekman type shown in Fig.69A. The Sverdrup type of streamline deflection also appears to be verified by the observations south of the Grand Banks shown in Fig.68. The pure Sverdrup type of streamline deflection can, however, also be obtained in a homogeneous ocean if the variation of the Coriolis parameter is taken into account and friction is neglected. If the variation of the Coriolis parameter is considered in addition to

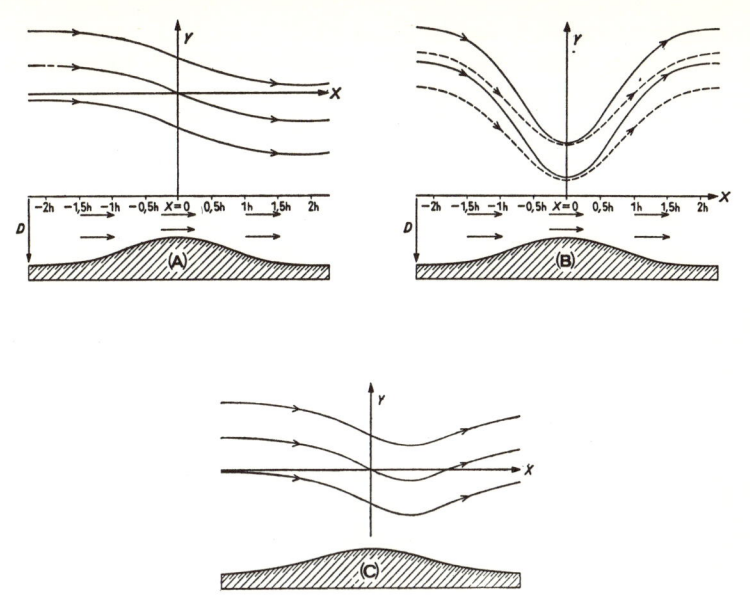

Fig.69A. Ekman type of streamline deflection for a flow over a submarine ridge in homogeneous water.
B. Sverdrup type of streamline deflection.
C. Mixed type of streamline deflection. The upper parts of the figures show the deflection in a horizontal plane (Northern Hemisphere). (After NEUMANN, 1960b.)

friction, a mixed type of streamline deflection results (NEUMANN, 1960b; see also SAINT-GUILY, 1962). These different types of streamline deflection are shown in Fig.69.

The planetary curl effect

The effect of the term W_ϕ appears more complicated, even when studied separately from the other terms. It contains the wind stress, τ_x, in the zonal direction, along with the term G_y. With $\tau_x = 0$, the sign of W_ϕ is given by the sign of G_y. Currents flowing towards the poles experience a negative curl effect (cum sole deflection), and currents flowing toward the equator experience a positive curl effect (contra solem deflection).

The combined effects of W_d and W_ϕ are examined for non-accelerated, frictionless currents. With $\tau = 0$, $B = 0$, and $b = kd$ it follows from eq.135 that:

$$G_x \frac{\partial d}{\partial x} + G_y \frac{\partial d}{\partial y} - \frac{G_y}{R_e \tan\phi} d = 0 \qquad (140)$$

or:

$$\frac{\partial \zeta}{\partial y} \frac{1}{d} \frac{\partial d}{\partial x} - \frac{\partial \zeta}{\partial x} \frac{1}{d} \frac{\partial d}{\partial y} = -\frac{\partial \zeta}{\partial x} \frac{1}{R_e \tan\phi} \qquad (141)$$

Eq.141 shows that a non-accelerated frictionless current follows the bottom contour lines only if $\partial f/\partial y = 0$, and in the case where d is constant, $\partial \zeta/\partial x$ must be zero. Thus, meridional geostrophic currents, G_y, are not possible in homogeneous water of constant depth on a rotating earth; only zonal currents of this type can exist. All current directions, however, are possible when d is a certain function of latitude (DEFANT, 1929a). If eq.141 is written in the form:

$$\frac{\partial \zeta}{\partial x}\left(\frac{1}{d}\frac{\partial d}{\partial y} - \frac{1}{R_e \tan\phi}\right) = \frac{\partial \zeta}{\partial y}\frac{1}{d}\frac{\partial d}{\partial x} \qquad (142)$$

it is seen that $\partial \zeta/\partial x$ can be unequal to zero even with $\partial d/\partial x = 0$, if:

$$\frac{1}{d}\frac{\partial d}{\partial y} = \frac{1}{R_e \tan\phi} \qquad (143)$$

This condition requires a change of the ocean depth, d, with latitude (or of the depth of no absolute current) according to an equation of the form:

$$d = d_0 \sin\phi + \text{constant} \qquad (144)$$

Ekman called a bottom slope that follows the depth distribution of eq.144 the "critical bottom" slope.

NEUMANN (1955) has shown that the average meridional slope of DEFANT's (1941) reference surface, or level of no motion in the Atlantic Ocean agrees with the basic concepts of eq.144 when the average meridional density stratification is considered. It is remarkable that in the zonal average the depth of Defant's

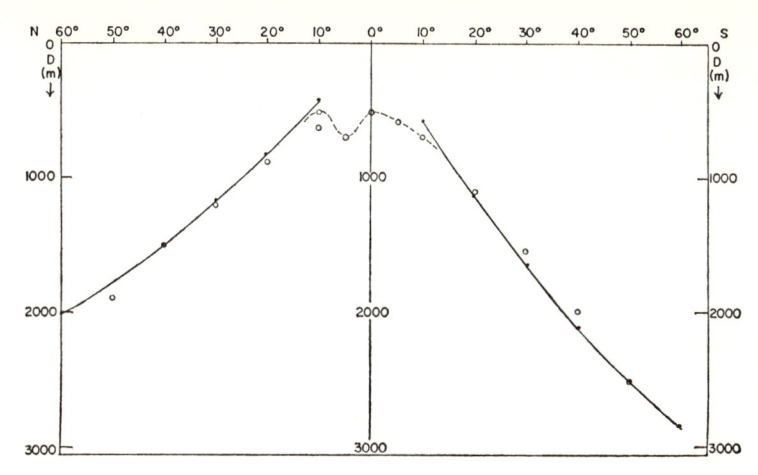

Fig.70. Zonal average depth of Defant's reference surface in the Atlantic Ocean shown by circles. The fully drawn curves represent the function $d = d_0 \sin\phi$. (After NEUMANN, 1955.)

reference surface is nearly proportional to the sine of the geographic latitude as expressed in eq.144. Fig.70 represents the depth of Defant's reference surface for zonal averages in the Atlantic Ocean by circles, and the fully drawn curves represent the function $d = d_0 \sin\phi$. The constant d_0 is different in the Northern and Southern Hemispheres, but the increase of d with latitude, ϕ, follows this function closely except in equatorial regions, where planetary vorticity effects are more and more overruled by relative vorticity effects. OSTAPOFF (1957) has shown that the "zero layer" in the Pacific Ocean as defined by DEFANT (1941) also changes its depth nearly proportional to the sine of the geographic latitude. Although these results apply to a baroclinic ocean, they are of interest with regard to Ekman's results obtained for a homogeneous ocean where the lower boundary is the solid bottom.

EKMAN (1932a) extended his studies to include nonlinear field accelerations. These accelerations have significant effects on the deep current. Although in general, the effects of the three terms on the right hand side of eq.137 are of the same type as before, they are no longer independent of each other. In particular, the topographic and planetary vorticity effects are interrelated in a complex manner. For example, in the case of a current flowing over increasing depth, the revised theory of EKMAN (1932a) indicates a tendency toward contra solem deflection not only where the depth increases in flow direction, but also where the depth is a maximum; and vice versa for decreasing depth where a tendency of cum sole deflection is also found on top of a submarine ridge. Thus, the streamline pattern shown in Fig.69A will be displaced to the right with respect to the bottom topography. Also, the streamline deflection was found to be dependent on the absolute depth of water in addition to the slope of the bottom.

The introduction of frictional forces resulting from lateral shearing stresses will modify Ekman's original results. When compared to vertical shearing stresses, DEFANT (1926) and ROSSBY (1936) pointed out that lateral shearing stresses associated with the horizontal exchange in eddies may play an equally important part in the general circulation of both atmosphere and ocean. MONTGOMERY (1939) and MONTGOMERY and PALMÉN (1940) indicated the importance of lateral stresses in the dynamic balance of oceanic current systems, and MUNK (1950) demonstrated most clearly the essential role of the lateral stress curl for the wind-driven circulation of a whole ocean.

Modern approaches to the problem of wind-driven currents

Considerable progress toward a better understanding of the average large-scale wind-driven oceanic circulation has been made since about 1946. Primarily, these investigations have been concerned with the general features of the horizontal current systems and their relationship to the distribution of the mean wind field over the oceans. It is interesting to follow the development of modern dynamic oceanography from the study of basic, linear theories to nonlinear approaches, and ultimately, to the attempt to combine wind-driven and thermohaline components in one model of the general ocean circulation (BRYAN and Cox, 1967). Although a complete mathematical analysis of the problem and successful consideration of all possible effects have not yet been achieved even for the stationary current system, progress since Ekman's classical theory is remarkable.

In the analysis of the total horizontal mass or volume transport by wind-driven ocean currents, most frequently the assumptions are retained that the currents are non-accelerated and that the coefficients of vertical (A_z) and horizontal (A_h) eddy viscosity are constant. Although these assumptions are very restrictive, a series of important and informative papers has emerged through their use.

In rectangular coordinates with the y-axis to the north, the x-axis to the east and the z-axis vertically upwards from a level surface, the equations of motion for horizontal non-accelerated flow with constant eddy viscosity coefficients are:

$$-f\rho v = -\frac{\partial p}{\partial x} + A_z \frac{\partial^2 u}{\partial z^2} + A_h \nabla^2 u$$

$$f\rho u = -\frac{\partial p}{\partial y} + A_z \frac{\partial^2 v}{\partial z^2} + A_h \nabla^2 v$$

(145)

where the operator $\nabla^2 = (\partial^2/\partial x^2 + \partial^2/\partial y^2)$. Integration of eq.145 between the lower boundary, $d(x, y)$, of the current system and the sea surface, $\zeta(x,y)$ yields:

$$-fS_y = -\frac{\partial}{\partial x}\int_d^\zeta p\,dz + p_\zeta \frac{\partial \zeta}{\partial x} - p_d \frac{\partial d}{\partial x} + \tau_x - (\tau_x)_d + \int_d^\zeta R_x\,dz$$

$$fS_x = -\frac{\partial}{\partial y}\int_d^\zeta p\,dz + p_\zeta \frac{\partial \zeta}{\partial y} - p_d \frac{\partial d}{\partial y} + \tau_y - (\tau_y)_d + \int_d^\zeta R_y\,dz$$

(146)

In eq.146:

$$S_x = \int_d^\zeta \rho u\,dz$$

$$S_y = \int_d^\zeta \rho v\,dz$$

(147)

are the x and y components, respectively, of the total horizontal water mass transport S. The wind stress components at the sea surface are τ_x and τ_y respectively, and $(\tau_x)_d$, $(\tau_y)_d$ represent possible tangential stresses exerted at the lower boundary, d, of the currents. In homogeneous water, and where the current reaches the bottom, bottom stresses should be included. In eq.146, the abbreviations:

$$R_x = A_h \nabla^2 u \\ R_y = A_h \nabla^2 v \tag{148}$$

are introduced in order to simplify the writing, and also because other investigators have used other forms for the effective internal lateral friction than those stated in eq.148.

In a stratified ocean where the current velocity decreases with depth and gradually vanishes at some distance above the solid bottom, the bottom stress components $(\tau_x)_d$ and $(\tau_y)_d$ either vanish or can be neglected. If a bottom stress is present, in some problems its effect can be thought of as being absorbed in a "virtual" (effective) internal friction.

Differentiation of the first equation in eq.146 with respect to y and of the second equation with respect to x, and subsequent subtraction of one equation from the other yields:

$$S_y \frac{\partial f}{\partial y} + \left(\frac{\partial \tau_x}{\partial y} - \frac{\partial \tau_y}{\partial x} \right) - \frac{\partial}{\partial x} \int_d^\zeta R_y \, dz \\ + \frac{\partial}{\partial y} \int_d^\zeta R_x \, dz = \frac{\partial p_d}{\partial y} \frac{\partial d}{\partial x} - \frac{\partial p_d}{\partial x} \frac{\partial d}{\partial y} \tag{149}$$

In arriving at eq.149, it is assumed that the pressure at the sea surface, p_ζ (the atmospheric pressure), is constant, and that the bottom stress vanishes or can be neglected. The pressure at the lower boundary of the current, p_d, is obtained from:

$$p_d = g\bar{\rho}(\zeta - d) \tag{150}$$

where:

$$\bar{\rho}(x, y) = \frac{1}{\zeta - d} \int_d^\zeta \rho(z) \, dz \tag{151}$$

In addition to eq.149, the divergence of the total horizontal mass transport:

$$\frac{\partial S_x}{\partial x} + \frac{\partial S_y}{\partial y} = 0 \qquad (152)$$

for stationary upper and lower boundaries of the current.[1] Substitution of eq.150 into eq.149 yields:

$$S_y \frac{\partial f}{\partial y} - \mathrm{curl}_z \tau - \frac{\partial}{\partial x} \int_d^\zeta R_y \mathrm{d}z + \frac{\partial}{\partial y} \int_d^\zeta R_x \mathrm{d}z$$

$$= gd \left(\frac{\partial \bar{\rho}}{\partial x} \frac{\partial d}{\partial y} - \frac{\partial \bar{\rho}}{\partial y} \frac{\partial d}{\partial x} \right) - g\bar{\rho} \left(\frac{\partial \zeta}{\partial x} \frac{\partial d}{\partial y} - \frac{\partial \zeta}{\partial y} \frac{\partial d}{\partial x} \right) \qquad (153)$$

if the factor $\zeta - d$ is replaced by $-d$. This is no serious restriction, since ζ, the deviation of the free surface from a level surface, is always much smaller than the depth d. The vertical component of the curl of the wind stress is $\mathrm{curl}_z \tau = \partial \tau_y/\partial x - \partial \tau_x/\partial y$.

The right hand side of eq.153 represents the effect of a variable ocean depth. For $\nabla \zeta \neq 0$ and $\nabla d \neq 0$, these terms disappear only if the cross products $\nabla \zeta \times \nabla d = 0$, $\nabla \bar{\rho} \times \nabla d = 0$, or if both terms balance each other and $g(\bar{\rho} \nabla \zeta - d \nabla \bar{\rho}) \times \nabla d = 0$. This means that either $\bar{\rho} \nabla \zeta$ is exactly equal to and opposite to $d \nabla \bar{\rho}$ or that the resultant vector $\bar{\rho} \nabla \zeta - d \nabla \bar{\rho}$ is exactly parallel to ∇d.

The vertically integrated equation of motion has been used to avoid some difficulties in the mathematical analysis of wind-driven ocean currents (STOCKMAN, 1946; SVERDRUP, 1947). Their use makes it possible to deal with the general case of a baroclinic ocean without detailed specification of the vertical density distribution. However, the integrated equations give no information about the vertical currents. Also, it has to be kept in mind that the depth of penetration of wind-driven currents (including slope currents) depends essentially on the

---

[1] In sperical coordinates, however, $\mathrm{div} S = -S_y \tan\phi/R_e$. The term containing $\tan\phi$ is in middle latitudes of the same order of magnitude as the term $\partial f/\partial y$ and should be included in large-scale studies of the general oceanic circulations (NEUMANN, 1955; GARNER et al., 1962; LONGUET-HIGGINS, 1965).

density structure of the ocean. With $\bar{\rho} =$ constant, eq.153 would correspond to Ekman's equation (eq.135) in a homogeneous ocean if the frictional terms were replaced by bottom friction and the currents between the upper and bottom layers of frictional influence were essentially geostrophic.

STOCKMAN (1946) first used vertically integrated equations for the total transport in a baroclinic ocean of constant depth. He employed friction in the form of eq.148; however, he neglected the important term $S_y(\partial f/\partial y)$. SVERDRUP (1947) neglected lateral friction ($R_x = 0$; $R_y = 0$) and assumed that the horizontal pressure gradient due to the slope of the sea surface and density stratification compensate at some depth $d =$ constant. Thus, eq.153 was employed by Sverdrup in the form:

$$S_y \frac{\partial f}{\partial y} = \text{curl}_z \tau \tag{154}$$

which simply expresses a balance between the wind stress curl and the planetary vorticity due to motion in the meridional direction. In addition, eq.152 must be satisfied. If the wind stress is predominantly in the zonal direction ($\tau_y = 0$), or if $\partial \tau_y/\partial x$ can be considered small enough, $S_y(\partial f/\partial y) + \partial \tau_x/\partial y = 0$. With consideration of eq.152 this yields:

$$\frac{\partial S_x}{\partial x} = \frac{\partial}{\partial y}\left[(\partial f/\partial y)^{-1} \frac{\partial \tau_x}{\partial y}\right] \tag{155}$$

or:

$$\frac{\partial S_x}{\partial x} = \frac{1}{2\omega \cos\phi}\left(\frac{\partial \tau_x}{\partial y} \tan\phi + \frac{\partial^2 \tau_x}{\partial y^2} R_e\right)$$

where R_e is the radius of the earth.

This equation can satisfy boundary conditions only at one meridional boundary. SVERDRUP (1947) introduced the meridional boundary condition that at the eastern coast the transport, S_x, perpendicular to the coast vanishes. Eq.155 has been successfully applied by SVERDRUP (1947) and R. O. REID (1948b) to the equatorial currents of the eastern Pacific Ocean.

STOMMEL (1957b) has given a clear illustration of the physical meaning of a simple Sverdrup-type solution in a homogeneous ocean of constant depth, bounded by an eastern coast, and acted upon by a zonal wind stress. The heavy, shaded arrows in Fig.71, hovering above the surface, represent winds blowing from the west and from the east, respectively (Westerlies and Trade winds in the Northern Hemisphere). Horizontal water transports by pure drift currents in the upper layer of frictional influence (Ekman layer) are indicated by open arrows. These transports are proportional to the wind stress, inversely proportional to the sine of the geographical latitude and directed 90° to the right of the wind. In the central part of the figure, between the maximum Westerlies and the maximum Trade winds, downward motion of water is imposed as a result of converging

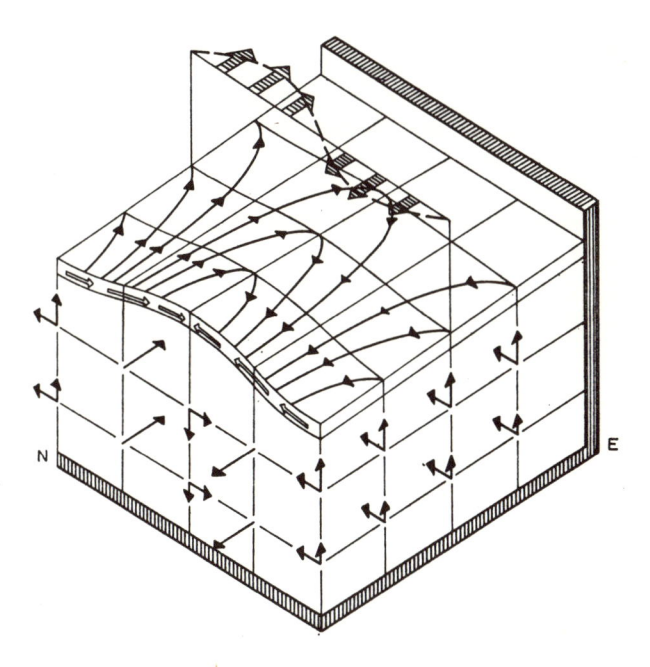

Fig.71. Sverdrup-type solution in a homogeneous ocean of constant depth, bounded by a meridional coastal wall on the east. Heavy, shaded arrows above the sea surface represent zonal winds. The curved lines are isobars and the arrowheads indicate direction of geostrophic horizontal flow. Solid arrows at subsurface levels show velocity components. (After STOMMEL, 1957b.)

currents in the Ekman layer. Outside of this zonal belt, surface water diverges and the vertical motion is upward. From the bottom of the Ekman layer to the bottom of the ocean, the vertical velocity component diminishes linearly to zero. At the latitudes of the maximum Westerlies and Trade winds, the vertical motion is zero. It is seen that a steady state geostrophic current system can be obtained which is consistent with the eastern coastal boundary condition and which matches by its field of divergence due to the meridional variation of the Coriolis parameter (eq.131) the field of divergence in the wind-driven layer of frictional influence (Ekman layer). The vertical transport of water in this model can be thought of as being absorbed by the divergence of the meridional component of the geostrophic current. Thus, at mid-latitudes (between the Westerlies and the Trades) where the vertical velocity is downward, the meridional component of geostrophic flow is southward. In Fig.71 the curved lines show contours of the sea surface topography, and the arrow heads at these lines indicate the direction of a geostrophic flow.

If the eastern boundary is removed, and the case of a zonal wind field is considered that acts upon a water belt all around the earth, zonal pressure gradients must vanish. This means that meridional components of a pure geostrophic current are impossible, and that without further consideration of other important physical terms (e.g., friction resulting from lateral shearing stresses), a steady state solution of ocean currents cannot be obtained. For a homogeneous ocean, Ekman introduced bottom friction to compensate for the anisobaric transport of water in the surface layer, and his model resulted in the "elementary current system" (Chapter IV).

It is also evident that boundary conditions at any coast to the west of the zonal water belt cannot be satisfied for the Sverdrup equation unless other processes are taken into account to provide for the dissipation of energy.

WELANDER (1959a) introduced the actual shape of ocean boundaries and wind stress data published by SCRIPPS INSTITUTION of OCEANOGRAPHY (1948) and by HIDAKA (1958) for transport computations in three oceans based on Sverdrup's equation. Since western boundary conditions could not be satisfied, no results could be obtained for the western boundary currents. In connection with his results, Welander discussed in detail the importance of terms neglected in the Sverdrup equation, such as lateral friction, incomplete compensation at the variable lower boundary of the currents, the effects of evaporation and precipitation, and local time variations. The complete theory, however, is very difficult to handle mathematically.

The important effect of the variation of the Coriolis parameter with latitude on the current in a closed rectangular ocean was clearly demonstrated by STOMMEL (1948a). In addition to a zonal wind stress acting at the sea surface, he introduced internal friction in a form comparable to the Guldberg–Mohn assumption. With this, four lateral boundary conditions in a closed (rectangular) ocean could be satisfied such that the boundaries are streamlines for the total horizontal water mass transport, and the transport component perpendicular to the coast vanishes. The ocean was still considered as a homogeneous layer of constant depth, d. If the components of frictional forces are taken as:

$$R_x = -\rho k u \qquad (156\text{A})$$
$$R_y = -\rho k v$$

where k is constant, eq.153 with d is constant yields:

$$S_y \frac{\partial f}{\partial y} + k\left(\frac{\partial S_y}{\partial x} - \frac{\partial S_x}{\partial y}\right) = \operatorname{curl}_z \tau \qquad (156\text{B})$$

By means of the equation of continuity (eq.152), a stream function, ψ, for the total horizontal mass transport can be introduced such that:

$$S_x = -\frac{\partial \psi}{\partial y}$$

$$S_y = \frac{\partial \psi}{\partial x} \qquad (157)$$

and eq.156B can be written in terms of the stream function:

$$\beta \frac{\partial \psi}{\partial x} + k\nabla^2 \psi = \text{curl}_z \tau \tag{158}$$

where $\beta = \partial f/\partial y$ is considered constant.

If a simple functional form of the wind stress is taken as:

$$\tau_x = -F \cos \frac{\pi y}{b}$$
$$\tau_y = 0 \tag{159}$$

eq.158 can be solved analytically with the boundary condition that the coast be a streamline. For a rectangular ocean basin of zonal length, l, and meridional width, b, the boundary conditions can be stated as in eq.160:

$$\psi(0,y) = \psi(l,y) = \psi(x,0) = \psi(x,b) = 0 \tag{160}$$

The solution to eq.158 with consideration of eq.159 and the boundary conditions, eq.160, was given by STOMMEL (1948a). His results are shown in Fig.72. Most striking is the intense crowding of streamlines toward the western border of the model ocean and the displacement of the center of the anticyclonic

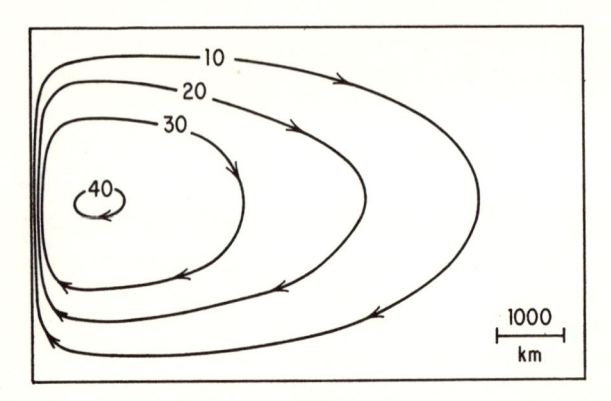

Fig.72. Westward intensification of wind-driven currents in a rectangular ocean of constant depth, according to STOMMEL (1948a).

gyre toward the west. This *westward intensification* of wind-driven ocean currents is the consequence of the variation of the Coriolis parameter with latitude. Eq.158 expresses a balance between planetary vorticity tendency, frictional vorticity tendency, and wind stress vorticity tendency. On the western side of the ocean where current speed and current shear are great, frictional and planetary vorticity tendencies are also great, whereas in the eastern part of the ocean where the currents are weak and diffuse, these terms are orders of magnitude smaller. However, the wind stress vorticity tendency is the same all over the ocean. The relative magnitude of the terms in eq.158 was estimated by STOMMEL (1952) as shown in Table III.

TABLE III

VORTICITY TENDENCIES IN AN ASYMMETRIC CIRCULATION

Vorticity tendency	North-flowing currents in the western part	South-flowing currents over remainder of ocean
Wind stress	− 1.0	−1.0
Frictional	+10.0	+0.1
Planetary	− 9.0	+0.9
Total	0.0	0.0

Stimulated by the results obtained by SVERDRUP (1947) and STOMMEL (1948a), a series of theoretical studies of the general wind-driven ocean circulation followed. The first were by HIDAKA (1949) and MUNK (1950). Munk succeeded in explaining many of the major features and some of the details of the general circulation on the basis of the mean wind stress distribution at the sea surface. In his analysis, more realistic wind stress data over a Northern Hemisphere ocean were used (R.O. REID, 1948b) and horizontal frictional forces in the form of eq.148 were introduced. With this, and for a constant depth, d, of the circulation system, eq.153 yields the basic vorticity tendency equation:

$$A_h \nabla^4 \psi - \beta \frac{\partial \psi}{\partial x} = - \text{curl}_z \tau \qquad (161)$$

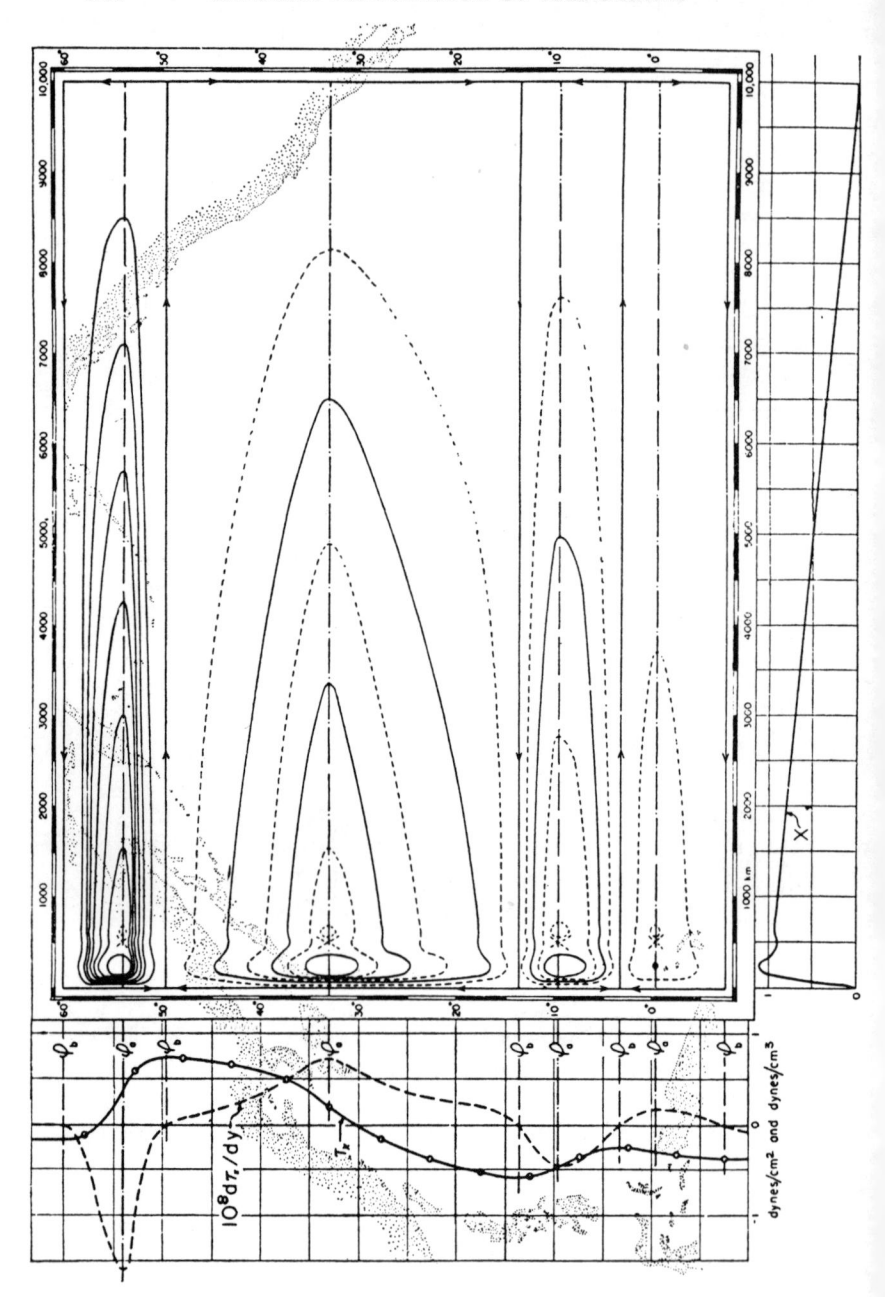

248   GENERAL CIRCULATION OF THE OCEANS

It is seen that Munk's equation (eq.161) as well as Stommel's equation (eq.158) reduce to Sverdrup's equation (eq.154) if the frictional vorticity tendency term is neglected. In addition to eq.161, eq.152 is satisfied, and for a solution the boundary conditions have been posed that the boundary itself is a streamline and that no slippage takes place against the boundary (MUNK, 1950). The operator:

$$\nabla^4 = \frac{\partial^4}{\partial x^4} + 2\frac{\partial^4}{\partial x^2 \partial y^2} + \frac{\partial^4}{\partial y^4}$$

in eq.161 is the biharmonic operator.

A graphic representation of the solution to eq.161 in terms of the stream function, ψ, for a rectangular basin of Pacific Ocean dimensions is shown in Fig.73. The mean annual *zonal* wind stress is considered the essential driving force. The result shows that the zonal wind system divides the circulation into a number of gyres. The dividing line between the gyres is found at latitudes where $\mathrm{curl}_z \tau = 0$, and the latitudinal axes of the gyres develop at latitudes where $\mathrm{curl}_z \tau$ has extreme values. The Sargasso Sea region of the Atlantic Ocean would be associated with the inflection point of the mean wind stress between Westerlies and Trade winds. The boundaries of the Equatorial Countercurrent are determined by the inflection points between the doldrums and the Northern and Southern Hemisphere Trade winds.

The Sverdrup type solution is closely approximated by Munk's solution in the central and eastern part of the ocean (for predominantly zonal wind stresses) where the planetary vorticity and $\mathrm{curl}_z \tau$ have opposite signs. In the western part planetary vorticity and wind stress curl have the same sign, and both must be balanced by the frictional vorticity tendency.

There are many other details in MUNK's (1950) solution for the streamline pattern that bear a remarkable resemblance to actual

---

Fig.73. Vertically integrated water volume transport lines (streamlines) for the wind driven circulation in a rectangular ocean. The mean annual zonal wind stress (over the Pacific Ocean) and its curl ($\partial \tau_x / \partial y$) are shown on the left. Transport between adjacent solid lines is 10^7 m³/sec. (After MUNK, 1950.)

ocean currents. For example, the countercurrent that develops to the east of the western boundary main current is a result which agrees with observations. Its transport amounts to about 17% of that of the main northward flowing boundary current.

MUNK (1950) also investigated theoretically the effect of meridional wind stress components on the general wind-driven circulation and arrived at an approximate solution. A schematic presentation of a possible circulation system given by Munk explains satisfactorily the dynamic balance of large-scale ocean currents by means of the vorticity tendency equation (eq.161) and realistic wind distributions over the oceans. Further work by HIDAKA (1950a,b), also dealing with an oceanic circulation of constant depth, contributed to a better understanding of the importance of the terms involved in the vorticity equation. For example, Hidaka found that the westward intensification of streamlines for the horizontal mass transport weakened with increased lateral friction. His mathematical treatment of the problem was different from Munk's by taking higher order terms into consideration, and in some instances, Hidaka used spherical coordinates. SARKISIAN (1954) obtained a numerical solution of the wind-driven circulation in a model shaped like the North Atlantic.

Application of the vorticity tendency equation (eq.161) to a triangular ocean basin of constant depth which approximates the North Pacific Ocean was made by MUNK and CARRIER

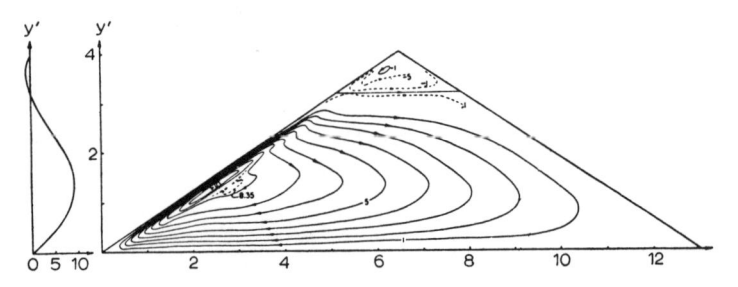

Fig.74. Streamlines of computed horizontal water mass transport in a triangular ocean. The transport between adjacent streamlines is $6 \cdot 10^6$ tons/sec in the direction of the arrows. (After MUNK and CARRIER, 1950.)

(1950). Fig.74 shows their analytical results for the horizontal mass transport represented by streamlines.

A more realistic approximation of actual coastal boundaries is made possible by numerical methods using electronic computing techniques. GARNER et al. (1962) examined the vorticity balance in the horizontal wind-driven volume transport of the whole Atlantic Ocean for the mean wind field of February. The shape of the continental boundaries is considered, along with a variable depth of the wind-driven circulation and appropriate sources and sinks on the open boundaries. The basic differential equation is shown in eq.162:

$$k\nabla^2\psi + \left(\frac{\partial f}{\partial y} - \frac{f}{d}\frac{\partial d}{\partial y} - \frac{k}{d}\frac{\partial d}{\partial x}\right)\frac{\partial \psi}{\partial x}$$

$$+ \left(\frac{f}{d}\frac{\partial d}{\partial x} - \frac{k}{R_e}\tan\phi - \frac{k}{d}\frac{\partial d}{\partial y}\right)\frac{\partial \psi}{\partial y}$$

$$= \operatorname{curl}_z \tau + \frac{\tau_y}{d}\frac{\partial d}{\partial x} - \tau_x\left(\frac{1}{d}\frac{\partial d}{\partial y} + \frac{\tan\phi}{R_e}\right) \quad (162)$$

In eq.162, d represents the variable depth of the currents, R_e the radius of the earth, ϕ the latitude, and k (sec^{-1}) the coefficient of internal friction (see eq.156A). Eq.162 is derived in spherical coordinates where $\partial x = R_e \sin\phi \, \partial\lambda$ ($\partial\lambda$ is an increment of longitude) and $\partial y = R_e \partial\phi$; ψ is the stream function of the vertically integrated horizontal volume transport, S. Eq.162 in spherical coordinates reduces to eq.158 in rectangular coordinates if the depth of the wind-driven currents is constant.

Coastal boundaries of the Atlantic Ocean were approximated by a series of straight lines joining a grid system of two by two degrees of longitude and latitude, nearest to the edge of the continental shelves of bordering land masses. The mean wind field for February and the approximation of coastlines by solid boundaries is shown in Fig.75A. The wind stress, τ, was obtained from the wind stress–wind relationship $\tau = 0.09\rho' w^{3/2}$, where ρ' is the air density and w the wind speed in cm/sec at a height of 9 m over the sea surface (NEUMANN, 1948a).

The fixed coastal boundaries were taken to be streamlines.

Constant values of the stream function, ψ, were assigned to each continuous section of coastal boundary. Their values were determined by the transport of water assumed to flow into or out from the field of integration. Values for the strengths of these sources and sinks were assumed as follows: Across the Drake Passage between Palmer Peninsula and the southern tip of South America a total eastward transport of $110 \cdot 10^6$ m³/sec was introduced (CLOWES, 1933). In the passage between the Antarctic Continent and Africa, the same amount of water was assumed to leave the South Atlantic. However, an additional source of $20 \cdot 10^6$ m³/sec was placed between the Cape of Good Hope and latitude $39°$S, representing a branch of the Agulhas Current entering the Atlantic Ocean. $14 \cdot 10^6$ m³/sec of this inflow was then carried back into the South Indian Ocean with the Antarctic Circumpolar Current, leaving the remainder of $6 \cdot 10^6$ m³/sec as a contribution to the Atlantic circulation. At the northern boundary along $55°$N it was assumed that the Labrador Current carries $4 \cdot 10^6$ m³/sec into the North Atlantic, and the North Atlantic Current between Iceland and the British Isles carries 10^7 m³/sec into the Norwegian Sea. These figures were chosen keeping in mind an estimated northward transport of $6 \cdot 10^6$ m³/sec in the Guiana Current along the northeast coast of South America (SVERDRUP et al., 1942).

With these assumptions about sources and sinks, the Atlantic area under consideration is balanced as far as the total net water transport of wind-driven currents across the open boundaries is concerned. The net gain of $6 \cdot 10^6$ m³/sec in the area north of

---

Fig.75A. Mean surface wind field over the Atlantic Ocean used in the computation of C. Isotachs in m/sec are shown as continuous lines. The arrows show the general wind direction.

B. Depth of the lower boundary of the wind-driven ocean circulation used in the computation of C. The numbers indicate depth in hundred of meters. (Based on DEFANT, 1941.)

C. Streamlines of the vertically integrated horizontal volume transport of the wind-driven circulation in the Atlantic Ocean corresponding to the driving wind pattern of A and the lower boundary topography of B. Flow is parallel to the streamlines in the direction indicated, and the volume of the flow between adjacent streamlines in millions of m³/sec is given by the difference of the indicated values. (After GARNER et al., 1962.)

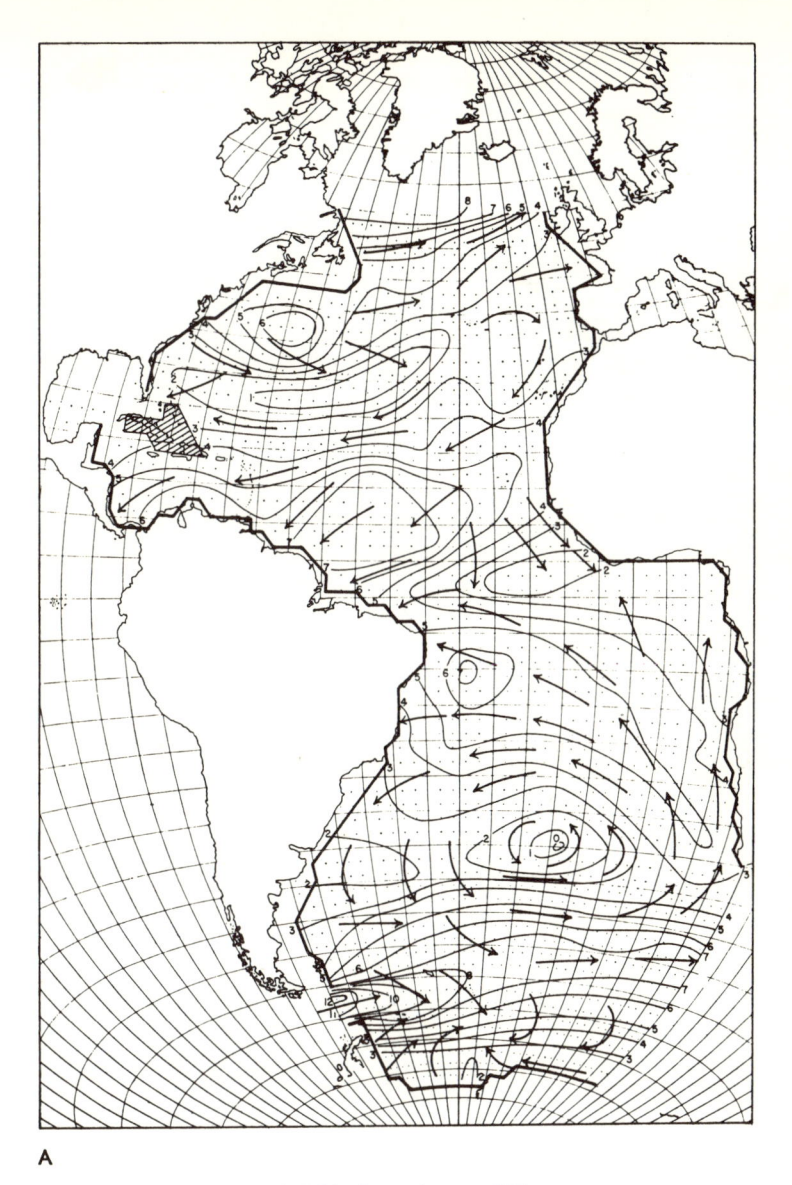

Fig.75A (legend see p.252).

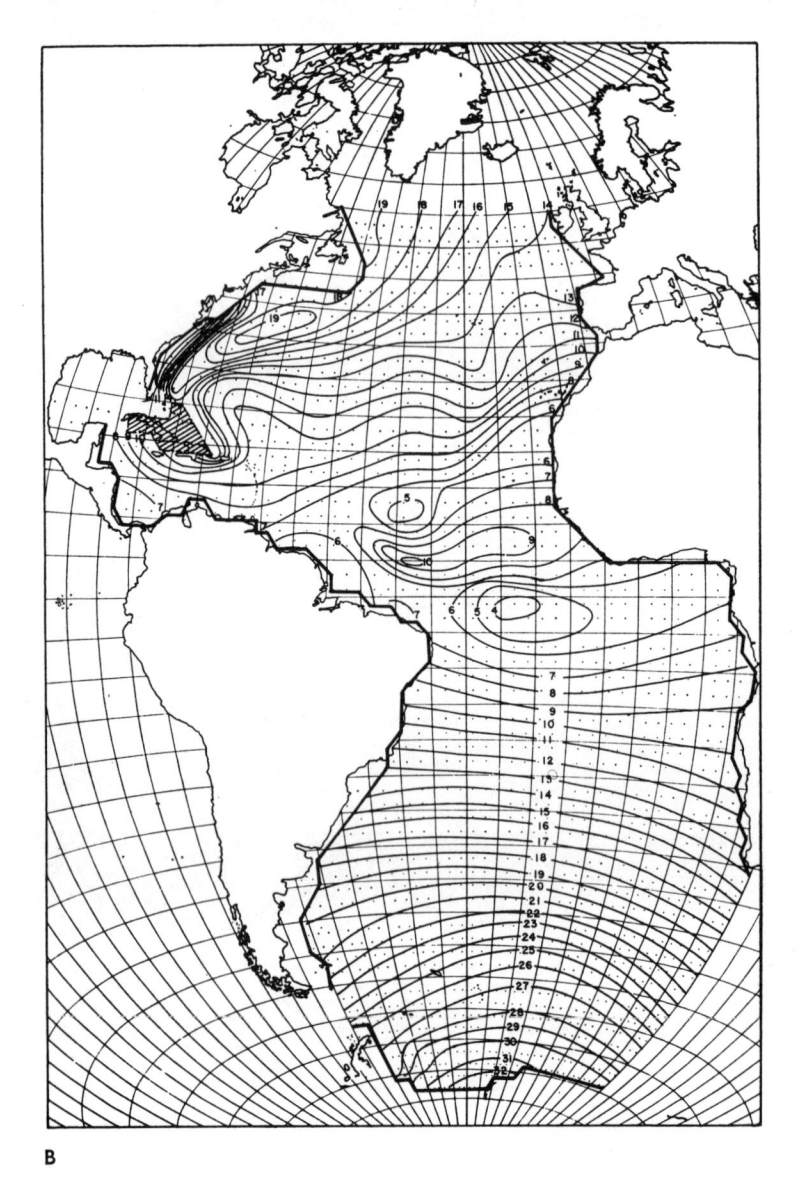

Fig.75B (legend see p.252).

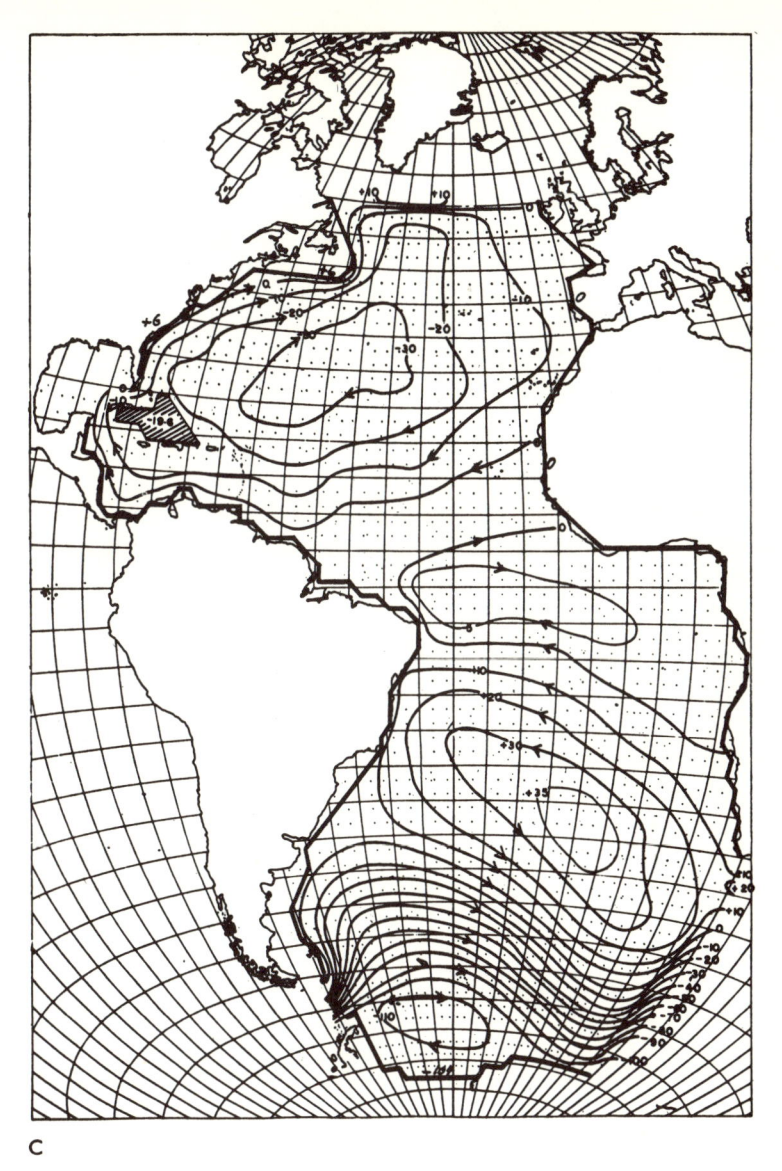

Fig.75C (legend see p.252).

55°N is supposed to return to the South Atlantic in deeper layers by thermohaline components of the general oceanic circulation.

A better approximation of the sources and sinks along the open boundaries shown in Fig.75 could probably be obtained on the basis of figures published by KORT (1959). Throughout the entire water depth across the Drake Passage, Kort estimates a net transport of $165 \cdot 10^6$ m³/sec. Between South Africa and Antarctica an east-going flow of $200 \cdot 10^6$ m³/sec was found between 38°S and 56°S. North of this, $25 \cdot 10^6$ m³/sec were found flowing from the Agulhas Current into the Atlantic Ocean, while 10^7 m³/sec flowed westward into the Atlantic between about 56°S and the Antarctic Continent. In this case, no net flow into the South Atlantic from the Southern Ocean remained. It is believed that the new figures would help to improve the results obtained in the southern South Atlantic.

Eq.162 includes consideration of a variable depth of the wind-driven circulation. There is evidence that this depth is shallow in equatorial regions and much deeper in higher latitudes. The assumed depth of the wind-driven circulation is shown in Fig. 75B, which is based on DEFANT (1941). This chart deviates from Defant's original chart essentially only in the southwestern part of the South Atlantic.

With consideration of these upper, lower, and lateral boundary conditions, the results of the numerical integration reveal an interesting picture of the total horizontal mass transport in the Atlantic Ocean. This is shown in Fig.75C. The circulation in the North Atlantic Ocean is dominated by a great anticyclonic gyre north of latitude 10°N. The center of this gyre is displaced to the west of the geographical center of the North Atlantic Ocean resulting in an intensification of the Gulf Stream branch relative to the more diffuse Canary Current branch. Between the western boundary ($\psi = +6$) and the streamline $\psi = -10$ nearly one half of the total transport of the North Atlantic gyre is found.

The total mass transport in the Gulf Stream system off the east coast of North America is close to $40 \cdot 10^6$ m³/sec. This can be compared with the figure of about $48 \cdot 10^6$ m³/sec derived by WÜST (1936a). SVERDRUP et al. (1942) estimated a total transport of about $55 \cdot 10^6$ m³/sec. For a section south of the Grand Banks, Soule (1939) derived a total transport of $40 \cdot 10^6$ m³/sec. These varying values can partly be explained by different assumptions about the lower boundary of the circulation and partly by seasonal or secular variations. There is some evidence that the flow of the Gulf Stream is greatest in July (Fuglister, 1951b).

A transport of nearly $26 \cdot 10^6$ m³/sec enters the Caribbean Sea from the North Equatorial Current and passes through the Yukatan Channel and the Straits of Florida. This amount is a result of the numerical computation. The boundary conditions of the idealized "Caribbean Island" as shown by the hatched area in Fig.75 was treated as a "floating" boundary. This means that its stream function value is obtained by numerical integration. It is determined by the geography and the dynamics of the situation as $\psi = -19.6$. Thus, about $26 \cdot 10^6$ m³/sec of water are passing through the Straits of Florida. This agrees with the amount estimated by Sverdrup et al. (1942). A weak water transport is indicated in the Antilles Current region. Sverdrup et al. (1942, p.676) estimated that about $12 \cdot 10^6$ m³/sec from the Antilles Current joins the Florida Current which leaves the Straits with a flow of about $26 \cdot 10^6$ m³/sec.

Northeast of the Grand Banks the solution shows a northgoing flow in an area that could be expected to be dominated by the Labrador Current. All inflow into the field of integration southwest of Greenland, imposed as a boundary condition, was turned abruptly eastward in the model to leave the field again in the North Atlantic Current. Indeed, there is little in the wind pattern used here to encourage the production of a Labrador Current that continues to flow southward in the region of the Grand Banks.

Although the total horizontal circulation in the eastern North Atlantic is broader than the circulation in the western part of the anticyclonic gyre, a significant transport along the northwest coast of Africa is indicated in the Canary Current region between about 30°N and 20°N.

Between the equator and north latitude 7° there is a distinct development of an Equatorial Countercurrent that extends across most of the Atlantic Ocean from about longitude 35°W into the Gulf of Guinea. This eastward flow is closely associated geographically with the doldrums belt in the applied wind field (Fig.75A).

The central South Atlantic is also dominated by a large anticyclonic gyre centered around 33°S, 10°W. Little or no westward displacement of the center of this anticyclonic gyre is evident. Some westward intensification of western boundary currents is indicated in the region of the Brazil Current. Separation of flow of the South Equatorial Current into the Guiana and Brazil Current branches occurs approximately at 10°S, south of Cape San Roque.

Streamlines that represent the Benguela Current off the Southwest African coast trend northwestward away from the coast between the Cape of Good Hope and 20°S. A large area of weak circulation is found between 20°S and the coast of Guinea. The general trend of water transport to the east-southeast or southeast between the equator and 10°S or 5°S agrees with results obtained during Equalant I expedition in February–April 1963 (WILLIAMS, 1966).

Unlike its Northern Hemisphere counterpart, the anticyclonic wind system over the South Atlantic has a simple cellular structure in the mean wind pattern for February (Fig.75A) with a center at about 35°S, 13°W that is close to the center of the corresponding current pattern in the ocean. The anticyclonic current gyre in the South Atlantic Ocean circulation is elongated, with its major axis extending southeastward into the Southern Ocean to a latitude of about 45°S. This configuration results in a marked southeastward trend of streamlines that represent the flow of the Antarctic Circumpolar Current. It seems likely that the greater part of this flow of the Circumpolar Current from the South Atlantic into the Indian Ocean would have developed in higher latitudes if boundary conditions along the meridian of 20°E would not have forced this current to be concentrated in the middle of the passage.

Further work in the theoretical study of wind-driven ocean currents between 1950 and 1960 was mainly directed toward a better understanding of the formation and dynamical explanation of intensified western boundary currents. CHARNEY (1955) and MORGAN (1956) suggested theoretical models by neglecting frictional (dissipative) terms altogether. Instead, they included nonlinear field accelerations. Their results have shown that a rather realistic theory of the development of a current like the Gulf Stream can be obtained by the action of acceleration terms (inertial theory). It appears that friction and field accelerations are dynamically of equal importance in producing intensified boundary currents.

More recently, attempts have been made to extend the inertial models into three-dimensional models, but the results are ambiguous. BLANDFORD (1965a) concluded that adding another moving layer to the models of Charney and Morgan leads to such unrealistic solutions that purely inertial steady models may be inadequate even for the growth regions of the stream. ROBINSON (1965) attempted a continuously stratified inertial model. His solutions are in the form of the first term of a series and in view of Blandford's work it is possible that a convergent series does not exist. However, the true relations between the two papers is not known.

FOFONOFF's (1954) studies of free, frictionless steady horizontal flow in a homogeneous ocean of constant depth are of abstract theoretical interest. No wind stress was applied to the sea surface. Fofonoff reached the conclusion that in an enclosed ocean a steady state circulation of this type cannot have slow, broad eastward currents. The eastward currents must rather occur as narrow streams of high velocity and high relative vorticity. Intensified currents are also present along eastern coasts. STEWART (1964) has indicated that a type of flow of this kind cannot exist in a real fluid.

The reader interested in theoretical work is referred to more recent studies on the subject of boundary currents by VERONIS (1963), MOORE (1963), and ILYIN and KAMENKOWICH (1963). An excellent survey article on the influence of friction on inertial models of the oceanic circulation was published by STEWART (1964). This article has also helped to clarify many confusing points in the present theoretical analysis of homogeneous models of the general oceanic circulation.

BRYAN's (1963) numerical time-integrations of the wind-driven circulation in a homogeneous ocean are of particular interest and practical meaning. He included nonlinear field accelerations and frictional, dissipative forces, although the friction terms were considered independent of depth. The solutions for the steady state appear quite realistic. Without time dependence the model and boundary conditions are equivalent to those of MUNK and CARRIER (1950). More recent work by BRYAN and COX (1967) on combined wind-driven and thermohaline circulations, and some of his recent results will be discussed in the section on wind-driven and thermohaline circulations.

Bottom topography effects as those represented on the right hand side of eq.153 can be extremely important where currents reach the ocean bottom. Western and eastern boundary currents flowing along the continental margins may "feel" the bottom at least in some regions of their course. South of Cape Hatteras, after leaving the Blake Plateau, the Gulf Stream flows partly over the continental slope and is guided by it. In this region, the Stream reaches the ocean bottom and its flow is directly influenced by the bottom topography (NEUMANN, 1960b). In higher latitudes where the density stratification is very weak, slope currents may reach the ocean bottom, and effects of bottom topography on currents have been observed. Evidence was provided by SVERDRUP et al. (1942) for the Antarctic Circumpolar Current and by DEFANT (1941) for the South Atlantic Ocean between about 30°S and 40°S. SCHUMACHER (1940, 1943) has also indicated that the currents in the central part of the Atlantic Ocean while crossing the Mid-Atlantic Ridge are affected by the topography of this bottom feature. The deflection of branches of the North Atlantic Current while crossing the Reykjanes Ridge southeast of Iceland is indicated in a current chart given by DIETRICH (1956c). (See also NEUMANN, 1960b.)

The inclusion of a term in the vorticity tendency equation containing the effect of a variable depth of the lower boundary on currents is certainly necessary if currents reach the ocean bottom. In non-homogeneous water, a lower boundary of variable depth for wind-driven currents can develop at some depth above the bottom. In lower latitudes, where the density stratification is strong, the depth of this lower boundary can be expected to be shallower than in the higher latitudes where the density stratification is weak. In higher northern latitudes and near the Antarctic Continent, slope currents may even reach the bottom and the term $(\partial p_a/\partial y)(\partial d/\partial x) - (\partial p_a/\partial x)(\partial d/\partial y)$ in eq. 149 cannot be omitted.

The effect of a variable lower boundary, d, on the wind-driven ocean circulation in a stratified ocean was also discussed by WELANDER (1959a). The question of whether or not the variable depth (other than the ocean bottom) plays a role in the dynamical analysis of wind-driven currents in a stratified ocean is coupled with the question of mass compensation. If the mass compensation is perfect, the variable lower boundary of wind-driven currents will not enter the problem. This means that friction is absent or insignificant in the deeper layers and the currents are in geostrophic equilibrium. However, it is probably unrealistic to carry the assumption of geostrophy so far in the dynamic analysis of world-wide wind-driven currents. Even minor deviations from geostrophy may play a more important part in the analysis of ocean currents than one would presently be inclined to admit.

WIND-DRIVEN AND THERMOHALINE CIRCULATIONS

Our present empirical knowledge of the three-dimensional structure of ocean currents is based on the analysis of results obtained by numerous expeditions. Indirect evidence drawn from the observed fields of temperature, salinity, oxygen, and other chemical factors has helped to construct a qualitative picture of what we believe represents a fairly good approximation of currents and water mass transports throughout the oceans. Direct current measurements in some parts of the ocean have substantiated this picture. Among the three oceans, the Atlantic Ocean appears to be the best surveyed. It is also an ocean which is more differentiated in its vertical structure of temperature, salinity, and oxygen than the Pacific Ocean and, probably, the Indian Ocean. There are several reasons for this fact. The exchange of water between the Atlantic and the Arctic Ocean is many times greater than the exchange between the Pacific and the Arctic Ocean. There are important sources for the formation of deep water in the northern North Atlantic that are missing in the northern North Pacific. In addition, the Atlantic is

262   GENERAL CIRCULATION OF THE OCEANS

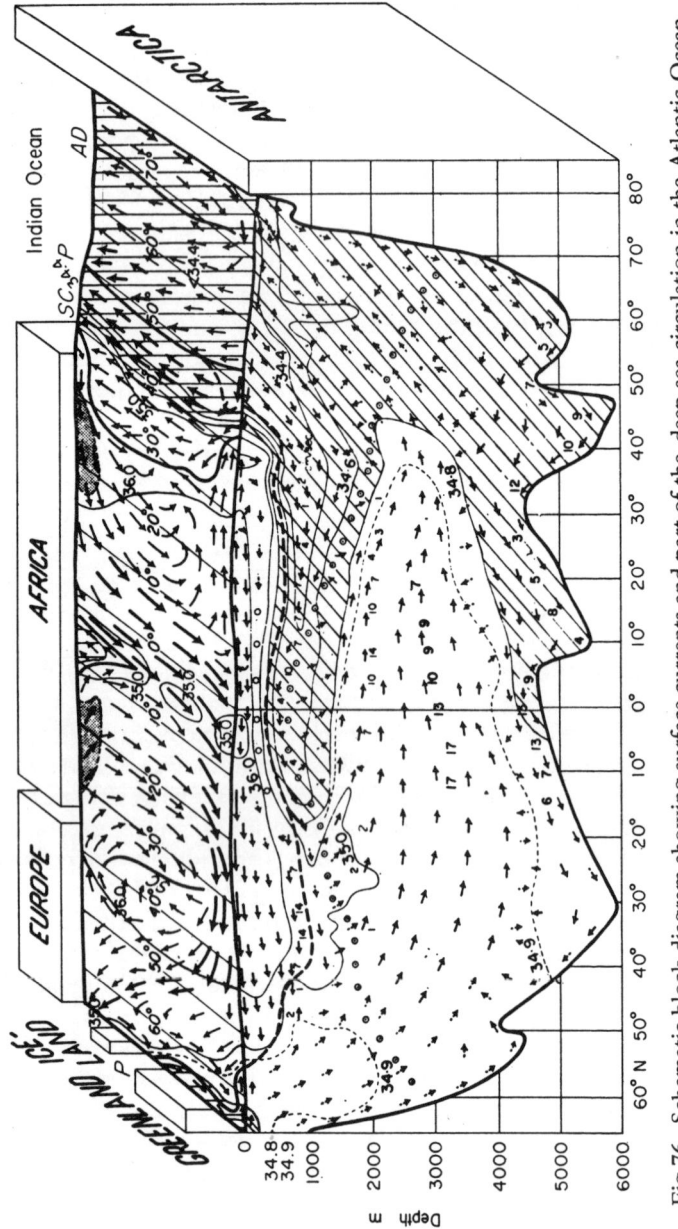

Fig. 76. Schematic block diagram showing surface currents and part of the deep sea circulation in the Atlantic Ocean. (After Wüst, 1950.)

bordered by a number of adjacent seas of which the Mediterranean Sea is the most important. It represents the source for another significant water mass, the Upper Deep Water, that characterizes the structure of Atlantic deep waters. Small, but continuous, lateral intrusions of the Mediterranean water of high salinity and relatively high temperature through the Strait of Gibraltar create a special "climate" in the Atlantic deep sea at mid-depths of about 1,000 m. This Mediterranean water spreads into the Atlantic Ocean, affecting its deep sea structure as far south as the Antarctic Circumpolar Current.

A schematic representation of surface currents and of deep sea circulations in the western part of the Atlantic Ocean is shown in Fig.76, according to Wüst (1950). This block diagram summarizes our present ideas about the general oceanic circulation in the Atlantic.

Similar results obtained from many expeditions and surveys in other oceans suggest that the driving forces of ocean currents are ultimately to be sought at the atmosphere–ocean interface. The wind stress or the transfer of momentum from the atmosphere to the ocean is doubtless the most important driving force for currents in the upper strata of the oceans. The exchange of heat (differential heating), and of water (evaporation and precipitation) across the air–sea boundary are next in importance. They change the density of sea water in contact with the atmosphere and lead to the development of thermohaline currents and the formation of characteristic water types. Spreading and mixing of these water types create water masses that dominate certain deep layers and produce a certain oceanic "climate" in the deeper strata of the oceans. In fact, almost all of our knowledge about the stratification and circulation of the deeper strata in the oceans (e.g., the summarized results shown in Fig.76 for the Atlantic Ocean) is derived from a study of the laminated water mass structure in the ocean. The core method, classical concepts about turbulence, diffusion and mixing of water masses, and the classical dynamic method based on the assumption of geostrophic flow have served as mainstays in arriving at this picture.

Wind-driven and thermohaline components of the general oceanic circulation are not independent of each other. A linear superposition of both can at the best serve only as a crude, qualitative approximation.

In a survey of ocean current theory, STOMMEL (1957b) gave a review of then existing theories for both wind-driven and thermohaline currents. His qualitative analysis and final synthesis of ocean currents are in accord with our present understanding of the general three-dimensional ocean circulation.

STOMMEL (1957b) considered a schematic ocean as shown in Fig.77 extending from pole to pole and bounded by coasts about 60° of longitude apart. This model roughly represents an ocean of Atlantic dimensions. The next assumption is that a level surface, L, divides this ocean into an upper and a lower layer. The depth of this layer is the same in equatorial and in polar regions (ca. 1,500–2,000 m deep). The upper layer, between the sea surface and the depth L, contains essentially the wind-driven currents, but also components of the thermohaline circulation. A more realistic assumption about the depth, L, which divides the vertical structure of the ocean into two parts, would have been a depth of L that is shallow (about 400 m deep) at the equator and deep (nearly reaching the ocean bottom) in polar regions.

Perhaps guided by observations in the Atlantic Ocean, Stommel introduced a sinking of surface water across the level surface, L, in subarctic regions (see Fig.76). This thermohaline process is indicated in Fig.77A.

Separated from this is a pure wind-driven circulation in the upper strata (above the level L). This wind-driven circulation is, essentially, based on the results obtained by MUNK (1950). Superposition of A and B in Fig.77 results in a qualitative picture as shown in C.

This schematic interpretation of the circulation of the Atlantic Ocean is based on a qualitative, linear superposition of wind-driven and (what is believed to be) a thermohaline circulation. It tries to explain why the Brazil Current as a western boundary current in the South Atlantic is not so well developed as the Gulf Stream.

STOMMEL's (1957b) model is a qualitatively reasonable model for the Atlantic Ocean. However, it is difficult to arrive at quantitative results. First of all, there is the question of whether

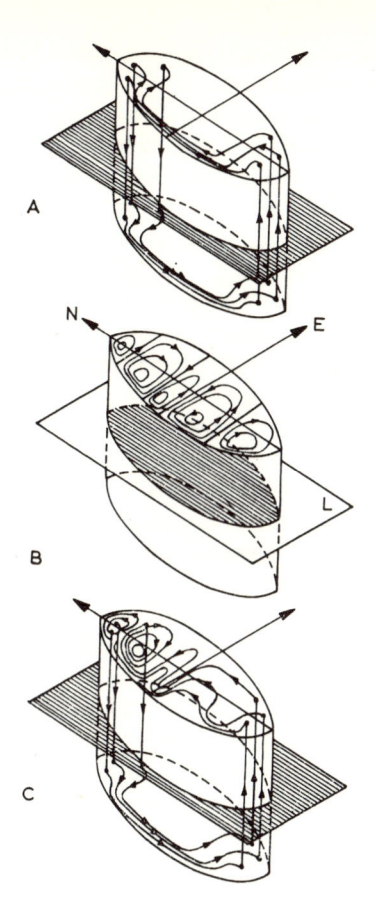

Fig.77. Schematic representation of the circulation in the Atlantic Ocean by superposition of a simple thermohaline mode of circulation (shown in A) with a flow across the level surface, L, at mid-depth, and a wind-driven mode of circulation in the layer above L (shown in B). Superposition of both modes is shown in C. (After STOMMEL, 1957b.)

all of the thermohaline components in this model are purely "thermohaline", particularly in high latitudes where wind-induced currents are deep and can even reach the ocean bottom in weakly stratified water. Secondly, few reliable quantitative speeds for thermohaline currents are available. Present estimates (Defant, 1936b, 1941; Wüst, 1936a, 1938, 1955, 1957) suggest relatively small current speeds. Unless quantitative results prove the contrary, it remains doubtful whether the small velocities of the thermohaline oceanic circulation can compensate more than one half of the wind-driven currents in the upper strata of the oceans. Stommel concludes from his model that the Gulf Stream is reinforced by the thermohaline circulation, and that the Brazil Current is so weakened that it almost disappears. Moreover, no significant thermohaline deep current seems to be present in the Pacific Ocean to be superimposed upon the wind-driven circulation.

More recently, some success has been achieved in constructing theories of the oceanic thermocline and the associated thermohaline circulation which predict major features of the observed density structure in the oceans (Lineikin, 1955a,b, 1962; Robinson and Stommel, 1959; Welander, 1959b; Stommel and Webster, 1962). The most complete model for the entire region of a whole ocean basin that deals with combined wind-driven and thermohaline currents has been studied by Bryan and Cox (1967). The analytic approach of previous studies of this kind was abandoned, and equilibrium solutions were sought by a direct numerical solution of an initial value problem. The mathematical model of Bryan and Cox includes nonlinear terms in the momentum equations for the horizontal components of velocity. The vertical diffusion coefficient is made a simple function of static stability. In regions of stable stratification, this coefficient is a constant, whereas in areas of unstable stratification it is set equal to infinity in order to model free convection regimes, and to allow complete overturning of water masses due to cooling at the surface in arctic and subarctic regions. Lateral mixing of momentum as a dissipative mechanism and lateral heat conduction are also included in the model.

Wind stress and temperature at the sea surface (upper boundary conditions) are specified as functions of latitude between about 10°N and 70°N (see Fig.78). The zonal wind stress pattern corresponds to Westerlies in mid-latitudes, to a Trade-Wind belt in subtropical latitudes, and to prevailing Easterlies in latitudes north of about 55°N. The sea-surface temperature

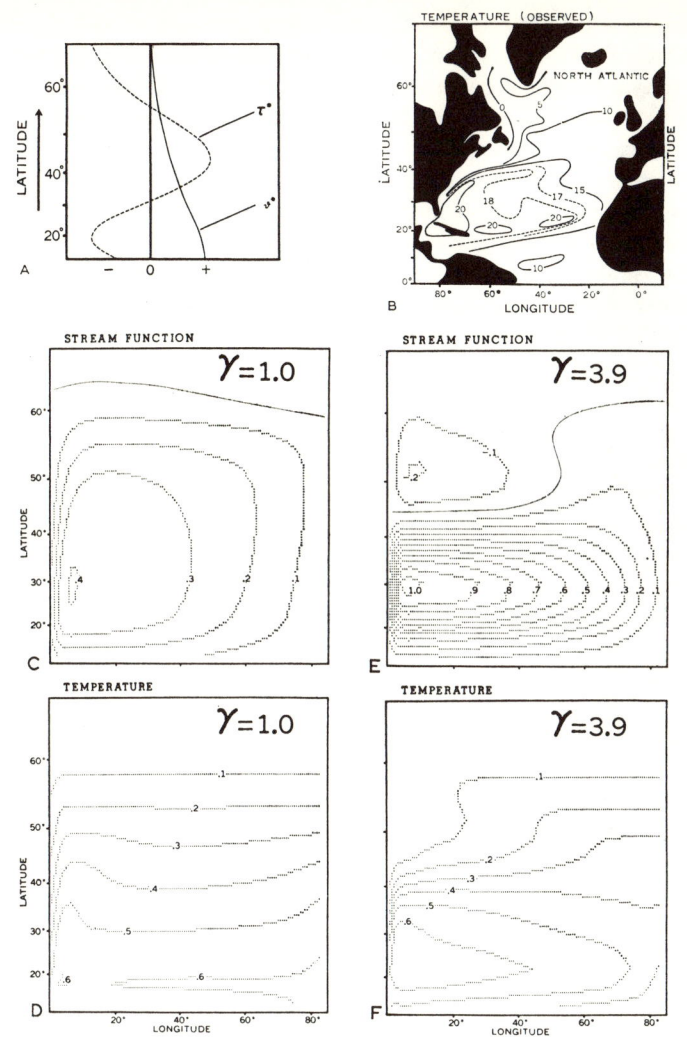

Fig.78A. Zonal wind stress (τ^*) and temperature (ϑ^*) specified at the sea surface.
B. Observed temperature pattern at 200 m depth in the North Atlantic Ocean, based on the "Meteor" atlas (WÜST and DEFANT, 1936).
C–F. Patterns of stream function and temperature at the level of about 200 m below the surface for cases $\gamma = 1$ and $\gamma = 3.9$. (After BRYAN and COX, 1967.)

distribution follows the average meridional decrease of temperature between tropical regions (10°N) and subpolar regions. Initial conditions for the numerical calculations were chosen to be a state of horizontally uniform water stratification and of an ocean at complete rest.

Solutions are obtained by direct numerical integration of the time-dependent equations using electronic computer techniques. Dimensional analysis of the basic equations indicates that the solution of the model for a closed ocean basin of planetary scale depends on five non-dimensional parameters. These parameters are varied, in turn, and the results are compared to a reference case ($\gamma = 1$) for which the parameters are chosen to correspond as closely as possible to the geophysical range. This reference case does not include the wind stress and, therefore, its equilibrium solution reproduces a purely thermal (thermohaline) mode of circulation. Computations including the wind stress show the interaction of the thermohaline and the wind-driven components of the large-scale circulation. The results are finally compared with observations in the North Atlantic Ocean.

Among the eight different cases studied by Bryan and Cox, the most interesting results can be obtained from two selected cases through a detailed comparison of the distribution of the stream function, and the horizontal and the vertical temperature distribution as shown in different parts of Fig.78 and 79. These two cases refer to the reference case ($\gamma = 1$) with no wind stress applied and to the case ($\gamma = 3.9$) where zonal wind stresses produce a wind-driven component in addition to the purely thermal circulation of the reference case. The zonal wind stress (τ^*) and the temperature (ϑ^*) specified at the sea surface are shown in Fig.78A.

The horizontal pattern of the stream function and of the temperature in Fig.78 corresponds to a depth of approximately 200 m in the real ocean. The pure thermal circulation ($\gamma = 1$) shows a single, large anticyclonic gyre with the typical westward displacement of the gyre center. The additional effect of the wind and of the wind-driven circulation ($\gamma = 3.9$) is to split this gyre into a strong anticyclonic subtropical gyre and a weak cyclonic subarctic gyre. The boundary between these two gyres coincides with the latitude of maximum wind stress in the

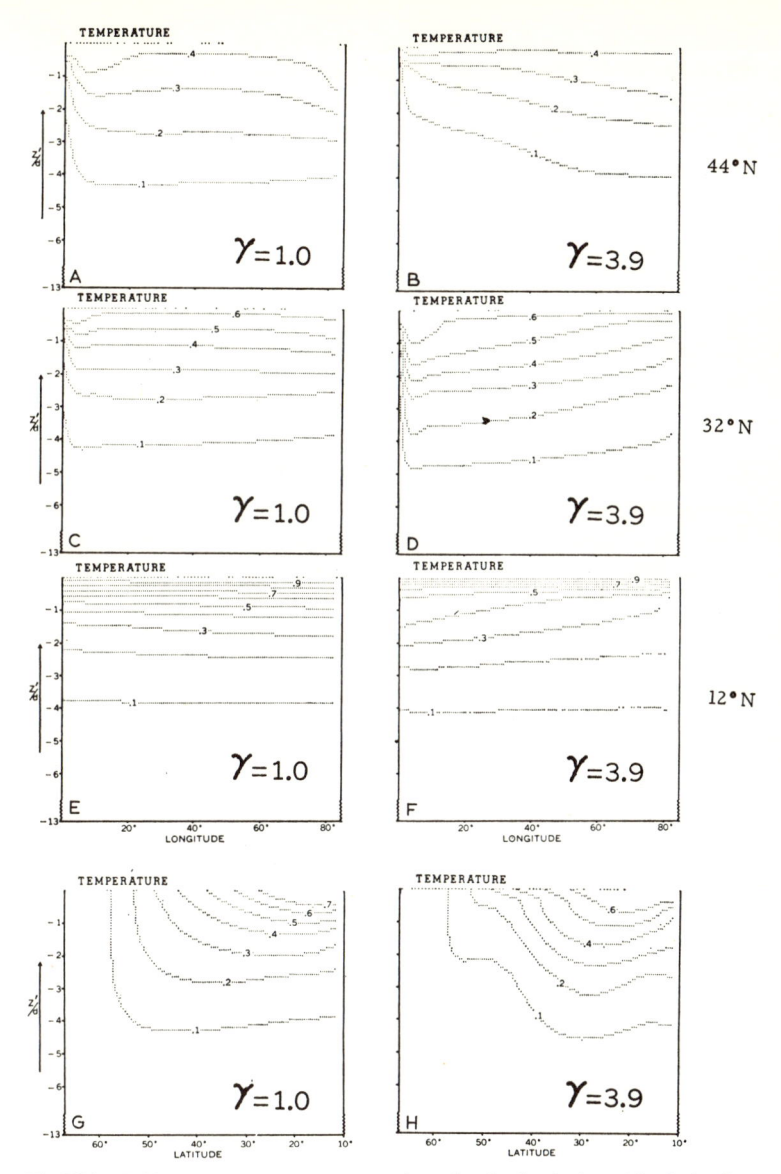

Fig.79A–F. East–west temperature sections for final solution. The latitudes of the sections are 44°N, 32°N and 12°N.
G,H. North–south section bisecting the basin.
(After Bryan and Cox, 1967.)

Westerlies. In the subtropical gyre, thermal and wind effects work in the same direction, whereas in the subarctic gyre they tend to oppose each other. This is the reason that the subarctic gyre is so much weaker.

One of the most striking features of the horizontal temperature distribution is the concentration of isotherms at mid-latitudes near the western boundary and the relatively weak horizontal temperature gradient at the same latitudes along the eastern boundary. This computed pattern reflects in an excellent way the observed temperature distribution at the 200 m level in the North Atlantic. The temperature chart shown in Fig.78B is based on the "Meteor" atlas by Wüst and DEFANT (1936).

The computed vertical temperature distribution is shown in the sections of Fig.79. Zonal sections are given for latitudes 44°, 32°, and 12°N, and one north–south section is shown for the center meridian of the basin. Comparison of the cases $\gamma = 3.9$ and $\gamma = 1.0$ indicates that the addition of wind brings the pattern into remarkably good agreement with observations, although some allowance must be made for the inflow of Mediterranean water at intermediate depths. This effect is not considered in the model. The section along 32°N latitude crosses directly through the subtropical gyre. Note the strong temperature gradient near the western boundary and the upward slope of isotherms from the center of the subtropical gyre towards the eastern boundary. A similar trend is indicated in the section along 12°N, at the south wall of the basin. However, detailed comparison with actual observations is not appropriate along this boundary because a possible water exchange between the North and South Atlantic is not included in the model. The northern section along 44°N, however, indicates the opposite slope of isotherms; that is, the eastern part of the ocean basin becomes warmer as one proceeds along subsurface levels from west to east across the ocean. These results agree remarkably well with our empirical knowledge of the Northern Hemisphere ocean stratification.

The vertical temperature section in a meridional plane is shown in the lower part of Fig.79. Even in the case of a purely thermal circulation there is a tendency for the isotherms to be bowed downward in mid-latitudes. The effect of wind, however, greatly accentuates this tendency and the thermocline becomes shallower in both low and high latitudes. The deep-reaching accumulation of warm water in subtropical latitudes is clearly shown in the case for $\gamma = 3.9$. Note also that the axis of this warm water accumulation tilts to the poleward side of the ocean with increasing depth. In the upper strata, the isotherms reach a maximum depth at a latitude of about 20° and in intermediate or deep layers the maximum depth is found near 30° latitude. This characteristic feature has also been discussed theoretically by PHILLIPS (1963). Compare this typical pattern qualitatively with the density section shown in Fig.57.

The results obtained by BRYAN and COX (1967) reproduce many observed features of the oceanic circulation and associated temperature stratification. It seems that there are no inherent difficulties in considering a better approximation of the shape of the continental boundaries, the irregular bottom topography, bottom friction, and other effects that may prove to be of importance. Continuation of this line of work will be of great value in dynamic oceanography.

Classical methods have helped to form an idea of the three-dimensional average structure of the oceans. Further progress in dynamic oceanography seems to depend strongly on a better understanding of air–sea interaction, turbulence, diffusion and mixing processes in the oceans. New approaches in experimental and theoretical work that provide a firmer basis for the air–sea interface problem, and for the measurement and study of turbulence in the oceanic space are needed.

SPECIAL CURRENT SYSTEMS AND CURRENT BRANCHES

The general circulation of the upper strata of the oceans divides into several anticyclonic and cyclonic gyres. Most conspicuous are the great, more or less elongated anticyclonic gyres centered around subtropical latitudes. In higher latitudes of both hemispheres, subpolar cyclonic gyres dominate the picture of the general ocean circulation which corresponds closely to the average climatological wind systems over the oceans.

The subpolar cyclonic gyre system in the North Atlantic is

formed by the North Atlantic Current, the Irminger Current, the East and West Greenland Currents and the Labrador Current. As a result of the irregular distribution of water and land areas, the circulation in the subpolar gyre of the North Atlantic appears more complicated than the corresponding circulation in its North Pacific counterpart, which is essentially composed of the North Pacific Current, the Alaskan Current, the Aleutian Current, the Kamchatka Current, and the Oya Shio. In the higher latitudes of the Southern Hemisphere, south of the Antarctic Circumpolar Current, very elongated cyclonic systems are indicated by the presence of the westward flowing polar current close to the Antarctic Continent. The geographical extent and major features of this polar current are not yet adequately explored. However, its existence seems fairly well established in most part of the Southern Hemisphere ocean (see the current chart in Fig.21).

A schematic division of ocean currents into cyclonic and anticyclonic gyres is shown in Fig.80. However, this circulation scheme presents a very generalized picture only.

The poleward flow on the western side of the great subtropical gyres in the North Atlantic and the North Pacific Oceans is fast and narrow as it moves in close proximity to the western ocean boundary. The Gulf Stream and the Kuroshio are two outstanding examples of *western boundary currents*. The development of these currents is associated with a westward displacement of the center of rotation of the subtropical gyres. Dynamic reasons for this westward intensification of ocean currents were given in the preceding sections of this chapter.

Also, western boundary currents in the Indian Ocean are outstanding for both their speed and concentration. One of the strongest ocean currents, the Somali Current, flows off the coast of East Africa (Tanganyika, Kenya, Somali) and crosses the geographical equator. This current changes its direction regularly during the year, following the regular change of prevailing winds in the tropical part of the Indian Ocean. During the Indian Southwest Monsoon period the Somali Current flows as fast as the Gulf Stream—and sometimes even faster. The Mozambique Current and the Agulhas Current are other outstanding western boundary currents in the Indian Ocean.

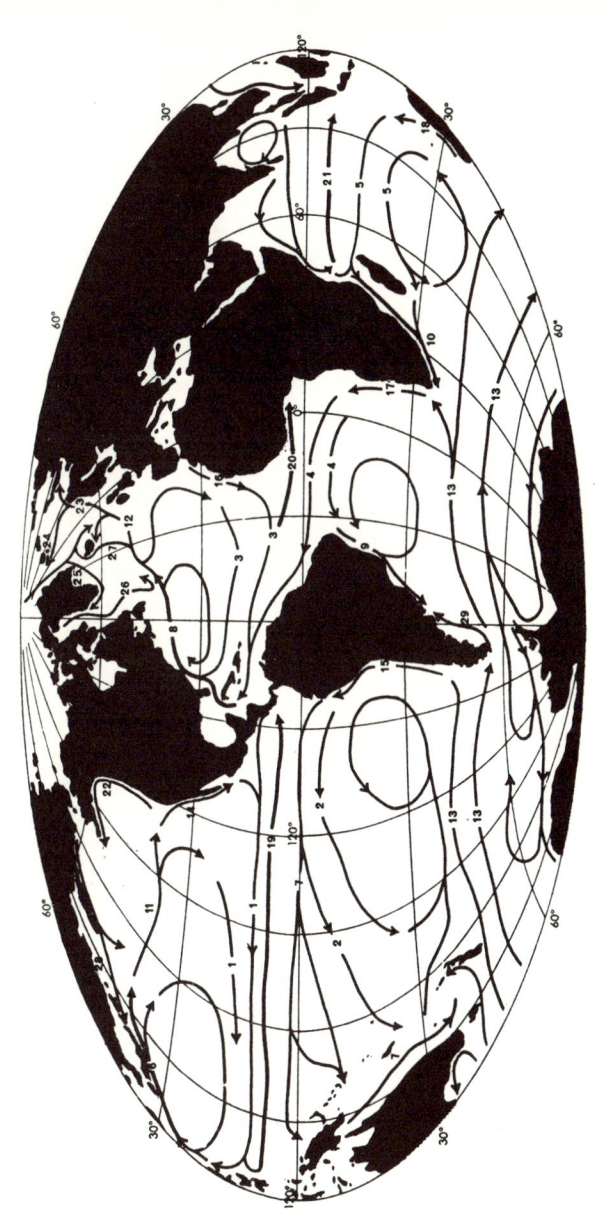

SPECIAL CURRENT SYSTEMS AND CURRENT BRANCHES 273

Fig.80. Schematic surface circulation pattern of the oceans. (After DIETRICH, 1963.)

1–5 = North and South Equatorial Currents; *6* = Kuroshio; *7* = East Australian Current; *8* = Gulf Stream; *9* = Brazil Current; *10* = Agulhas Current; *11* = North Pacific Current; *12* = North Atlantic Current; *13* = Antarctic Circumpolar Current; *14* = California Current; *15* = Peru Current; *16* = Canary Current; *17* = Benguela Current; *18* = West Australian Current; *19–21* = Equatorial Countercurrents; *22* = Alaskan and Aleutian Currents; *23* = Norwegian Current; *24* = West Spitsbergen Current; *25* = East Greenland Current; *26* = Labrador Current; *27* = Irminger Current; *28* = Oya Shio; *29* = Falkland Current.

Some westward intensification of ocean currents in the subtropical gyre of the South Atlantic seems to be indicated by surface current observations in the region of the Brazil Current. (See current chart, Fig.25.) However, the density structure off the coast of Brazil between the sea surface and the bottom does not support any significant poleward mass transport (SVERDRUP et al., 1942; NEUMANN, 1958; FISHER, 1965). This western boundary current is by far not as well developed as its Northern Hemisphere counterpart, the Gulf Stream. An eastern boundary current in the South Atlantic, the Benguela Current, appears to be comparable to the Brazil Current.

Also, the circulation in the South Pacific Ocean seems to behave in an anomalous way. The East Australian Current (HAMON, 1965) is a western boundary current but it does not reveal any outstanding characteristics when compared to its Northern Hemisphere counterpart, the Kuroshio. On the contrary, an eastern boundary current, the Peru Current (WOOSTER and CROMWELL, 1958), appears to be even more significant, although, in general, eastern boundary currents are relatively shallow and often indicate the presence of numerous eddies and countercurrents. Surface currents shown in Fig.25 indicate that the

TABLE IV

WESTERN AND EASTERN BOUNDARY CURRENTS ASSOCIATED WITH SUBTROPICAL GYRES IN THE OCEANS

Boundary currents	Atlantic Ocean north	Atlantic Ocean south	Pacific Ocean north	Pacific Ocean south	Indian Ocean
Western	Gulf Stream	Brazil Current	Kuroshio	East Australian Current	Somali Current Mozambique Current Agulhas Current
Eastern	Canary Current	Benguela Current	California Current	Peru Current	West Australian Current

center of rotation of the anticyclonic gyre in the South Pacific Ocean is displaced to the east rather than to the west.

Western and eastern boundary currents associated with subtropical gyres in the oceans are given by their geographical names in Table IV.

The currents in the equatorial region of the oceans flow mainly in zonal direction and are maintained by the driving forces of the prevailing winds. Near continental boundaries, the currents are forced to follow the coastlines. In all three oceans complicated current systems develop along the tropical western and eastern boundaries.

The equatorial current system

The circulation and stratification of water masses in the upper layers of the oceans in tropical and subtropical regions is dominated by the North and South Equatorial Currents, Equatorial Countercurrents, the Equatorial Undercurrent, and in some parts of the ocean by Monsoon currents. The equatorial current system is very sensitive to changes in the prevailing winds, and strongly pronounced seasonal variations are observed as a result of seasonal changes of the wind system. DEFANT (1961) called the system of equatorial currents the "backbone of the circulation".

The wind system in tropical ocean regions is governed by the Trade winds, the doldrums, and in some areas by Monsoon winds. Without Monsoon winds, there are two major cases to be considered besides the case that the Trade winds are symmetrical to the equator: either the southeast Trade winds reach over the equator into the Northern Hemisphere or the northeast Trade winds reach over the equator into the Southern Hemisphere while changing their directions clockwise or counterclockwise, respectively, when crossing the equator. However, in both cases a more or less broad belt of calms (or light variable winds) develops between the Trade winds. This belt is called the doldrums.

The doldrums of the Atlantic Ocean are, on the average, located in the Northern Hemisphere. Only during late winter or early spring do the Atlantic doldrums move close to the equator in the western part of this ocean. During this season, they may even extend a little into the Southern Hemisphere close to the South American coast.

In the eastern Pacific, the doldrums belt is also found in the Northern Hemisphere throughout the year, although significant variations of its geographical extent can be observed in different years. However, in the western Pacific Ocean (west of about 180°), the doldrums or a belt of light variable winds, are more regularly found during northern winter months in the Southern Hemisphere between about 10° or even 20° and the equator.

The Indian Ocean is known for most thorough changes of the prevailing wind system during the year. During northern winter, northeast winds of the Northern Hemisphere reach over the equator into the Southern Hemisphere (Northeast Monsoon) and a doldrums belt develops at about 5°S to 10°S between weak northerly winds and the southeast Trade winds, south of 10°S. During the northern summer months, doldrums or light, variable winds are the rule in the central and eastern Indian Ocean, approximately at the equator, while southeast winds from the Southern Hemisphere in the western part of the Indian Ocean gradually turn into southerly and southwesterly winds after crossing the equator. This is the period of the Indian Southwest Monsoon. Southwest winds can be especially strong near the Somali coast and in the Arabian Gulf.

These wind systems and their seasonal changes are responsible for the changes in the circulation of the equatorial current system in the oceans. Variations of the sea surface circulation during the course of the year in the East Asian archipelago between southeast Asia and Australia as a result of the changing Monsoon winds have been described by WYRTKI (1957).

Among the outstanding branches of the equatorial current system are the *Equatorial Countercurrents*. The most impressive Countercurrent is the North Pacific Equatorial Countercurrent. The surface speed is 35–60 cm/sec, except in March and April when it decreases to 20 cm/sec or less. Direct observations in the North Equatorial Countercurrent have shown that the maximum speed of this narrow eastward flowing stream is often found somewhat below the sea surface, sometimes 50–100 m deep (KNAUSS, 1961). The North and South Equatorial Currents have, in general, streaks of maximum speed near their southern and northern edges, respectively.

In the western Pacific Ocean, an eastward flowing South Equatorial Countercurrent has been found and described by J. L. REID (1959). This current has not previously been recognized. It is much weaker than the North Equatorial Countercurrent; however, its existence seems fairly well established between latitudes of about 10°S and 14°S. An oceanographic description of the equatorial current system in the western Pacific was given by TSUCHIYA (1961).

The Equatorial Countercurrent of the Atlantic Ocean is a more complex branch of the equatorial current system. During most of the year it is divided into a "western" and an "eastern" Countercurrent (SCHUMACHER, 1940, 1943). Only during July through September (or October) do these two branches join and form a continuous Equatorial Countercurrent between about 55°W and the inner part of the Gulf of Guinea. The eastern part of the Atlantic Equatorial Countercurrent is also called the Guinea Current.

The Guinea Current reaches its maximum speed during August–September close to the Guinea coast (Cape Palmas, Cape Three Points). The latitudinal extent of the North Atlantic Equatorial Countercurrent is between about 5° and 10°N. Recently, J. L. REID (1964) also pointed to the possible existence of a weak South Atlantic Equatorial Countercurrent between 5°S and 12.5°S at 14°W longitude embedded in the general westward flow of the South Equatorial Current. Results obtained during Equalant Expeditions (1963–1964) revealed a very complicated flow pattern south of latitude 5°S or 10°S (NEUMANN, 1965). This region, previously thought to be occupied by a rather uniform slow westward flow, appears to be characterized by a number of irregular current branches, although the general trend of water movements is westward (WILLIAMS, 1966).

It is of interest to note that in both the Atlantic and Pacific Oceans, the South Equatorial Current is on the average stronger than the North Equatorial Current. The South Equatorial Current in the Atlantic Ocean often has two maximum streaks of its velocity; one is found north of the equator at about 2° latitude and the other 3°–5° south of the equator.

In the Indian Ocean, during the time of the Northeast Monsoon, a strong Equatorial Countercurrent is found in the Southern Hemisphere between 2°S and 8°S. This location corresponds to the location of the doldrums belt (or thermal equator) which is always south of the equator during this season. In the Northern Hemisphere, the North Equatorial Current (Monsoon drift) flows almost due west, with strongest speeds to the south and southwest of Ceylon. An anticyclonic gyre develops in the Bay of Bengal. Near the African coast, the Somali Current flows

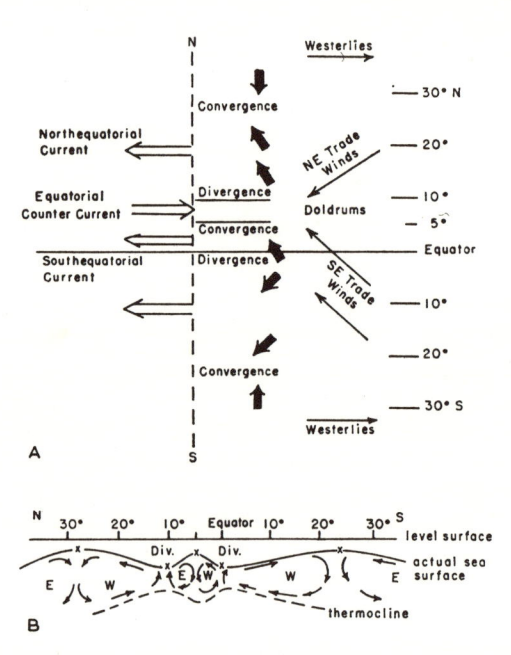

Fig.81A. Schematic representation of an asymmetric wind distribution with respect to the equator showing the doldrums between 5°N and 10°N. The Ekman transports of water in the upper layer of frictional influence are indicated by short, heavy arrows.
B. The system of surface water divergences and convergences leads to a characteristic slope of the sea surface in meridional direction and to the appearance of an Equatorial Countercurrent (E) embedded in the westward flowing North and South Equatorial Currents. See also p.206. (After NEUMANN, 1947, 1965.)

southwestward, where at about 7°S the Equatorial Countercurrent starts rather abruptly.

The German research vessel "Meteor" surveyed this area during the Northeast Monsoon period in 1964 (Indian Ocean Expedition). Results have not yet been published, but the data will be most interesting for a study of current changes in the western tropical Indian Ocean associated with changes of the Monsoon winds.

During the period of the Indian Southwest Monsoon (Northern Hemisphere summer) currents have reversed their direction in the tropical western and Northern Hemisphere part of the Indian Ocean. The Equatorial Countercurrent in the Southern Hemisphere has disappeared and the South Equatorial Current has become stronger than during the winter months. The Somali Current flows north or northeastward as fast as, or sometimes even faster than, the Gulf Stream (WARREN, 1965). This remarkable current feeds its water masses into the eastward flowing Southwest Monsoon Current. This current has its maximum speeds to the south and southeast of Ceylon where the width of the northern Indian Ocean between the equator and land (at about 80°E) is smallest.

Theoretical analyses of currents in the equatorial regions of the oceans by HIDAKA (1951), YOSHIDA et al. (1953) who used R. O. REID's (1948a) model for the mean density distribution and the mean zonal wind field (Pacific Ocean), have essentially confirmed the fact that the basic circulation in the upper strata of the oceans is the result of the typical wind systems in equatorial regions.

A schematic picture and a qualitative explanation of the tropical circulation system for the case where the doldrums occupy latitudes between 5°N and 10°N can be obtained from Fig.81. As a result of wind-driven water transports in the Ekman layer of frictional influence, convergences and divergences of surface water transport lead to a piling up or depression of the sea surface, respectively. Such convergences and divergences are shown in Fig.81A. An idealized meridional profile along the dashed line north–south in Fig.81B shows characteristic slopes of the sea surface. These slopes produce horizontal pressure gradients, and currents that superpose the pure wind-driven currents which ultimately caused the slopes. In the Northern Hemisphere, according to the slopes, westward flowing currents must occur between about 25°N and 10°N and between 5°N and the equator. In the Southern Hemisphere westward flow must be found between the equator and about 25°S. These currents represent the North and South Equatorial Currents. However, the reversed slope in the doldrums belt between 10°N and 5°N requires a narrow, swift eastward flowing current, the Equatorial Countercurrent. In Fig.81B, these currents are indicated by W or E, respectively, together with a qualitative indication of lateral and vertical circulation patterns (arrows).

An explanation of the driving forces for the Equatorial Countercurrent does not necessarily require a west to east slope as was assumed by MONTGOMERY and PALMÉN (1940). North-south slopes of the sea surface in the central parts of the oceans caused by the asymmetrical distribution of the Trade wind-doldrums system with respect to the equator seem to provide the necessary horizontal pressure gradient to drive this current (NEUMANN, 1947). The zonal slope of the sea surface that can develop in all oceans as a result of a westward water transport in the region of the North and South Equatorial currents has, nevertheless, important consequences on the dynamics of the equatorial current system.

Our present knowledge indicates that a downward sea surface slope along the equator from west to east is responsible for the existence of a most remarkable current, the *Equatorial Undercurrent*. Oceanographic surveys in equatorial regions during recent years have established the worldwide existence of this current. Especially, observations during the Indian Ocean Expedition and during Equalant Expeditions (International Cooperative Investigations of the Tropical Atlantic) have helped greatly to further our knowledge of this most significant branch in the equatorial current system which was hardly mentioned some fifteen years ago.

The Equatorial Undercurrent was first extensively observed and studied in the central and eastern part of the Pacific Ocean (CROMWELL et al., 1954). Further observations and systematic measurement by KING et al. (1957) and KNAUSS (1960, 1961, 1962, 1966) are outstanding efforts in the exploration of this current. The Equatorial Undercurrent of the Pacific Ocean is called *Cromwell Current* after its first explorer.

After the discovery of the Cromwell Current, not much attention was paid to the possible existence of a similar phenomenon in the Atlantic Ocean. NEUMANN (1960c) provided some evidence for an Equatorial Undercurrent in the Atlantic Ocean. This evidence was based partly on some direct older observations and partly on results obtained in 1946 from an analysis of oceanographic stations between 10°N and 10°S in the Atlantic. In the years after 1960, direct observations clearly demonstrated this current in the central Atlantic Ocean (VOIGT, 1961; METCALF et al., 1962; PONOMARENKO, 1963; SCHEMAINDA et al., 1964). The first report on systematic observations of the Indian Equatorial Undercurrent is, probably, by KNAUSS and TAFT (1963), although MONTGOMERY (1962) indicated a world-wide occurrence of this current.

Results obtained during the Indian Ocean Expedition for the period June through September 1962 have shown that the Equatorial Undercurrent was missing. Although strong currents near the equator in the region of the thermocline were measured, the steady, narrow eastward flow and the characteristic spreading of the thermocline were not observed. It was also found that the surface of the Indian Ocean at the equator sloped down toward the west (KNAUSS and TAFT, 1963). However, during Northern Hemisphere winter when the winds on both sides of the equator blow towards the west, the slope of the sea surface and of isobaric surfaces at the Undercurrent core was downward from west to east, and during this season, the Equatorial Undercurrent in the Indian Ocean seems to be clearly developed (SWALLOW, 1964, 1965a; TAFT, 1965). As in the Atlantic Ocean, the Equatorial Undercurrent was also associated with a core of high salinity water centered on the equator.

Two vertical sections in Fig.82 for the Atlantic and Pacific Oceans illustrate the vertical and lateral extent and speed of this current. A characteristic feature in the core of the Undercurrent is the spreading (weakening) of the thermocline. This feature was first observed in the Pacific Equatorial Undercurrent (WOOSTER and JENNINGS, 1955) and since then it has been found in the Atlantic as well as Indian Oceans.

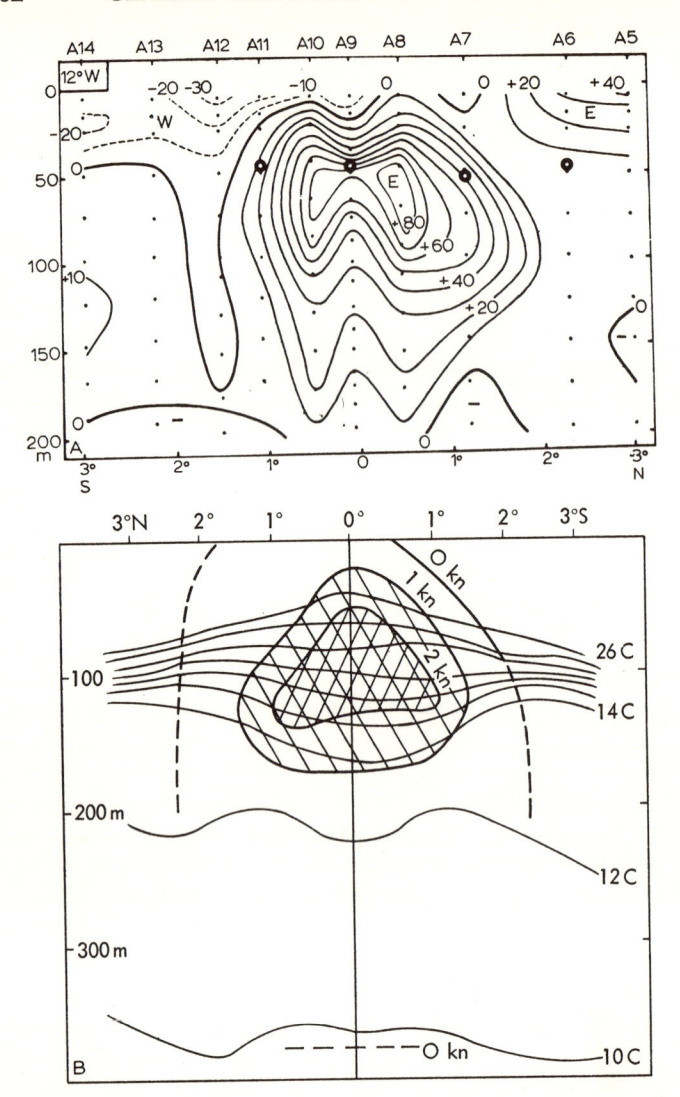

Fig.82. Velocity cross sections of the Equatorial Undercurrent in the Atlantic Ocean (A) and Pacific Ocean (B). The Pacific section also shows the associated temperature field. The spreading (weakening) of the thermocline is characteristic for the undercurrent in all three oceans. The heavy circles in A show the depth of salinity maxima. A is after STURM and VOIGT (1966) and B after MONTGOMERY (1962).

The width, depth, and speed of the Equatorial Undercurrent can vary, depending on the ocean and, probably, the season. On the average, the Equatorial Undercurrent is a narrow, shallow, subsurface current of high speed. Its width is about 200–300 km, centered on the equator. The depth occupies about 150–300 m with maximum speeds of 100–150 cm/sec at depth between about 50 and 150 m. Total water volume transports of this narrow, swift currents are of the order of 20 to 30 · 10^6 m^3/sec or more (KNAUSS, 1960.)

Not much is known about the termination of this strong current flowing from west to east along the equator. It can be expected that complicated, three-dimensional currents develop at the "termination" of the strong eastward flowing branches of the equatorial current system. Very detailed and closely spaced current observations would be needed to answer this question (RINKEL et al., 1966).

The Equatorial Undercurrent of the Atlantic is characterized by a maximum salinity content of the water in its core. This fact was first mentioned by METCALF et al. (1962). This salinity maximum, associated with the Equatorial Undercurrent seems to be indicated also in the Indian Ocean during the Northern Hemisphere winter season. However, it is absent or at least not well developed in its Pacific counterpart (MONTGOMERY and STROUP, 1962). In the eastern Pacific a weak isolated high salinity core associated with the Cromwell Current was described by KNAUSS (1966). The source for the high-salinity water in the core of the Atlantic Undercurrent is mainly the western part of the South Atlantic Ocean (METCALF et al., 1962), although intrusions of high-salinity water from the Northern Hemisphere are indicated (WILLIAMS, 1966).

The high salinity core of the Atlantic Equatorial Undercurrent, with salinities up to 36.2‰ and more could be traced as far as 8°E into the Gulf of Guinea (RINKEL et al., 1966). Parachute drogue measurements in and around the maximum salinity core in the immediate vicinity of the equator have helped to follow the course of the eastern end of the Atlantic Equatorial Undercurrent. It was found that near the island of São Tomé, the core of the Undercurrent turned southeastward. Its momentum and water mass characteristics are probably lost in a number of eddies south of 2°S close to the African coast.

Although the Equatorial Undercurrent seems to be a typical phenomenon bound to the equator, observations have shown that its course can fluctuate. Sometimes it flows a little north, sometimes a little south of the equator. This is not surprising if one considers that the Coriolis force is the stabilizing factor for the course of this current, acting together with a downward slope of the sea surface from west to east in the oceans. External effects, for example, wind effects, can cause deviations from the "stable" course due east. Some evidence for longer period fluctuations in the direction of the Equatorial Undercurrent in the Atlantic Ocean was found by NEUMANN and WILLIAMS (1965) who traced parachute drogues at about 15°W for a little longer than 2 days, and 3 h (see Fig.8). Although this time of tracking is not long enough to establish a "free" period of about two and a half days with fair accuracy, it is interesting to note that this approximate period of north–south fluctuations of the Equatorial Undercurrent is close to theoretical results. FEDEROV (1965) has dealt analytically with the problem of possible free oscillations of the Equatorial Undercurrent and arrived at the result that the period of such oscillations can be of the order of magnitude of three days.

Western and eastern boundary currents

The general theory of wind-driven currents suggests that on the eastern sides of the oceans the vorticity balance is achieved mainly by the wind stress curl and the planetary vorticity with little effect of lateral stress vorticity and inertial terms, which play a dominant role in western boundary currents. The important difference in the advection of vorticity in western and eastern boundary currents has been clearly demonstrated (CARRIER and ROBINSON, 1962).

Some of the western boundary currents such as the Gulf Stream, the Kuroshio, the Somali Current, and to some degree, the Mozambique–Agulhas Currents are outstanding current branches. They are swift, narrow and extend to great depth, in contrast to the eastern boundary currents which are relatively slow, broad and shallow. However, great differences between western boundary currents have been observed in corresponding branches in the Northern and Southern Hemispheres. Neither the Brazil Current nor the East Australian Current have the outstanding features of their Northern Hemisphere counterparts, the Gulf Stream and the Kuroshio, respectively.

The total water volume transport of the Brazil current was estimated by SVERDRUP et al. (1942) as $17 \cdot 10^6$ m³/sec. This figure includes all the southward transport along the western part of the subtropical gyre in the South Atlantic. On the opposite side, the Benguela Current including the diffuse eastern branches of the subtropical gyre transports about $23 \cdot 10^6$ m³/sec northward. These transports are distributed over nearly the same zonal width in both the western and eastern South Atlantic.

A comparison with the North Atlantic Ocean shows the remarkable difference in the total transport values as well as in their zonal distribution. The transport chart in Fig.83A according to SVERDRUP et al. (1942) suggests that $55 \cdot 10^6$ m³/sec flows northward with the Gulf Stream. This transport is mainly concentrated in a narrow western boundary current of about 100 km width, whereas the total return flow of $49 \cdot 10^6$ m³/sec to the south is distributed over the width of about 4,000 km with only a slight indication of some concentration in the Canary Current region where about $16 \cdot 10^6$ m³/sec flows southward.

The total volume transport of the Kuroshio has been calculated between about 30 and $60 \cdot 10^6$ m³/sec with a maximum in the spring and fall and a minimum in winter and early summer (ICHIYE, 1965b). This is in contrast to seasonal changes in the Gulf Stream northeast of Cape Hatteras. ISELIN (1940) computed transport figures for the period 1937-1940 and arrived at an annual variation of about $15 \cdot 10^6$ m³/sec with a maximum in the summer and a minimum in the fall and early spring. On the average, it appears that the Kuroshio transports about two-thirds of the water masses when compared with the Gulf Stream transport.

In the North Pacific Ocean, the northward transport with the Kuroshio is also concentrated in a narrow, swift and deep stream. As in the North Atlantic Ocean, the return flow to the south is diffuse and distributed over nearly the entire width of the North Pacific. This is shown in the transport chart Fig.83B.

The transport in the California Current is about 10 to 13 · 10⁶ m³/sec (WOOSTER and REID, 1963).

A comparison of the South Pacific Ocean with the North Pacific Ocean offers a similar striking difference between western

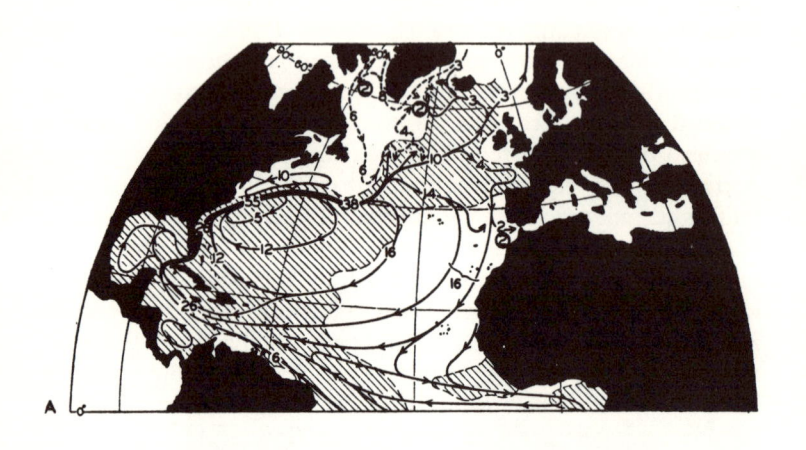

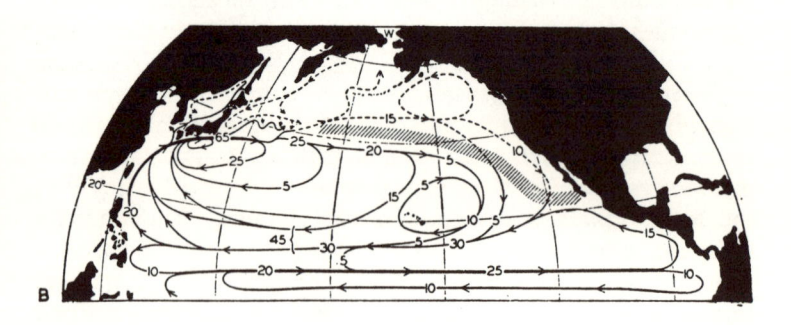

Fig.83A. Transport chart of Central Water and Subarctic Water in the North Atlantic. Inserted numbers indicate transport volumes in millions of m³/sec. Areas of positive temperature anomaly are shaded.

B. Transport chart of currents of the North Pacific Ocean. This represents, essentially, the transport above 1,500 m. The inserted numbers indicate transport volumes in millions of m³/sec. Dashed lines indicate cold currents, full-drawn lines warm currents.

Both charts from SVERDRUP et al. (1942).

and eastern boundary currents as the Atlantic Ocean. The East Australian Current is often shown on current atlases as a relatively strong, narrow current close to the edge of the continental shelf. However, this average picture seems to be misleading. Recent work by HAMON (1965) on the structure of the East Australian Current based on results of eight cruises between 1960 and 1964 revealed a very complicated flow pattern. The volume transport above 1,300 m is in the range of 12 to 43 · 10^6 m³/sec, with an indication that the East Australian Current is strongest between December and April. The current if often U-shaped, with a southward branch near the continental shelf and a northward countercurrent further offshore. The current speed decreases to half its surface value at a depth of 250 m. There is evidence that the current can reach as deep as 2,000 m.

Although the East Australian Current has some features in common with the Gulf Stream and the Kuroshio, it appears to differ in other respects. HAMON (1965) finds that the narrow, swift southward flowing branch near the continental shelf between 27°S and 32°S is one feature in common, whereas the equally narrow northward or northeastward return branches of the whole system offer the greatest contrast with the Northern Hemisphere counterparts. Hamon also points out that if the East Australian Current is considered as the western boundary current for the subtropical South Pacific Ocean, its volume transport would be expected to be larger than the Gulf Stream transport. Actually, the transport of the East Australian Current is about one half of the transport of the Gulf Stream.

The northward or northwestward flowing current in the eastern South Pacific is called the Humboldt Current. Where it is close to the Chilean and Peruvian coasts the name Peru Current is most commonly applied. The Peru Current was first studied in greater detail by GUNTHER (1936) and later by WOOSTER and GILMARTIN (1961) and WYRTKI (1963). GUNTHER (1936) for the first time, distinguished between a narrow Peru Coastal Current and a broader offshore branch, the Peru Oceanic Current. Both branches flow in the same general direction. They are separated from each other by a south-flowing Peru Countercurrent which reaches the sea surface only occasionally (WYRTKI, 1963). This countercurrent is also indicated in the sections published by Wooster and Gilmartin (1961). It should be well distinguished from the weaker Peru Undercurrent which flows southward close to the coast.

The Benguela Current is the major oceanographical feature of the southeast Atlantic Ocean and its effects spread widely over the eastern side of the South Atlantic basin (Currie, 1965). In the Benguela Current system, a narrow coastal current and a broader offshore oceanic current have also been described by Hart and Currie (1960). The Benguela Current and the Peru Current have many features in common. In fact, all of the better known eastern boundary currents have certain common characteristics which have been examined by Wooster and Reid (1963) on a comparative basis.

Poleward undercurrents near the west coasts of the continents beneath the equatorward flowing eastern boundary currents have been observed in the Peru–Chile region (Gunther, 1936), in the Benguela region (Hart and Currie, 1960), along the North American west coast (Sverdrup and Fleming, 1941; J. L. Reid et al., 1958), and in the region of the Canary Current (Montgomery, 1938b). Along the coast of California and Oregon, north of 35°N, this countercurrent is known as the Davidson Current. These countercurrents seem to be closely connected with the process of *upwelling* which is another outstanding phenomenon associated with eastern boundary currents in subtropical regions. Along the North American west coast, during the entire season of upwelling, a countercurrent flows close to the coast northward at depths below 200 m. When upwelling ceases, the countercurrent also develops in the surface layers (Sverdrup et al., 1942).

The exact dynamic nature of the relationship between upwelling and countercurrents is not yet fully understood. The phenomenon of coastal upwelling has been studied theoretically by Defant (1952), Hidaka (1954), Yoshida and Mao (1957), and Yoshida (1958). The earlier qualitative works by Thorade (1909), McEwen (1912), Sverdrup (1930, 1938), Gunther (1936), Defant (1936a), and Sverdrup and Fleming (1941) have added significantly to the knowledge of this process. It has been shown that coastal upwelling is largely confined to a narrow band of probably less than 100 km width close to the coast.

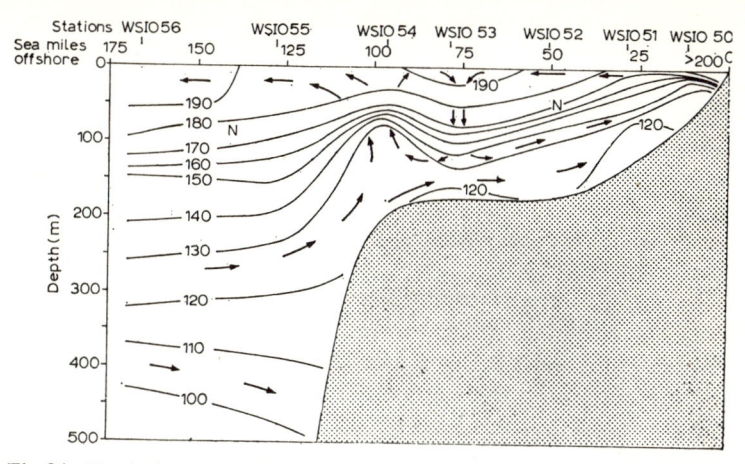

Fig.84. Vertical section of the anomaly of specific volume off the west coast of Africa in latitude 28°30′S in September, 1950. The arrows show the pattern of lateral and vertical circulation on the assumption that the distribution of mass is maintained by the stress of the wind on the sea surface. The north flowing branches of the Benguela Current are indicated by the letter N in the region of counterclockwise circulation. (After CURRIE, 1965.)

The ascending motion of water, called upwelling, is usually restricted to a relatively shallow depth of about 100–200 m, or occasionally, 300 m.

A typical vertical section showing the anomaly of the specific volume off the west coast of Africa at latitude 28°30′S, and the probable pattern of vertical circulation is shown in Fig.84, according to CURRIE (1965). The characteristic uplift of isosteric lines (and isotherms) and indications of offshore water transport in the surface layer in regions of upwelling are clearly indicated. The north-flowing branches of the Benguela Current are found in the region of counterclockwise lateral and vertical circulation (indicated by N in Fig.84). Near the edge of the continental shelf where the circulation in the vertical plane is clockwise, a southflowing countercurrent may be present or a series of eddies may develop close to the surface convergence.

Besides in the divergence at the coast there is also evidence of upward movement and upwelling at the edge of the continental shelf. This offshore divergence probably corresponds to the divergence observed by DEFANT (1936), and the vertical circulation pattern shown in Fig.84 agrees well with the theoretical model suggested by DEFANT (1952). At a depth of 200–300 m, CURRIE (1965) finds some evidence for a southerly movement along the edge of the continental slope. Estimates of speeds of vertical motion in regions of upwelling water are from 10–20 m/month (McEWEN, 1929) to about 80 m/month (SAITO, 1951; HIDAKA, 1954).

In connection with eastern boundary currents a small, and under normal conditions, rather insignificant current in the eastern Pacific Ocean should at least briefly be mentioned. In January to March the Peru Current turns west a few degrees south of the equator, and the North Equatorial Countercurrent sends a warm branch southward close to the west coast of South America. This small current branch is called El Niño. However, occasionally, the warm El Niño Current spreads its waters abnormally far south, causing a rapid rise of sea surface temperatures in normally cool water regions. Such disturbances can extend as far as 15°S along the coast of Peru. The anomalous shift of currents and the high surface temperatures lead to instabilities in the overlying atmosphere which cause excessive rainfall and damaging floods and erosion in regions where the normal precipitation is very small. The anomalous condition also has disastrous effects on marine life, from plankton to fish and birds. Often the fish are killed to such an extent that decomposing organic matter creates foul conditions in the water and air. Hydrogen sulfide in the water liberated by the decomposition may even blacken the paint of ships.

It is possible that intensity fluctuations of the eastern part of the South Equatorial Countercurrent (J. L. REID, 1959; WOOSTER, 1961) are also responsible for the development of abnormal El Niño conditions. WOOSTER (1960) has given a more recent description of this interesting branch of eastern boundary currents.

The best explored *western boundary currents* are the Gulf Stream and the Kuroshio. A comprehensive summary of the exploration, description and dynamical explanation of the Gulf Stream (and of the western North Atlantic) has been given by STOMMEL

(1958, 1965). This book serves at the same time as a survey of our knowledge of the general dynamics of outstanding western boundary currents. The three-dimensional structure and the dynamics of the Gulf Stream and of the Kuroshio are in many ways similar. Knowledge of both currents has considerably increased during the past 20 years and has, by far, surpassed the exploration of western boundary currents in other parts of the oceans. Only recently, during the Indian Ocean Expeditions has the Indian Ocean, and particularly its western part, received more attention. Such remarkable currents as the Somali Current, the Mozambique Current and the Agulhas Current are, presently, very much in the focus of oceanographic exploration.

The systematic exploration of the Gulf Stream started in 1931 with the famous research vessel "Atlantis" of the newly founded Woods Hole Oceanographic Institution. The first comprehensive account of the circulation of the western North Atlantic in general and of the Gulf Stream in particular was published by ISELIN (1936). This study was essentially based on the numerous, detailed "Atlantis" sections made during the early thirties across the Gulf Stream. Further exploration of the Gulf Stream system is almost exclusively credited to oceanographers from the Woods Hole Oceanographic Institution.

At the same time, Japanese oceanographers—notably UDA (1930), SIGEMATSU (1933), KISINDO (1934) and KOENUMA (1939)—advanced our knowledge of the Kuroshio. Modern studies of this current (HIDAKA, 1951; UDA, 1951; SAITO, 1952; ICHIYE, 1954; and others) have shown that not only the large scale features of the surface circulation but also the detailed structure of the Kuroshio are very similar to those of the Gulf Stream. One of the more recent accounts is the paper by UDA (1964) on the nature of the Kuroshio, its origin and meanders.

The current system in the western North Atlantic, the Gulf Stream system (ISELIN, 1936), is composed of the Florida Current, the Gulf Stream between Cape Hatteras and the tail of the Grand Banks, and the North Atlantic Current. A rather complicated current pattern seems to develop in the region to the south and southeast of the Grand Banks between the Gulf Stream and the western end of the North Atlantic Current. It is possible that this pattern which led WORTHINGTON (1965) to a "two gyre hypothesis" is related to effects of bottom topography. The strong southward deflection of the Gulf Stream near longitude 50°W when approaching a submarine ridge and the extreme northward deflection on the eastern side of this ridge was discussed in connection with Fig.68.

The Kuroshio system (WÜST, 1936a) also consists of three major divisions. From east of Luzon (Philippines) the Kuroshio flows north and later northeastward close to Japan as far as 35°N. Here it turns nearly to the east to form the Kuroshio Extension and, further east, the North Pacific Current. A branch of the Kuroshio enters the Japan Sea and flows west of the Japanese Islands to the north. This branch, the Tsushima Current, has no counterpart in the Gulf Stream system.

The Gulf Stream and the Kuroshio are usually sharply defined along their continental side (see Gulf Stream sections in Fig.35). This oceanic front marks a transition between the warm water of high salinity of the central ocean and the cool, somewhat fresher coastal water that spreads in a wedge-like form underneath the warmer ocean water. An outstanding feature of the Gulf Stream and the Kuroshio is their intense baroclinicity. Vertical density sections reveal horizontal density differences by as much as two units of σ_t in the upper 500-m layer when crossing the frontal zone.

Another essential fact is that the Gulf Stream, in its southern part, flows over the continental slope and is guided by it. When the stream reaches the ocean bottom it is directly affected by the bottom topography, particularly by the continental slope where the stream is restricted to shallower depths on the continental side and extends to much greater depth on the Sargasso Sea side (NEUMANN, 1960a). After leaving Cape Hatteras, the Gulf Stream crosses the bottom isobaths of the continental slope at an angle. WARREN (1963) has suggested that the topographic curl effect produces a meander pattern which depends on the crossing angle between isobaths and the current. Warren's calculations agree well with the observed meander pattern of the Stream. The Gulf Stream seems to "feel" the bottom, at least along parts of its track, as far east as the tail of the Grand Banks. Gulf Stream meanders as a driving agency for the current were theoretically discussed by WEBSTER (1951).

The lower boundary of the Gulf Stream, even if the Stream is not guided by the continental slope, probably increases from left to right when one faces in the direction of the current. The velocity section shown in Fig.85 is computed from the mass distribution and the relative horizontal pressure field using DEFANT's method (1941). In Fig.85 it is assumed that the lower boundary of the Gulf Stream is about 600 m deep on the continental side and about 2,000 m deep on the Sargasso Sea side. The main axis of the Stream (maximum speed) at different depths slopes to the right (Sargasso Sea) with increasing depth. This agrees with results obtained by VON ARX (1952b) by means of simultaneous current measurements at the surface and 100 m

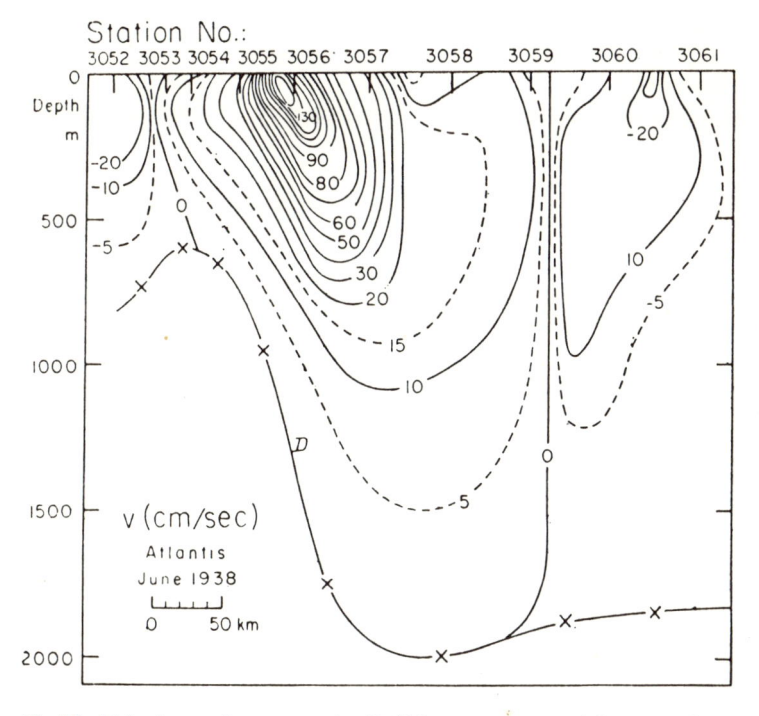

Fig.85. Velocity section across the Gulf Stream computed from the distribution of mass with the assumption that the flow is geostrophic. The lower boundary, D, of the stream is shown by the line ×−×−×. (After NEUMANN, 1956.)

depth. Fig.86 according to VON ARX (1952b) illustrates three important facts: (*1*) a sharply peaked distribution of speed across the Gulf Stream, (*2*) an asymmetrical horizontal current shear on the two sides of the current maximum, and (*3*) a displacement to the right of the velocity profile at 100 m depth as compared with the surface.

If a lower boundary of the Gulf Stream as shown in Fig.85 is accepted, another fact about the density distribution in the moving layer can be obtained. Although the horizontal density gradient in level surfaces at different depths across the Stream is extremely strong, the mean density for vertical columns of water between the surface and the lower boundary of the Stream is practically constant (NEUMANN, 1956; MARTINEAU, 1958). The Gulf Stream tends to approach an equivalent-barotropic system.

Direct current observations with Swallow floats, and hydrographic stations in the anticyclonic shear region of the Gulf Stream (SWALLOW and WORTHINGTON, 1957, 1961) have shown

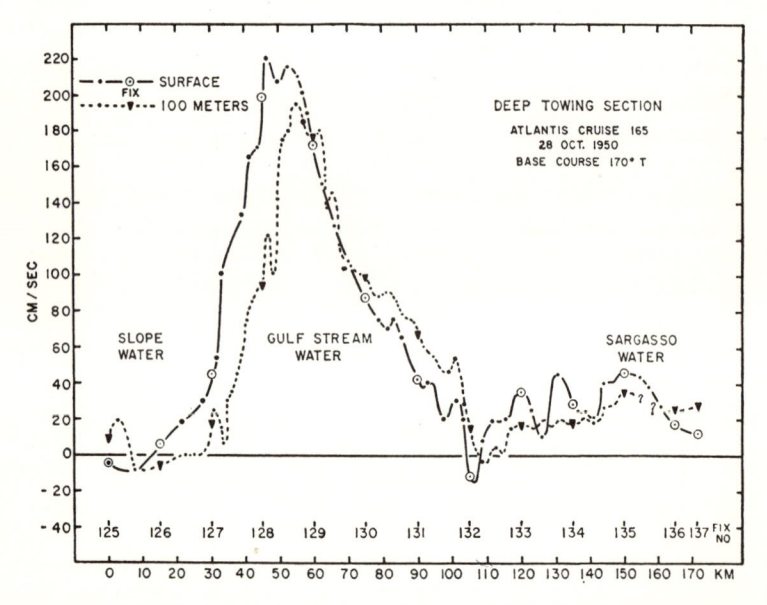

Fig.86. Transverse profile of GEK velocities observed in the Gulf Stream at the sea surface and 100-m levels (After VON ARX, 1952b.)

that at a latitude of about 33°N the lower boundary of the Stream was between 1,500 and 2,000 m. Below this depth a southward moving countercurrent with speeds of 5–10 cm/sec was observed. The existence of such a countercurrent underneath the Gulf Stream was theoretically deduced by STOMMEL (1955) as part of the thermohaline circulation. Japanese measurements in 1960 also suggested the existence of a deep counterflow below the Kuroshio. However, later observations in the western Atlantic (VOLKMANN, 1962; FUGLISTER, 1963) raised some doubt as to the significance of the earlier measurements, and the question remains whether the deep countercurrent under the Gulf Stream is a characteristic feature of the system. It may not always be present. Deep current observations between 36° and 37°N by KNAUSS (1965) gave further evidence that the Gulf Stream (at least at the time when the observations were made) extended to the bottom without a significant change in direction.

Quasi-synoptic observations in parts of the Gulf Stream have helped to explore the fine structure of this current in great detail. One of the earliest examples depicting the detailed temperature structure of the Gulf Stream is represented in Fig.87A which is deduced from quasi-synoptic temperature observations. Fig.87B indicates the formation of an eddy at about 39°N, 61°W according to FUGLISTER and WORTHINGTON (1951). Such eddies can form, detach from the main currents, and disappear. FUGLISTER and VOORHIS (1965) have recently tracked such eddies for longer periods of time and obtained evidence that they can have a lifetime of possibly several weeks.

The first quasi-synoptic "Operation Cabot" in 1950 was followed by a multiple ship operation "Gulf Stream '60" (FUGLISTER, 1963). In 1964, FUGLISTER and VOORHIS (1965) observed eddies to shift toward the east with translational speeds of 5–6 cm/sec. This survey also suggested that the deep and bottom currents of the Gulf Stream must occasionally follow a path different from that of the currents in the upper strata. Fig.88 shows the position of the strongest horizontal temperature gradient at the 200 m level in June 1964. The path of the Gulf Stream could rapidly be followed by towing a temperature sensor at 200 m depth and following the 15°C isotherm downstream. Fuglister and Voorhis note that, according to UDA

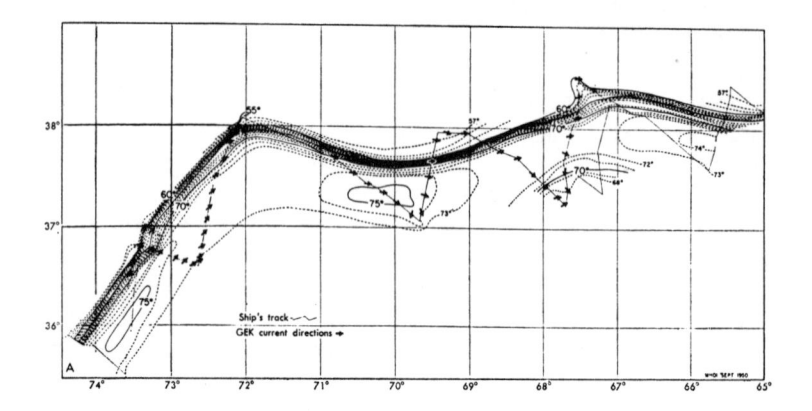

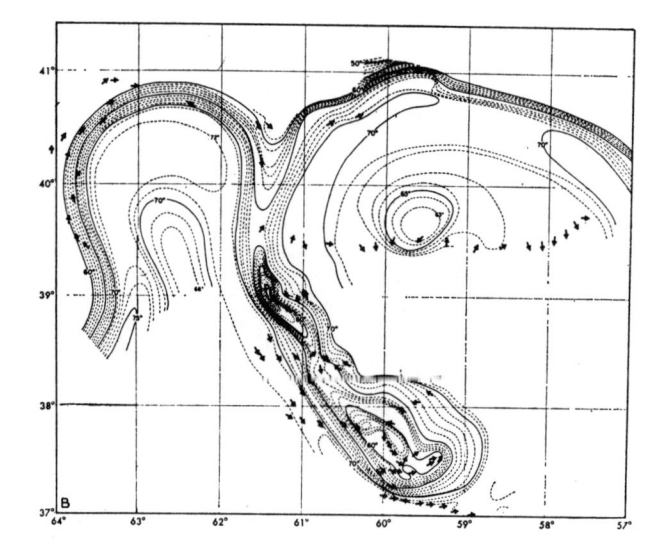

Fig.87. Details of the structure of the Gulf Stream observed during the multiple ship survey of 1950.
A. Isotherms (°F) in the upper 200-m layer of the Gulf Stream and observed currents (GEK measurements).
B. Detailed map of an eddy before the time of its separation from the main current. The temperature in °F is shown for the depth of 200 m. Arrows show GEK measurements. (After FUGLISTER and WORTHINGTON, 1951.)

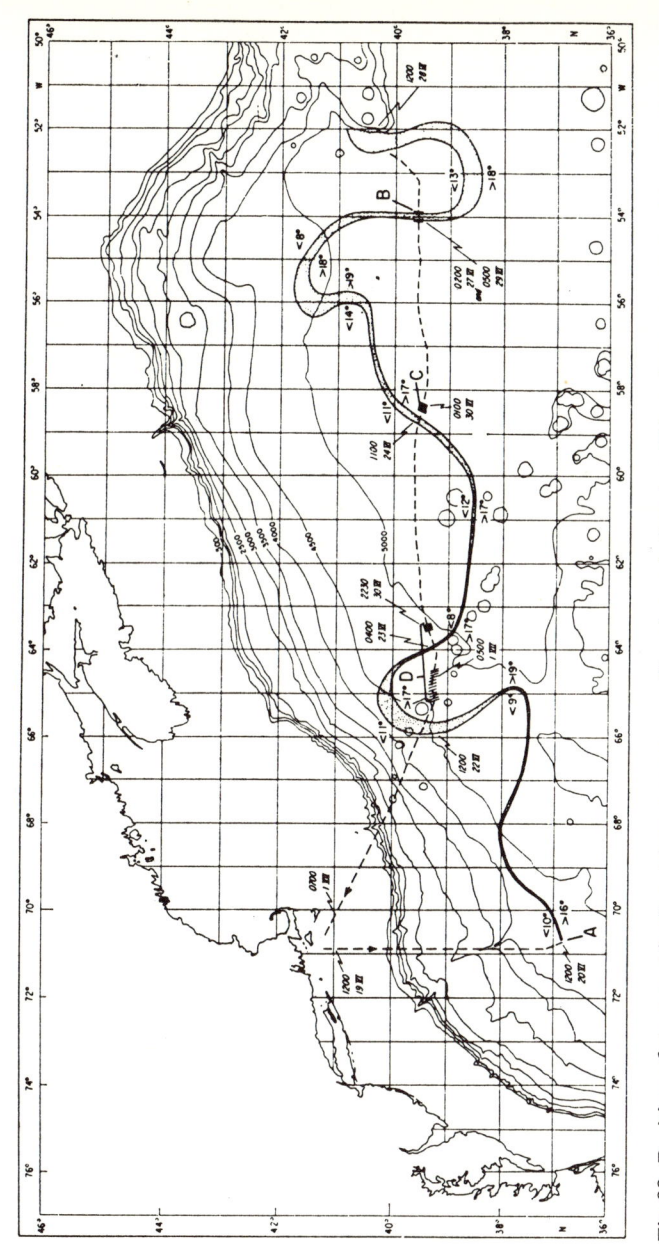

Fig.88. Position of strongest horizontal temperature gradient (oceanic front) at 200 m depth. "Crawford" cruise 110, June 1964. A ship following this gradient downstream may be in or near the maximum surface current; however, the major portion of the deep Stream is to the left of the front shown in the chart. (After FUGLISTER and VOORHIS, 1965.)

(1964), the 15° isotherm at the 200 m level can be considered to be a good indicator of the main stream axis of the Kuroshio. The meandering of the Kuroshio is very similar to the Gulf Stream.

The Gulf Stream study in 1960 (FUGLISTER, 1963) indicated a nearly stationary meander pattern during April and May. In 1964 a changing pattern not only between cruises (e.g., November and December) but to some degree during each cruise was clearly observed (FUGLISTER and VOORHIS, 1965).

A countercurrent is sometimes found in the region east of the Gulf Stream. Its width is roughly one half of the width of the Gulf Stream and speeds can be as high as 1.5 knots.

Southflowing countercurrents over the continental shelf are often observed along the western edge of the Gulf Stream. Charts by BUMPUS and LAUZIER (1965) show that these coastal countercurrents are most conspicuous north of Cape Hatteras.

The Antarctic Circumpolar Current and the Antarctic Ocean

The circulation around the Antarctic Continent is mainly in zonal direction from west to east. This current is called the Antarctic Circumpolar Current or "West Wind Drift". In a narrow zone around most of the continent a westward flowing Polar Current or "East Wind Drift" is observed. Both currents are largely due to the frictional stress exerted by the prevailing winds. Their flow is almost unobstructed by land barriers. The only exception is in the narrow Drake Passage between South America and the Palmer Peninsula projecting north from the Antarctic Continent.

The Antarctic Circumpolar Current is a deep current in contrast to the wind-driven currents in tropical and subtropical regions. In some parts of the Antarctic Ocean, the currents reach the ocean bottom at 3,000–5,000 m depth and are significantly affected by the bottom topography. The surface speeds are only up to 15–20 cm/sec, but as the result of the great depth, the water volume transports are higher than that of any other current system in the oceans. SVERDRUP et al. (1942) estimated a transport of $90 \cdot 10^6$ m³/sec through the Drake Passage and $150 \cdot 10^6$ m³/sec between Antarctica and Tasmania. More recent figures based on observations made during the Soviet Marine Antarctic Expedition 1956–1958 were obtained by KORT (1959). These figures suggest higher transports and do not show such great variation for different sections. KORT (1962) found the following average values: between Antarctica and Africa, $190 \cdot 10^6$ m³/sec; between Antarctica and Tasmania, $180 \cdot 10^6$ m³/sec; and between Antarctica and South America, $150 \cdot 10^6$ m³/sec.

The volume of $190 \cdot 10^6$ m³/sec is carried by the Antarctic Circumpolar Current from west to east into the Indian Ocean mainly between 56°S and 38°S. However, $25 \cdot 10^6$ m³/sec were found to be transported westward by the Agulhas Current between 38°S and Africa, and another $10 \cdot 10^6$ m³/sec were carried westward between 56°S and Antarctica by the Polar Current (Western Coastal Current, according to Kort). The total balance of water transports across the three major sections is difficult to estimate. For example, it appears from KORT's (1962) figures that most of the transport of the Agulhas Current and of the Polar Current are returned with the eastward flow of the Antarctic Circumpolar Current into the Indian Ocean. $10 \cdot 10^6$ m³/sec are then diverted into the circulation of the Indian Ocean. Again, an exact water balance is difficult to obtain, and in trying to arrive at such balances, KORT (1962) refers to other branches of the ocean circulation, for example, to the flow through the Indonesian Straits, and to river discharges, ice melting and evaporation–precipitation differences over the world oceans.

It is also to be expected that the water transport of currents in the Antarctic Ocean is subject to considerable seasonal fluctuations and to fluctuations from year to year. Such fluctuations are well known in the Gulf Stream system (MONTGOMERY, 1938a; ISELIN, 1940; HELA, 1952). In the Antarctic Ocean their investigation has not yet even been started.

The dynamics of the Antarctic Circumpolar Current offers some problems. MUNK and PALMÉN (1951) considered the Antarctic Circumpolar Current as a wind-driven eastward flow on a plane tangential to the earth at the South Pole. The essential balance of forces is obtained from wind stress and lateral friction. With an average surface wind stress of 2 dynes/cm², a lateral eddy viscosity coefficient $K_h = 10^8$ cm²/sec, Munk and Palmén arrived at a total eastward transport of $5 \cdot 10^{10}$ m³/sec between 70°S and 45°S while the observed transport is about $150 \cdot 10^6$ m³/sec. This result is not essentially changed by taking spherical coordinates or by introducing a variation of wind speed with latitude. To obtain a better agreement with the observed transport it is necessary to increase the value of K_h by a factor of 100. Since this seems unreasonable, Munk and Palmén suggested additional bottom friction.

An estimate of the water transport in a pure zonal flow can also be obtained from eq.149. With $S_y = 0$, and a zonal wind field where $\tau_y = 0$, it follows that:

$$\frac{\partial \tau_x}{\partial y} + \frac{\partial}{\partial y} \int_d^\zeta R_x \, dz = 0$$

if the lower boundary of the current, d, is constant. A simple assumption about effective internal friction for vertically integrated horizontal transport in wind-driven ocean currents is that the frictional force is proportional to the current speed. Thus, if one uses $R_x = -\rho k u$ (eq.156A), the essential balance of forces in the Antarctic Circumpolar Current is given by $\partial \tau_x / \partial y - k \, \partial S_x / \partial y = 0$, where S_x is the vertically integrated transport to the east across a section of 1 cm width. By integration over the meridional extent (y) of the current it follows that $S_x = (\bar{\tau}_x/k) + C$ where C is an integration constant and $\bar{\tau}_x$ is an average wind stress. As long as the current is purely wind-driven, it seems reasonable to assume that $S_x = 0$ for $\bar{\tau}_x = 0$; thus $C = 0$. A reasonable value for k is $3 \cdot 10^{-6}$ sec⁻¹ (NEUMANN, 1955). Although the physical meaning of this "friction coefficient" k is somewhat obscure (as a matter of fact, this is also true of the meaning of a constant lateral eddy viscosity coefficient, K_h), it was shown in Chapter IV, pp.166–167) that for the open ocean $k \approx 3 \cdot 10^{-6}$ is a good average value. It has been applied successfully in other problems of the general wind-driven circulation (GARNER et al., 1962).

From $S_x = \bar{\tau}_x/k$, with $\bar{\tau}_x = 2$ dynes/cm² and $k = 3 \cdot 10^{-6}$ sec⁻¹ it follows that $S_x = 6.7 \cdot 10^5$ g/cm/sec for the mean mass transport of the Antarctic Circumpolar Current across a vertical section of 1 cm width. Between 70°S and 45°S, the total transport is $6.7 \cdot 10^5 \cdot 2.8 \cdot 10^8 = 18.8 \cdot 10^{13}$ g/sec, or about $188 \cdot 10^6$ m³/sec. This is the right order of magnitude for the water volume transport in the Antarctic Circumpolar Current.

Such estimates do not contribute much to the understanding and explanation of the dynamics of this magnificent current. Hydrodynamic solutions have been offered by HIDAKA and TSUCHIYA (1953) and by TAKANO (1955a). In a practical evaluation of their theories, again unreasonably high lateral eddy viscosity coefficients are required to obtain agreement with the observed values.

An essentially different approach to the problem of the dynamics of the Antarctic Circumpolar Current was made by STOMMEL (1957b) who introduced the important meridional barrier in the Drake Passage. Most of the energy dissipation and other disturbances are assumed to occur or originate at this part of the current system. STOMMEL's (1957b) explanation seems to suggest a more realistic analysis of the dynamics of the Antarctic Circumpolar Current than preceding theories. An exact quantitative explanation of the dynamics of the Circumpolar Current has not yet been offered.

The surface structure of the currents around the Antarctic Continent is fairly well known. DEACON's (1963) summary of the oceanography of the Antarctic Ocean gives an account of our present knowledge of this remote part of the earth's hydrosphere.

The distribution of temperature and salinity in vertical sections across the Antarctic Ocean suggests a natural division into two zones which can be described as antarctic and subantarctic. The transition zone between antarctic and subantarctic water, the Antarctic Convergence, can be very sharp. This was first shown by MEINARDUS (1923) and SCHOTT (1926). WÜST (1928) and DEFANT (1928) used the name Oceanic Polar Front. The voyage of R.V. "Discovery II" in 1932 has shown that this "front" is continuous all around the Antarctic Continent (DEACON, 1937b; MACKINTOSH, 1946).

The origin and the dynamical character of the Antarctic Convergence has been the subject of discussions by SVERDRUP (1934b) and DEACON (1934). Essentially, Deacon puts more emphasis on the development of a thermohaline circulation in the region of the convergence, whereas Sverdrup stresses the wind-driven circulation pattern and associated vertical motions. Probably both are involved, and a satisfactory explanation can be given by a combination of wind-driven and thermohaline circulations (DEFANT, 1938). Other aspects of a dynamic character concerned with the formation of this front and the sinking of surface waters in its region have been discussed by WYRTKI (1960). Near the Antarctic Convergence between about 45°S and 55°S is the origin of the Antarctic Intermediate Water.

Fig.89 according to DEACON (1963) shows the probable meridional circulation of water in a vertical section between 30°S and 60°S along the meridion of 30°W (see also Fig.76). The Antarctic Intermediate Current and the Antarctic Bottom Current are two outstanding northward flowing branches deduced from the distribution of temperature, salinity and oxygen in the deep sea. The Warm Deep Current as it is called by Deacon, spreads southward while gradually rising. It contains some water that has sunk into greater depth from the surface in the North Atlantic Ocean, and some water of Mediterranean

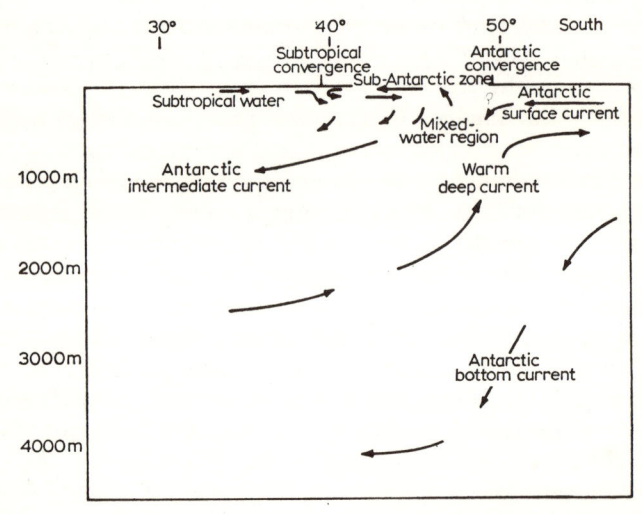

Fig.89. The circulation of water in a meridional section from 30°S to 60°S in the Atlantic Ocean. (After DEACON, 1963.)

Sea origin. The latter mass is warm and has a higher salinity.
Closer to the Antarctic Continent the Antarctic Polar Current flows westward. Between the Polar Current and the Antarctic Circumpolar Current a surface divergence can be expected. KOOPMANN (1953) called this divergence the Antarctic Divergence. Based on a detailed analysis of "Discovery II"-sections he suggests for the South Atlantic Ocean another divergence zone which is located between the Antarctic Convergence and the Antarctic Divergence. Because in the sections along 1°E this divergence was close to the latitude of the Bouvet Island, Koopmann suggested the name Bouvet Divergence.

Two other convergences are found in middle and lower latitudes, the Subtropical and Tropical Convergences. They are not as clearly indicated in the horizontal temperature distribution as the Antarctic Convergence. Their location is mainly deduced from the distribution of sea surface currents. Both the Antarctic Convergence and the Subtropical Convergence are shown in the current chart of Fig..25 The Arctic Convergence, or Polar Front, and the Subtropical Convergence are also shown for the North Atlantic and North Pacific Oceans.

Deep-sea circulation and bottom currents

WÜST (1936b) recognized five major water masses in the Atlantic Ocean: the Intermediate Water, the Upper Deep Water, the Middle Deep Water, the Lower Deep Water, and the Bottom Water. Fig.90 represents in a schematic pattern the spreading of these water masses in the western Atlantic Ocean deduced from the core method. The figure also shows some of the major characteristics of these water masses. The spreading of these water types and water masses is often considered to be the result of currents such that the main axis of a current is placed in the core of a water mass. This may be true for some water masses, and in some parts of the oceans, but in general, the spreading of water types cannot easily be identified with currents. The reason for this was discussed in Chapter IV, pp. 174– 177.

Bottom water of the oceans is formed in polar regions. The most important source for bottom water seems to be in the Weddell Sea. Here, in winter, with cooling and ice formation

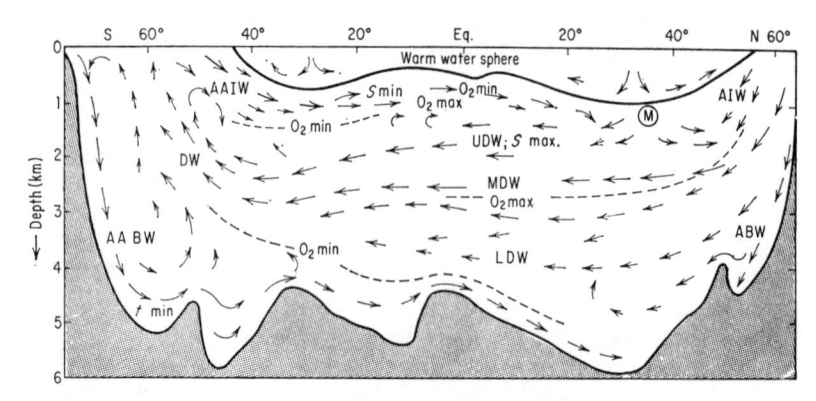

Fig.90. Schematic diagram showing the spreading and major characteristics (maximum or minimum salinity, oxygen or temperature) of Atlantic water masses. Abbreviations are used for water masses, e.g., $AABW$ = Antarctic Bottom Water; $AAIW$ = Antarctic Intermediate Water; DW = Deep Water; UDW = Upper Deep Water; ABW = Arctic Bottom Water; AIW = Arctic Intermediate Water.

the density of the water increases to such high values that it can sink along the slopes of the continental margin to the bottom from where it spreads northward and eastward. The distribution of Antarctic Bottom Water and of the water masses over the Bottom Water is influenced by the deep-reaching, wind-driven Antarctic Circumpolar Current. Around the Antarctic Continent, Antarctic Bottom Water mixes with the overlying water and continuously becomes warmer.

In the Atlantic Ocean, the Mid-Atlantic Ridge imposes a separate spreading in the western and eastern troughs. Since the Walfisch Ridge in the Eastern Atlantic Trough at about 30°S blocks the northward spread, most of the Antarctic Bottom Water spreads northward in the Western Atlantic Trough (Wüst, 1936b) where it can be traced as far north as 30° or 40°N.

There is some indication of Arctic Bottom Water formation in the Greenland Sea between Spitzbergen and Greenland where potential temperatures of $-1.4°C$ are observed at depths of more than 2,000 m (Wüst, 1941). Dietrich (1956b) has shown that this cold water can occasionally overflow the passages in the Iceland–Färöer Ridge and enter the Eastern Atlantic Trough. An overflow of this kind can, at times, also occur over the

Greenland–Iceland Ridge and supply the Western Atlantic Trough with cold bottom water. Another, probably minor, source of Arctic Bottom Water is the region south and southeast of Greenland (Wüst, 1943).

Other sources of cold bottom water in the Southern Hemisphere are indicated by Deacon (1937a) on the basis of "Discovery II" expeditions between the South Orkney and South Shetland Islands. The British–Australian–New Zealand Expedition supported evidence for bottom water formation in the South Indian Ocean (Sverdrup, 1940a).

The Antarctic Intermediate Water (see Fig.89, 90) is formed at the sea surface near the Antarctic Convergence. In this region, precipitation exceeds evaporation and this water type is characterized by a low salinity (see the TS-diagram in Fig.12). It can be traced between 700 and 1,000 m depth along the South American coast into the North Atlantic to about 25°N. The high oxygen content of the Antarctic Intermediate Water is also an important indicator of this water mass which allowed Wüst (1936b) to determine the vertical extent of the Antarctic Intermediate Water.

Arctic Intermediate Water in the Atlantic Ocean seems to be restricted to some areas east of the Grand Banks. It is rather unimportant when compared to the North Pacific Ocean where both Arctic and Antarctic Intermediate Water seem to be developed almost equally on both sides of the equator, at least in the western part of the ocean.

The Deep Water masses in the Atlantic Ocean spreading southward are sandwiched between the Antarctic Bottom Water and the Antarctic Intermediate Water. This is clearly shown in Fig.76 according to Wüst (1950). The upper strata of this deep water are characterized by a high salinity of more than 35‰ in the central eastern North Atlantic Ocean. Fig.76 shows this feature clearly for 30°N at about 1,500 m depth. This component of the Atlantic Deep Water was called Mediterranean Water by Wüst. It spreads at depths between 1,000 and nearly 3,000 m southward. This water type originates in the eastern Mediterranean Sea (Wüst, 1936b, 1961) and overlies the Middle and Lower Deep Water of the North Atlantic.

The Lower Deep Water is not so clearly separated from the

Middle Deep Water and the underlying Bottom Water as is the Upper Deep Water separated from the Middle Deep Water. The Middle Deep Water is characterized by its high oxygen content and the Lower Deep Water by a slight secondary increase of oxygen content. The latter, however, seems to merge with the rudimentarily developed Arctic Bottom Water. The Middle and Lower Deep Water masses from the North Atlantic fill the Atlantic Ocean between about 2,500 and 4,000 m depth.

The meridional water exchange in the deep sea of the Atlantic Ocean is, essentially, an exchange between the North and the South Atlantic, where Antarctic Intermediate Water and Bottom Water spread northward and the North Atlantic Deep Water spreads southward. Superimposed on this major water exchange is a meridional exchange in the South Atlantic where Antarctic Intermediate Water and Antarctic Bottom Water mix with the southward moving Deep Water and partly return into the water body of the Antarctic Circumpolar Current. The Antarctic Circumpolar Water flowing around the Antarctic Continent is most important for the deep sea stratification of the Pacific Ocean.

A zonal section approximately following the region of the Antarctic Convergence shows a well developed salinity maximum at about 1,500 m depth in the southern Atlantic and an indication of a second salinity maximum at about 2,000 m depth in the southern Indian Ocean (SVERDRUP et al., 1942). These salinity maxima nearly coincide with minima in the oxygen distribution, and their origin seems related to the southward spreading Deep Water which is characterized by a salinity maximum and an oxygen minimum. The "regeneration" of the subsurface salinity maximum in the southern Indian Ocean suggests the admixture of higher salinity water to the Antarctic Circumpolar Water at mid-depth. Such water has, indeed, been found to flow from the Red Sea into the Indian Ocean over the sill of the Strait of Bab el Mandeb (VERCELLI, 1925; MÖLLER, 1929; CLOWES and DEACON, 1935; THOMPSON, 1939a,b). The effect of Red Sea Water on the stratification of the Indian Ocean Deep Water is, probably, very similar to that of the Mediterranean Water in the Atlantic. Red Sea Water spreads southward between 1,000 and 2,000 m depth in the western Indian Ocean.

SVERDRUP et al. (1942) suggest an independent circulation in the southern Indian Ocean, such that Antarctic Intermediate Water and Antarctic Bottom Water mix with the Deep Water of higher salinity and that the return flow to the south takes place between 1,000 and 4,000 m depth. The Antarctic Intermediate Water is well developed in the Indian Ocean.

From the Indian Ocean, Antarctic Circumpolar Water with its components of Atlantic and Indian Ocean origin enters the Pacific Ocean. The Pacific Deep and Bottom Waters between about 2,500 m and the bottom are apparently supplied entirely by the Antarctic Circumpolar Current. Its water masses spread north at 160°W to 180°W and, after crossing the equator, branch northwestward and northeastward into the North Pacific. As in the Atlantic and Indian Oceans, the flow is, probably, strongest in the western part of the ocean.

There is no Deep Water formation in the North Pacific which is comparable to that of the North Atlantic. Adjacent seas of the type of the Mediterranean Sea (and the Red Sea) are missing. Deep reaching convection as a result of surface cooling that produces the Middle and Lower Deep Water masses of the North Atlantic do not occur in the North Pacific. Only in the Okhotsk Sea, during winter, cold bottom water is formed. This water of subarctic origin can overflow the submarine boundaries of the Okhotsk Sea, but does not contribute to any significant North Pacific bottom water formation.

As a result of these conditions, the deep water of the Pacific is very uniform. However, it forms the largest water mass in the world's oceans which MONTGOMERY (1958) called the Common Water. The greatest uniformity is found in the North Pacific Ocean which is by far the poorest in oxygen content of the deep layers. In the Atlantic deep sea values are between 3 and 6.5 ml of oxygen per liter of water (ml/l), whereas in the Pacific Ocean they range between 0.5 (or less) to 4.5 ml/l.

The weak stratification of the Pacific deep sea may also suggest that the thermohaline component of the general oceanic circulation in this ocean is much less important than, for example, in the Atlantic Ocean. This, however, does not mean that strong currents in the deep sea are absent. During transient stages of the wind-driven circulation, strong currents can even be observed

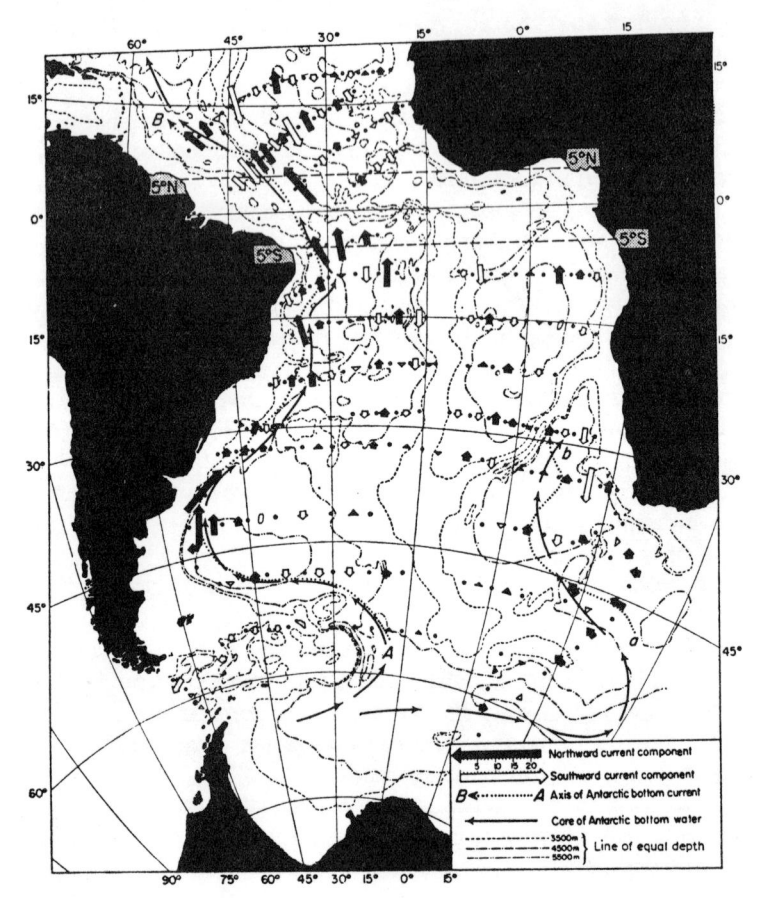

Fig.91. Bottom water velocities in the Atlantic Ocean according to WÜST (1957).

at the ocean bottom. One reason for this was given in Chapter IV, pp.216–218 and explained with the help of Fig.66.

For the Bottom Water of the Atlantic Ocean, WÜST (1955, 1957) determined maximum velocities of 10–15 cm/sec. These values, however, were obtained by the classical method of dynamic computations which assumes a geostrophic flow. They can probably be considered as a good "average" speed for stationary conditions. Results of Wüst's computations are shown in Fig.91.

Direct current measurements near the ocean bottom in recent years were obtained by with SWALLOW (1955), SWALLOW and WORTHINGTON (1961), and VOLKMANN(1962), and continue until the present (KNAUSS, 1965). These measurements have shown that strong and often irregular currents can occur at the ocean bottom. For example, Swallow observed a current speed of 42 cm/sec at 4,000 m depth (ocean bottom depth 4,600 m) 200 miles west of Bermuda with his neutrally buoyant float.

Relatively strong bottom currents in the deep sea have also been inferred by marine geologists. HEEZEN and HOLLISTER (1964) have given an excellent summary of our present knowledge of deep sea current evidence from abyssal sediments. This evidence is based on ocean bottom photographs and sediment cores. The majority of ocean bottom photographs of high and steep topographic prominences, such as seamounts, escarpments and the crests of major ridges, show ripples in the sediment, scour marks and rock outcrops. Current scour and ripples in sediment are also observed beneath the Antarctic bottom current in the western South Atlantic, in the Drake Passage and other parts of the oceans, including the Pacific Ocean. Estimated current speeds up to 60 cm/sec are, probably, not unusual. Although, at present, reliable direct current measurements in the deep sea and near the ocean bottom are rare, they seem to indicate, together with the results obtained from ocean bottom photography, that even the abyssal regions of the oceans are, at least sometimes, moving at significant speeds.

As early as 1650, Bernhard Varenius wrote in his *Geographia Generalis* the words: "Cum pars Oceani movetur, totus Oceanus movetur". This statement has greater importance in connection with the general ocean circulation than earlier oceanographers have, apparently, recognized. Today, we know that we have to abandon the idea of a deep sea being devoid of strong currents. At the same time, we also have to admit that we do not really understand the dynamics of the deep and bottom currents in the general, three-dimensional ocean circulation and that they offer many unsolved problems.

REFERENCES

ANONYMOUS, 1963. Pegram current meter. *U.S. Army Beach Erosion Board, Interim Rept.*
ARTHUR, R. S., 1960. A review of the calculation of ocean currents at the equator. *Deep-Sea Res.*, 6(4): 287-297.
BASCOM, W. N., 1957. Deep sea moored buoy instrument stations. *Aspects of Deep-Sea Research—Natl. Acad. Sci.—Natl. Res. Council, Comm. Undersea Warfare, Publ.*, 473: 105-111.
BELKNAP, G., 1885. Narrative I. *Rept. Sci. Results Voyage "Challenger" 1873-1876*, 1.
BJERKNES, V., 1916. Über thermodynamische Maschinen, die unter der Mitwirkung der Schwerkraft arbeiten. *Abhandl. Sächs. Akad. Wiss. Leipzig, Math. Naturw. Kl.*, 35(1).
BJERKNES, V., 1921. On the dynamics of the circular vortex with applications to the atmosphere and atmospheric vortex and wave motions. *Geofys. Publikasjoner, Norske Videnskapsakad. Oslo*, 2(4).
BJERKNES, V. and SANDSTRÖM, J. W., 1910. Dynamic meteorology and hydrography. 1. Statics. *Carnegie Inst. Wash. Publ.*, 88: 146 pp.
BJERKNES, V. und SANDSTRÖM, J. W., 1912. *Statik der Atmosphäre und der Hydrosphäre. Mit Anhang: Hydrographische Tabellen.* Vieweg, Braunschweig.
BJERKNES, V., BJERKNES, J., SOLBERG, H. S. und BERGERON, T., 1933. *Physikalische Hydrodynamik.* Springer, Berlin, 787 pp.
BLANDFORD, R., 1965a. Inertial flow in the Gulf Stream. *Tellus*, 17: 69-76.
BLANDFORD, R., 1965b. Notes on the theory of the thermocline. *J. Marine Res.*, 23(1): 18-29.
BÖHNECKE, G., 1937. Bericht über die Strommessungen auf. der Ankerstation 369 der 1. Teilfahrt der D.A.E. "Meteor", Februar/Mai 1937. *Ann. Hydrograph. Maritimen Meteorol., September-Beih.*, 1937: 14-16.
BÖHNECKE, G., 1955. The principles of measuring currents. *Assoc. Oceanog. Phys., Publ. Sci.*, 14: 28 pp.
BOWDEN, K. F., 1950. Processes affecting the salinity of the Irish Sea. *Monthly Notices Roy. Astron. Soc., Geophys. Suppl.*, 6: 63.
BOWDEN, K. F., 1953. Measurement of wind currents in the sea by the method of towed electrodes. *Nature*, 171(4356): 735-737.
BOWDEN, K. F., 1954. The direct measurement of subsurface currents in the oceans. *Deep-Sea Res.*, 2(1): 33-47.
BOWDEN, K. F., 1956. The flow of water through the Straits of Dover

related to wind and differences in sea level. *Phil. Trans. Roy. Soc. London, Ser. A*, 248(953): 517–551.
BOWDEN, K. F., 1962a. Measurements of turbulence near the sea bed in a tidal current. *J. Geophys. Res.*, 67(8): 3181–3186.
BOWDEN, K. F., 1962b. Turbulence. In: M. N. HILL (General Editor), *The Sea*. Interscience, New York, N.Y., 1: 802–825.
BROECKER, W., 1963. Radioisotopes and large-scale ocean mixing. In: M. N. HILL (General Editor), *The Sea*. Interscience, New York, N.Y., 2: 88–107.
BRYAN, K., 1963. A numerical investigation of a nonlinear model of a wind driven ocean. *J. Atmospheric Sci.*, 20(6): 594–606.
BRYAN, K. and COX, D., 1965. A numerical investigation of the oceanic general circulation. *Tech. Rept. Geophys. Fluid Dyn. Lab.*, *ESSA*,
BRYAN, K. and COX, D., 1967. A numerical investigation of the oceanic general circulation. *Tellus*, 19(1): 54–80.
BUCHANAN, J. Y., 1886. On similarities in the physical geography of the great oceans. *Proc. Roy. Geograph. Soc.*, 8: 753–770.
BUMPUS, D. F. and LAUZIER, L. M., 1965. *Serial Atlas of the Marine Environment. 7. Surface Circulation on the Continental Shelf off Eastern North America between Newfoundland and Florida*. Am. Geograph. Soc., New York, N.Y.
BURT, W. V. and WYATT, B., 1964. Drift bottle observations of the Davidson Current off Oregon. *Studies Oceanog., Hidaka Vol.*, pp.156–165.

CARRIER, G. F. and ROBINSON, A. R., 1962. On the theory of the wind-driven ocean circulation. *J. Fluid Mech.*, 12: 49.
CARRUTHERS, J. N., 1926. A new current measuring instrument for the purpose of fishery research. *J. Conseil, Conseil Perm. Intern. Exploration Mer*, 1: 127–139.
CARRUTHERS, J. N., 1928. New drift bottles for the investigation of currents in connection with fishery research. *J. Conseil, Conseil Perm. Intern. Exploration Mer*, 3: 194–205.
CARRUTHERS, J. N., 1935. A "vertical log" current meter. *Hydrograph. Rev.*, 12(2): 62–76.
CARRUTHERS, J. N., 1939. The Lowestoft crossbow float. *Nautical Mag.*, 141: 511–519.
CARRUTHERS, J. N., 1954a. A new automatic current float. *Water Sanit. Engr.*, 5(1): 400.
CARRUTHERS, J. N., 1954b. A fisherman's current meter. *Arch. Meteorol., Geophys. Bioklimatol., Ser. B*, 1/2 (Defant Heft).
CARRUTHERS, J. N., 1954c. New approaches in current-measuring. *Commun., Assoc. Océanog. Phys., Rept. Abstr.*
CARRUTHERS, J. N., 1957. A discussion of "A determination of the relation between wind and sea-surface drift". *Quart. J. Roy. Meteorol. Soc.*, 83(356): 276–277.
CARRUTHERS, J. N., 1962. Measurement of ocean bed currents. A new cheap method and a suggested plan. *Nature*, 195(4845): 976–981.

REFERENCES 313

CARRUTHERS, J. N., 1964. Various desiderata in current-measuring and a new instrument to meet some of them. *Studies Oceanog.*, *Hidaka Vol.*, pp.296–301.
CARRUTHERS, J. N., LAWFORD, A. L. and VELEY, V. F. C., 1951. Water movement at the North Goodwin Light Vessel. *Marine Observer*, 1951 (January): 36–46.
CASTENS, G., 1931. Strömung und Isolinienform. *Ann. Hydrograph. Maritimen Meteorol.*, 59: 41–46.
CHARNEY, J. G., 1955. The Gulf Stream as an inertial boundary layer. *Proc. Natl. Acad. Sci. U.S.*, 41: 731–740.
CLOWES, A. J., 1933. Influence of the Pacific on the circulation in the southwest Atlantic Ocean. *Nature*, 131: 189–191.
CLOWES, A. J. and DEACON, G. E. R., 1935. The deep-water circulation of the Indian Ocean. *Nature*, 136: 936–938.
COCHRANE, J. D., 1963. Equatorial Undercurrent and related currents off Brazil in March and April 1963. *Science*, 14(3593): 669–671.
CORIOLIS, G., 1835. Mémoire sur les équations du mouvement relatif des systèmes de corps. *J. École Rech. Polytech.*, 15: 142.
CORRSIN, S., 1959. Outline of some topics in homogeneous shear flow. *J. Geophys. Res.*, 64: 2134.
CORRSIN, S., 1962. Theories of turbulent dispersion. In: *Mécanique de la Turbulence, Marseille, 1961*, pp.27–82.
CRAIG, H. and GORDON, L. I., 1966. Isotopic study of the formation of deep water masses. *Trans. Am. Geophys. Union*, 47(1): 112 (abstr.).
CREASE, J., 1962. Velocity measurements in the deep water of the western North Atlantic. *J. Geophys. Res.*, 67(8): 3173–3176.
CROMWELL, T., MONTGOMERY, R. B. and STROUP, E. D., 1954. Equatorial Undercurrent in Pacific Ocean revealed by new methods. *Science*, 119: 648–649.
CURRIE, R. I., 1965. The oceanography of the Southeast Atlantic. *Anais Acad. Brasil. Cienc.*, *Suppl. Vol.*, 37: 11–22.

DEACON, G. E. R., 1934. Die Nordgrenzen antarktischen und subantarktischen Wassers im Weltmeer. *Ann. Hydrograph. Maritimen Meteorol.*, 62(4): 129–136.
DEACON, G. E. R., 1937a. The hydrology of the southern ocean. *Discovery Rept.*, 15: 1–124.
DEACON, G. E. R., 1937b. Note on the dynamics of the southern ocean. *Discovery Rept.*, 15: 125–152.
DEACON, G. E. R., 1938. A general account of the hydrology of the South Atlantic Ocean. *Discovery Rept.*, 7: 190.
DEACON, G. E. R., 1963. The Southern Ocean. In: M. N. HILL (General Editor), *The Sea*. Interscience, New York, N.Y., 2: 281–296.
DEFANT, A., 1925. Gezeitenprobleme des Meeres in Landnähe. *Probl. Kosmischen Physik*, 6: 80 pp.
DEFANT, A., 1926. Die Austauschgrösse der atmosphärischen und ozeanischen Zirkulation. *Ann. Hydrograph. Maritimen Meteorol.*, 54: 12.

DEFANT, A., 1927. Triftströme bei geschichtetem Wasser. *Z. Geophys.*, 1927: 310.
DEFANT, A., 1928. Die systematische Erforschung des Weltmeeres. *Z. Ges. Erdkunde Berlin, Jubiläum-Sonderband*, pp.459–505.
DEFANT, A., 1929a. Dynamische Ozeanographie. Einführung in die Geophysik. 3. *Naturwiss. Monographien Lehrbücher*, 9(3): 222 pp.
DEFANT, A., 1929b. Stabile Lagerung ozeanischer Wasserkörper und dazu gehörige Stromsysteme. *Veröffentl. Inst. Meereskunde, Univ. Berlin, A*, 19: 33 pp.
DEFANT, A., 1931. Bericht über die ozeanographischen Untersuchungen des Vermessungsschiffes "Meteor" in der Dänemarkstrasse und in der Irmingersee. Zweiter Bericht. *Sitzber. Preuss. Akad. Wiss., Phys. Math. Kl., Sonderausgabe*, 19: 3–17.
DEFANT, A., 1935. Zur Dynamik des Antarktischen Bodenstromes im Atlantischen Ozean. *Z. Geophys.*, 1935: 50–55.
DEFANT, A., 1936a. Das Kaltwasserauftriebsgebiet vor der Küste Südwestafrikas. *Länderkundliche Forsch., Festschr. Norbert Krebs*, pp.52–66.
DEFANT, A., 1936b. Die Troposphäre des Atlantischen Ozeans. *"Meteor"-Werk*, 6(1): 289–411.
DEFANT, A., 1936c. Ausbreitungs- und Vermischungsvorgänge im Antarktischen Bodenstrom und im Subantarktischen Zwischenwasser *Deut. Atlantische Expedition "Meteor", 1925–1927, Wiss. Ergeb.*, 6(2): 55–96.
DEFANT, A., 1937. C. G. Rossby, Dynamik stationärer ozeanischer Ströme im Lichte der experimentellen Stromlehre. Nebst einigen Bemerkungen. *Ann. Hydrograph. Maritimen Meteorol.*, 65: 58.
DEFANT, A., 1938. Aufbau und Zirkulation des Atlantischen Ozeans. *Sitzber. Preuss. Akad. Wiss., Phys. Math. Kl.*, 14: 29 pp.
DEFANT, A., 1940a. Scylla und Charybdis und die Gezeitenströmungen in der Strasse von Messina. *Ann. Hydrograph. Maritimen Meteorol.*, 68: 145–157.
DEFANT, A., 1940b. Die Lage des Forschungsschiffes "Altair" auf der Ankerstation 16–20 Juni 1938 und das auf ihre gewonnene Beobachtungsmaterial. *Ann. Hydrograph. Maritimen Meteorol.*, 68: 35 pp.
DEFANT, A., 1940c. Die ozeanographische Verhältnisse während der Ankerstation des "Altair" am Nordrand des Hauptstromstriches des Golfstromes nördlich der Azoren. *Ann. Hydrograph. Maritimen Meteorol.*, 68 (November Beih.): 35 pp.
DEFANT, A., 1941. Die absolute Topographie des physikalischen Meeresniveaus und der Druckflächen, sowie die Wasserbewegungen im Atlantischen Ozean. *"Meteor"-Werk*, 6(2): 191–260.
DEFANT, A., 1950. Reality and illusion in oceanographic survey. *J. Marine Res.*, 9(2): 120–138.
DEFANT, A., 1952. Theoretische Überlegungen zum Phänomen des Windstaus und des Auftriebs an ozeanischen Küsten. *Deut. Hydrograph. Z.*, 5(2/3): 69–80.
DEFANT, A., 1961. *Physical Oceanography*. Pergamon, Oxford, 1: 745 pp.; 2: 606 pp.

REFERENCES

DELAND, R. J. and LIN, Y-J., 1967. On the movement and prediction of travelling planetary-scale waves. *Monthly Weather Rev.*, 95(1): 21–31.
DELEONIBUS, P. S., 1966. Momentum flux observations from an ocean tower. *Rept. U.S. Naval Oceanog. Office*, in press.
DIETRICH, G., 1935. Aufbau und Dynamik des südlichen Agulhasstromgebietes. *Veröffentl. Inst. Meereskunde, Univ. Berlin, A*, 27: 79 pp.
DIETRICH, G., 1937a. Die Lage der Meeresoberfläche im Druckfeld von Ozean und Atmosphäre. *Veröffentl. Inst. Meereskunde, Univ. Berlin, A*, 33: 52 pp.
DIETRICH, G., 1937b. Die "dynamische" Bezugsfläche, ein Gegenwartsproblem der dyn. Ozeanographie. *Ann. Hydrograph. Maritimen Meteorol.*, 65: 506–519.
DIETRICH, G., 1946. *Unveröffentlichter wissenschaftlicher Bericht*. Deut. Hydrograph. Inst., Hamburg.
DIETRICH, G., 1956a. Beitrag zu einer vergleichenden Ozeanographie des Weltmeeres. *Kieler Meeresforsch.*, 12(1): 3–24.
DIETRICH, G., 1956b. Schichtung und Zirkulation der Irminger See im Juni 1955. *Ber. Deut. Wiss. Komm. Meeresforsch.*, 15: 1935.
DIETRICH, G., 1956c. Überströmung des Island-Färöer-Rückens in Bodennähe nach Beobachtungen mit dem Forschungsschiff "Anton Dohrn" 1955/1956. *Deut. Hydrograph. Z.*, 9: 78–89.
DIETRICH, G., 1963. Die Meere. In: *Die grosse illustrierte Länderkunde. II—Die Grosse Bertelsmann Lexikon Bibliothek*, 13: 1523–1606.
DIETRICH, G. und KALLE, K., 1957. *Allgemeine Meereskunde. Eine Einführung in die Ozeanographie*. Bornträger, Berlin, 492 S.
DOODSON, A. T., 1940. A current meter for measuring turbulence. *Hydrograph. Rev.*, 17(1): 79–100.
DUNCAN, C. P., 1965. Disadvantages of the Olson drift card, and description of a newly designed card. *J. Marine Res.*, 23(3): 233–236.
DUROCHE, J., 1953. The BBT-Neypric currentograph. *Intern. Hydrograph. Rev.*, 30(1): 159–166.
DURST, C. S., 1924. The relationship between current and wind. *Quart. J. Roy. Meteorol. Soc.*, 50: 113.

EKMAN, V. W., 1902. Om jordrotationens inverkan på vindströmmar i hafvet. *Nyt. Mag. Naturv.*, 20: 20 pp.
EKMAN, V. W., 1905. On the influence of the earth's rotation on ocean currents. *Ark. Math., Astron. Fysik.*, 2(11): 1–53.
EKMAN, V. W., 1906. Beiträge zur Theorie der Meeresströmungen. *Ann. Hydrograph. Maritimen Meteorol.*, 34: 423, 472, 527, 566.
EKMAN, V. W., 1908. Zur Frage der Ablenkung von Meeresströmungen. *Ann. Hydrograph. Maritimen Meteorol.*, 36: 429–482.
EKMAN, V. W., 1909. Nochmals zur Frage von der Ablenkung der Triftströmungen. *Ann. Hydrograph. Maritimen Meteorol.*, 37: 77–80.
EKMAN, V. W., 1923. Über Horizontalzirkulation bei winderzeugten Meeresströmungen. *Ark. Math., Astron. Fysik.*, 17(26): 1–74.
EKMAN, V. W. 1926a. Können Verdunstung und Niederschlag im Meere

merkliche Kompensationsströme hervorrufen? *Ann. Hydrograph. Maritimen Meteorol.*, 54: 261.

EKMAN, V. W., 1926b. On a new repeating current meter. *Conseil Perm. Intern. Exploration Mer, Publ. Circonstance*, 91: 27 pp.

EKMAN, V. W., 1928a. Eddy viscosity and skin friction in the dynamics of winds and ocean currents. *Mem. Roy. Soc. (London)*, 2(20): 161–172.

EKMAN, V. W., 1928b. A survey of some theoretical investigations on ocean currents. *J. Conseil, Conseil Perm. Intern. Exploration Mer*, 3(3): 295–327.

EKMAN, V. W., 1929. Über die Strommenge der Konvektionsströme im Meer. *Lunds Univ. Årsskr., Avd.2*, 25(6): 15 pp.

EKMAN, V. W., 1932a. Studien zur Dynamik der Meeresströmungen. *Gerlands Beitr. Geophys.*, 36: 385–438.

EKMAN, V. W., 1932b. An improved type of current meter. *J. Conseil, Conseil Perm. Intern. Exploration Mer*, 7(1): 3–10.

EKMAN, V. W., 1953. Studies on ocean currents. *Geofys. Publikasjoner, Norske Videnskapsakad. Oslo*, 19(1):

EKMAN, V. W. and HELLAND-HANSEN, BJ., 1931. Measurement of ocean currents. *Kgl. Fysiograf. Sällskap. Lund, Forh.*, 1(1): 7 pp.

ELLISON, T. H., 1954. On the correlation of vectors. *Quart. J. Roy. Meteorol. Soc.*, 80: 93–96.

EWING, M., WOOLLARD, G. P., VINE, A. and WORZEL, J. L., 1946. Photography of the ocean bottom. *J. Opt. Soc. Am.*, 36(6): 307–321.

EXNER, F., 1917. *Dynamische Meteorologie*. Springer, Berlin, 308 pp.

FARADAY, M., 1832. Bakerian lecture, experimental researches in electricity. *Phil. Trans. Roy. Soc. London*, 1: 163–177.

FEDEROV, K. N., 1965. Equatorial seiches. *Oceanology (English Transl.)*, 5(1): 37.

FELSENBAUM, A. I., 1956. An extension of Ekman's theory to the case of a non-uniform wind and an arbitrary bottom relief in a closed sea. *Dokl. Akad. Nauk S.S.S.R.*, 109: 299–302.

FISHER, A., 1965. *The Circulation and Stratification of the Brazil Current*. Thesis, Dept. Meteorol. Oceanog., N.Y. Univ., New York, N.Y.

FJELDSTAD, J. E., 1929. Ein Beitrag zur Theorie der winderzeugten Meeresströmungen. *Gerlands Beitr. Geophys.*, 23: 237.

FJELDSTAD, J. E., 1936. Results of tidal observations. *Norweg. N. Polar Expedition "Maud" 1918–1925, Sci. Res.*, 4(4): 88 pp.

FOFONOFF, N. P., 1954. Steady flow in a frictionless homogeneous ocean. *J. Marine Res.*, 13(3): 254–262.

FOFONOFF, N. P., 1962. Dynamics of ocean currents. In: M. N. HILL (General Editor), *The Sea*. Interscience, New York, N.Y., 1: 323–395.

FOMIN, L. M., 1964. *The Dynamic Method in Oceanography*. Elsevier, Amsterdam, 212 pp.

FORCH, C., KNUDSEN, M. und SÖRENSEN, S. P. L., 1902. Berichte über die Konstantenbestimmungen zur Aufstellung der Hydrographischen Tabellen. *Kgl. Danske Videnskab. Selskab, Skr., Raekke Naturv. Mat., Afd. 12*, 6(1): 151 pp.

FRANTZ JR., D. H., 1957. The use of recording and telemetering buoys in deep sea research. In: *Aspects of Deep Sea Research—Natl. Acad. Sci.- Natl. Res. Council, Publ.*, 473: 137–142.
FUGLISTER, F. C., 1951a. Multiple currents in the Gulf Stream. *Tellus*, 3: 230–233.
FUGLISTER, F. C., 1951b. Annual variations in current speeds in the Gulf Stream system. *J. Marine Res.*, 11: 119–127.
FUGLISTER, F. C., 1963. Gulf Stream '60. *Progr. Oceanog.*, 1: 265–373.
FUGLISTER, F. C. and VOORHIS, A. D., 1965. A new method of tracking the Gulf Stream. *Limnol. Oceanog., Redfield Vol., Suppl.*, 10: R115–R124.
FUGLISTER, F. C. and WORTHINGTON, L. V., 1951. Some results of a multiple ship survey of the Gulf Stream. *Tellus*, 3: 1–14.
FULTZ, D., 1951. Experimental analogies to atmospheric motions. In: *Compendium of Meteorology*. Am. Meteorol. Soc., Boston, Mass., pp. 1235–1248.

GARNER, D. M., 1961a. Results of a treatise on decaying inertia currents. *Lecture-Problems, Dept. Meteorol. Oceanog., N.Y. Univ.*, unpubl. manuscript.
GARNER, D. M., 1961b. Notes on the structure of the oceanic upper mixed layer. In: *Contribution to the Problem of Oceanic Circulation*. N.Y. Univ., Coll. Eng., Res. Div., New York, N.Y.—U.S. Navy Hydrograph. Office, Washington, D.C., pp.48–91.
GARNER, D. M., 1962. Some studies on the ocean circulation. *N.Y. Univ., Coll. Eng., Res. Div., Tech. Rept.*, 285(03): 182 pp.
GARNER, D. M., NEUMANN, G. and PIERSON, W. J., 1962. The average horizontal wind driven mass transport of the Atlantic for February as obtained by numerical methods. *Proc. Symp. Math. Hydrodyn. Methods Phys. Oceanog., Hamburg, 1961*, pp.297–319.
GEHRKE, J., 1909. Beitrag zur Hydrographie des Finnischen Meerbusens. *Finn. Hydrograph. Biol. Untersuch.*, 3.
GOLDSBROUGH, G. R., 1933. Ocean currents produced by evaporation and precipitation. *Proc. Roy. Soc. (London), Ser.A.*, 141: 512–517.
GOLDSBROUGH, G. R., 1935. On ocean currents produced by wind. *Proc. Roy. Soc. (London), Ser.A*, 148: 47–58.
GRANT, H. L., STEWART, R. W. and MOILLIET, A., 1962. Turbulence spectra from a tidal channel. *J. Fluid Mech.*, 12: 241–268.
GULDBERG, C. M. et MOHN, H., 1876 (revised 1883–1885). *Études sur les Mouvements de l'Atmosphère*. Christiania. Transl. by C. Abbe: *Smithsonian Inst., Misc. Collections*, 1910, 3: 122–248.
GUNTHER, E. R., 1936. A report on oceanographical investigations in the Peru Coastal Current. *Discovery Rept.*, 13: 107–276.
GUSTAFSON, T. und KULLENBERG, B., 1933. Trägheitsströmungen in der Ostsee. *Medd. Göteborgs Oceanograf. Inst.*, 5: 10.
GUSTAFSON, T. und KULLENBERG, B., 1936. Untersuchungen von Trägheitsströmungen in der Ostsee. *Svenska Hydrograf. Biol. Komm. Skrifter, Ny Ser.: Hydrograf.*, 13: 28 pp.

HADLEY, G., 1735. Concerning the cause of the general trade-winds. *Phil. Trans. Roy. Soc., London*, 39: 58.
HAMON, B. V., 1965. The East Australian Current. *Deep-Sea Res.*, 12(6): 899–921.
HANSEN, W., 1938. Amplitudenverhältnis und Phasenunterschied der harmonischen Konstanten in der Nordsee. *Ann. Hydrograph. Maritimen Meteorol.*, 66: 429–443.
HANSEN, W., 1951. Winderzeugte Strömungen im Ozean. *Deut. Hydrograph. Z.*, 4: 161–172.
HANSEN, W., 1952. Einige Bemerkungen zum Golfstromproblem. *Deut. Hydrograph. Z.*, 5: 80–94.
HANSEN, W., 1956. Theorie zur Errechnung des Wasserstandes und der Strömungen in Randmeeren nebst Anwendungen. *Tellus*, 8(3): 287–300.
HART, T. J. and CURRIE, R. I., 1960. The Benguela Current. *Discovery Rept.*, 31: 123–298.
HAURWITZ, B., 1941. *Dynamic Meteorology*. McGraw-Hill, New York, N.Y., 365 pp.
HAURWITZ, B. and PANOFSKY, H. A., 1950. Stability and meandering of the Gulf Stream. *Trans. Am. Geophys. Union*, 31(5): 723–731.
HEEZEN, B. C. and HOLLISTER, C., 1964. Deep-sea current evidence from abyssal sediments. *Marine Geol.*, 1: 141–174.
HELA, I., 1952. The fluctuation of the Florida Current. *J. Marine Sci. Gulf Caribbean*, 1(4): 241–248.
HELLAND-HANSEN, B., 1916. Nogen hydrografiske metoder. *Forh. Skand. Naturforsker Møte*, 16: 357–359.
HELLAND-HANSEN, B. and EKMAN, V. W., 1931. Measurements of ocean currents. (Experiments in the North Atlantic.) *Kgl. Fysiograf. Sällskap. Lund, Forh.*, 1(1): 7 pp.
HERSEY, J. B., 1952. Acoustic instrumentation as a tool in oceanography. In: J. D. ISAACS and C. O'D. ISELIN (Editors), *Oceanographic Instrumentation—Natl. Acad. Sci.—Natl. Res. Council, Publ.*, 309: 101–112.
HIDAKA, K., 1933. Non-stationary ocean currents. I. *Mem. Imp. Marine Obs., Kobe*, 5(3): 141–265.
HIDAKA, K., 1940. Absolute evaluation of ocean currents in dynamical calculations. *Proc. Imp. Acad. Tokyo*, 15(8): 391.
HIDAKA, K., 1949. Mass transport in ocean currents and lateral mixing. *J. Marine Res.*, 8(2): 132–136. Also *Tokyo Univ., Geophys. Notes*, 2(3):1–4.
HIDAKA, K., 1950a. Drift currents in an enclosed ocean. 1. *Tokyo Univ., Geophys. Notes*, 3(23): 1–23.
HIDAKA, K., 1950b. Drift currents in an enclosed ocean. 2. *Tokyo Univ., Geophys Notes*, 3(38): 1–10.
HIDAKA, K., 1951. Drift currents in an enclosed ocean. 3. *Tokyo Univ., Geophys. Notes*, 4(3): 1–19.
HIDAKA, K., 1954. A contribution to the theory of upwelling in coastal currents. *Trans. Am. Geophys. Union*, 35(3): 431–444.
HIDAKA, K., 1955. A theoretical study on the general circulation of the Pacific Ocean. *Pacific Sci.*, 9(2): 183–220.

HIDAKA, K., 1958. Computation of the wind stress over the oceans. *Tokyo Univ., Geophys. Notes*, 11(1): 77–123.
HIDAKA, K., 1961. A contribution to the computation of three-dimensional ocean currents by high speed computers. *Records Oceanog. Works Japan*, 6(1): 16–28.
HIDAKA, K. and TSUCHIYA, M., 1953. On the Antarctic Circumpolar Current. *J. Marine Res.*, 12(2): 214–222.
HUGHES, P., 1956. A determination of the relation between wind and sea surface drift. *J. Roy. Meteorol. Soc.*, 82: 494–502.

ICHIYE, T., 1949. On the theory of drift currents in an enclosed sea. *Oceanog. Mag.*, 1: 128–132.
ICHIYE, T., 1950. A note on the friction terms in the equation of ocean currents. *Oceanog. Mag.*, 2: 49–52.
ICHIYE, T., 1951. On the variation of oceanic circulation. *Oceanog. Mag.*, 4(6): 79–82, 89–96.
ICHIYE, T., 1954. On the variation of oceanic circulation. *Oceanog. Mag.*, 6(1): 185–217.
ICHIYE, T., 1965a. Diffusion experiments in coastal waters using dye techniques. *Symp. Diffusions Oceans Fresh Waters, 1964, Lamont Geol. Obs. Columbia Univ., Palisades, N.Y.*, pp.54–67.
ICHIYE, T., 1965b. The Kuroshio system. *Oceanus*, 12(1): 18–21.
IDRAC, P., 1927. Sur un apparail enregistrateur pour l'étude océanographique des courants de profondeur. *Compt. Rend.*, 184: 1472–1473. Also *Hydrograph. Rev.*, 1928, 2: 155–158.
IDRAC, P., 1933. Appareil Idrac pour le mesure de courants verticaux sousmarins. *Bull. Inst. Océanog.*, 637.
IDRAC, P., 1935. Idrac apparatus for measuring vertical submarine currents. *Hydrograph. Rev.*, 12(1): 158.
ILYIN, A. N. and KAMENKOWICH, V. M., 1963. On the influence of friction on ocean currents. *Dokl. Akad. Nauk S.S.S.R.*, 150: 1274–1277.
ISAACS, J. D., 1963. Deep-sea anchoring and mooring. In: M. N. HILL (General Editor), *The Sea*. Interscience, New York, N.Y., 2: 516–527.
ISELIN, C. O'D., 1936. A study of the circulation of the western North Atlantic. *Papers Phys. Oceanog. Meteorol.*, 4(4): 101 pp.
ISELIN, C.O'D., 1939. The influence of vertical and lateral turbulence on the characteristics of the waters at mid-depths. *Trans. Am. Geophys. Union*, 3: 414–417.
ISELIN, C.O'D., 1940. Preliminary report on long-period variations in the transport of the Gulf Stream system. *Papers Phys. Oceanog. Meteorol.*, 8(1): 40 pp.

JACOBSEN, J. P., 1913. Beitrag zur Hydrographie der Dänischen Gewässer. *Medd. Komm. Havundersøg., Ser. Hydrograph.*, 2(2): 94.
JACOBSEN, J. P., 1915. Hydrographical investigations in Faroe waters. *Medd. Komm. Havundersøg., Ser. Hydrograph.*, 4(4).

JACOBSEN, J. P., 1918. Hydrographische Untersuchungen im Randersfjord. *Medd. Komm. Havundersøg., Ser. Hydrograph.*, 7.
JEFFREYS, H., 1923. The effect of a steady wind on the sea level near a straight shore. *Phil. Mag.*, 46: 115–125.
JEFFREYS, H., 1925. On fluid motions produced by differences of temperature. *Quart. J. Roy Meteorol. Soc.*, 51.
JOHNS, B., 1965. Inertia currents. *Deep-Sea Res.*, 12: 825–830.
JOSEPH, J., 1954. The paddle wheel current meter and the bifilar current meter of the German Hydrographical Institut. *Commun. Assoc. Oceanog. Phys., Rept. Abstr.*, 1.

KING, J. E., AUSTIN, T. S. and DOTY, M. S., 1957. Preliminary report on Expedition Eastropic. *U.S. Fish Wildlife Serv., Spec. Sci. Rept., Fisheries*, 201: 155 pp.
KIRWAN, A. D., 1963. Circulation of the Antarctic Water deduced through isentropic analysis. *Texas A. M. Res. Found., A. M. Proj. 261*, 63-34F:
KIRWAN, A. D., 1965. On the use of the Rayleigh–Ritz method for calculating the eddy diffusivity. *Symp. Diffusion Oceans Fresh Waters, 1964, Lamont Geol. Obs. Columbia Univ., Palisades, N.Y.*, pp.86–92.
KIRWAN, A. D., ADELFANG, S. I. and MCNALLY, J. G., 1966. Recent results in turbulence studies on air–sea interaction. *Conf. Marine (Oceanic) Meteorol., Virginia Beach, 1966, Papers*.
KISINDO, S., 1934. Methods of ocean current observations now used by the Hydrographical Department in Tokyo, and some results obtained to date. *Hydrograph. Rev.*, 11(1): 4 pp.
KNAUSS, J. A., 1960. Measurements of the Cromwell Current. *Deep-Sea Res.*, 6(4): 265–286.
KNAUSS, J. A., 1961. The structure of the Pacific Equatorial Countercurrent. *J. Geophys. Res.*, 66(1): 143–155.
KNAUSS, J. A., 1962. Recent measurements of the Cromwell Current. *J. Geophys. Res.*, 67: 3571–3572 (abstr.).
KNAUSS, J. A., 1963. Drogues and neutral-buoyant floats. In: M. N. HILL (General Editor), *The Sea*. Interscience, New York, N.Y., 2: 303–305.
KNAUSS, J. A., 1965. A technique for measuring deep ocean currents close to the bottom with an unattached current meter, and some preliminary results. *J. Marine Res.*, 23(3): 237–245.
KNAUSS, J. A., 1966. Further measurements and observations on the Cromwell Current. *J. Marine Res.*, 24(2): 205–240.
KNAUSS, J. A. and TAFT, B. A., 1963. Measurements of currents along the equator in the Indian Ocean. *Nature*, 198: 376–377.
KNAUSS, J. A. and TAFT, B. A., 1964. Equatorial undercurrent of the Indian Ocean. *Science*, 143: 354.
KNUDSEN, M., 1899. *Beretning fra Commissionen for Videnskabelig Undersøgelse af de Danske Farvande*. 2(2): 40–44.
KNUDSEN, M., 1900. Ein hydrographischer Lehrsatz. *Ann. Hydrograph. Maritimen Meteorol.*, 28: 316–320.

KNUDSEN, M., 1901. *Hydrographical Tables*. Bianco Luno, Kopenhagen, 63 pp.; 2nd ed.: 1931.
KOCZY, F. F., KRONENGOLD, M. and LOEWENSTEIN, J. M., 1963. A doppler current meter. In: R. D. GAUL (Editor), *Marine Sciences Instrumentation*. Plenum Press, New York, N.Y., 2: 127–134.
KOENUMA, K., 1939. On the hydrography of the southwestern part of the North Pacific and the Kuroshio. *Mem. Imp. Marine Obs.*, 7: 41.
KOOPMANN, G., 1953. Entstehung und Verbreitung von Divergenzen in der oberflächennahen Wasserbewegung der antarktischen Gewässer. *Deut. Hydrograph. Z., Ergänzungsheft, A*, 2.
KORT, V. G., 1959. New data on the Antarctic water mass transport. *Intern. Oceanog. Congr., 1st, New York, Am. Assoc. Advan. Sci., Preprints.*
KORT, V. G., 1962. The Antarctic Ocean. *Sci. Am.*, 1962 (September): 11 pp.
KORT, V. G., 1964. New data on the transport of Antarctic waters. In: *Soviet Antarctic Expedition*. Elsevier, Amsterdam, 1: 358–361.
KRÜMMEL, O., 1907. *Handbuch der Ozeanographie*. Engelhorn, Stuttgart, 1: 537 pp.
KRÜMMEL, O., 1911. *Handbuch der Ozeanographie*. Engelhorn, Stuttgart, 2: 766 pp.
KUHLBRODT, E., 1941. Zur Meteorologie des Seeweges Kanal—New York. *Ann. Hydrograph. Maritimen Meteorol.*, 69: 233–238.

LA FOND, E. C., 1951. Processing oceanographic data. *U.S. Navy Hydrograph. Office, Publ.*, 614: 1–114.
LA FOND, E. C., 1962. Deep current measurement with the bathyscaph "Trieste". *Deep-Sea Res.*, 9: 115–116.
LAMB, H., 1932. *Hydrodynamics*, 6th ed. Cambridge Univ. Press, London, 738 pp.
LAPLACE, P. S., 1775. Recherches sur plusieurs points du système du monde. 1. *Mem. Acad. Roy. Soc.*, 88: 75–1812.
LAPLACE, P. S., 1776. Recherches sur plusieurs points du système du monde. 2. *Mem. Acad. Roy. Soc.*, 89: 177–267.
LARSEN, K. D., 1960. Study of velocity measurements in water by the use of thermistors. *Davidson Lab., Stevens Inst. Tech., Tech. Mem.*, 128.
LAWFORD, A. L., 1956. The effect of wind upon the surface drift in the Northeastern Atlantic and the North Sea. *Weather*, 11(5): 155–161.
LENZ, E., 1847. Bericht über die ozeanischen Temperaturen in verschiedenen Tiefen. *Bull. Class. Hist. Phil. Acad. Sci., Petersburg, Suppl.*, 3: 11.
LINEIKIN, P. S., 1955a. Wind-driven currents in the baroclinic layer of the sea. *Tr. Gos. Okeanogr. Inst.*, 29(41).
LINEIKIN, P. S., 1955b. On the determination of the thickness of the baroclinic layer in the sea. *Dokl. Akad. Nauk S.S.S.R.*, 101: 461–464.
LINEIKIN, P. S., 1957. On the dynamics of the baroclinic layer in the ocean. *Dokl. Akad. Nauk S.S.S.R.*, 177: 971–974.
LINEIKIN, P. S., 1962. Some new research on the dynamics of ocean currents. *Oceanog. Soc., Japan, 20th Anniv. Vol.*, pp.448–457.

LONGUET-HIGGINS, M. S., 1947. The electric field induced in a channel of moving water. *Admiralty Res. Lab., R.*, 2/102: 22
LONGUET-HIGGINS, M. S., 1949. The electrical and magnetic effects of tidal streams. *Monthly Notices Roy. Astron. Soc., Geophys. Suppl.*, 5(8): 295–307.
LONGUET-HIGGINS, M. S., 1953. Mass transport in water waves. *Phil. Trans. Roy. Soc. London, Ser. A*, 245(90): 535–581.
LONGUET-HIGGINS, M. S., 1965. Some dynamical aspects of ocean currents. *Quart. J. Roy. Meteorol. Soc.*, 91(390): 425–451.
LONGUET-HIGGINS, M. S. and BARBER, N., 1948. Water movements and earth currents: electrical-magnetic effects. *Nature*, 161: 192–193.
LONGUET-HIGGINS, M. S., STERN, M. E. and STOMMEL, H., 1954. The electrical field induced by ocean currents and waves with applications to the method of towed electrodes. *Papers Phys. Oceanog. Meteorol.*, 13(1): 1–37.
LUKASIK, S. J. and GROSCH, C. E., 1963. Pressure–velocity correlations in ocean swell. *J. Geophys. Res.*, 68(20): 5689–5699.

MACKINTOSH, N. A., 1946. The Antarctic convergence and the distribution of surface temperatures in Antarctic waters. *Discovery Rept.*, 23: 177–212.
MACLAURIN, C., 1740. De causa physica fluxus et refluxus maris. *Pièces qui ont remporté le prix de l'Académie des Sciences, Paris.*
MALKUS, W. V. R., 1953. A recording bathypitotmeter. *J. Marine Res.*, 12(1): 51–59.
MALKUS, W. V. R. and STERN, M. E., 1952. Determination of ocean transports by electromagnetic effects. *J. Marine Res.*, 11(2): 97–105.
MARGULES, M., 1906. Über Temperaturschichtung in stationär bewegter und ruhender Luft. *Meteorol. Z.*, 1906: 241–244.
MARTINEAU, D., 1958. *The Gulf Stream as an Equivalent-Barotropic System.* Ph. D. Thesis, Dept. Meteorol. Oceanog., N. Y. Univ., New York, N.Y.
MAZZARELLI, G., 1938. I vortici, i tagli e altri fenomeni delle currenti dello stretto di Messina. *Atti Real. Accad. Peloritana*, 40.
MCEWEN, G. F., 1912. The distribution of ocean temperatures along the west coast of North America deduced from Ekman's theory of upwelling of cold water from adjacent ocean depth. *Intern. Rev. Ges. Hydrobiol. Hydrographie*, 5: 243–286.
MCEWEN, G. F., 1929. A mathematical theory of the vertical distribution of temperature and salinity in water under the action of radiation, conduction, evaporation and mixing due to the resulting convection. *Bull. Scripps Inst. Oceanog., Univ. Calif.*, 2: 197–306.
MEINARDUS, W., 1923. Die zonale Verteilung der Luft- und Wassertemperatur. *Deut. Südpol Expedition*, 3: 528–546.
MERZ, A., 1925. Die Deutsche Atlantische Expedition auf dem Vermessungs- und Forschungsschiff "Meteor". *Sitzber. Akad. Wiss., Berlin, Math. Phys. Kl.*, 31: 562–586.
MERZ, A., 1929. The Alfred Merz apparatus for measuring the flow of strong and weak currents. *Hydrograph. Rev.*, 6(1): 158–162.
METCALF, W. G., VOORHIS, A. D. and STALCUP, M. C., 1962. The Atlantic Equatorial Undercurrent. *J. Geophys. Res.*, 67(6): 2499–2508.

MEYER, H. H. F., 1923. Die Oberflächenströmungen des Atlantischen Ozeans im Februar. *Veröffentl. Inst. Meereskunde, Univ. Berlin, Reihe A*, 11: 35 pp.

MICHAELIS, G., 1923. Die Wasserbewegung an der Oberfläche des Indischen Ozeans im Januar und Juli. *Veröffentl. Inst. Meereskunde, Univ. Berlin, Reihe A*, 8: 32 pp.

MIDDLETON, F. H., 1955. An ultrasonic current meter for estuarine research. *J. Marine Res.*, 14(2): 176–186.

MÖLLER, L., 1929. Die Zirkulation des Indischen Ozeans. Auf Grund von Temperatur- und Salzgehaltstiefenmessungen und Oberflächenstrombeobachtungen. *Veröffentl. Inst. Meereskunde, Univ. Berlin, Geograph. Naturwiss. Reihe*, 21: 48 pp.

MONTGOMERY, R. B., 1938a. Fluctuations in monthly sea level on eastern U.S. coast as related to dynamics of western North Atlantic Ocean. *J. Marine Res.*, 1: 165–185.

MONTGOMERY, R. B., 1938b. Circulation in upper layers of the southern North Atlantic deduced with use of isentropic analysis. *Papers Phys. Oceanog. Meteorol.*, 6(2): 1–55.

MONTGOMERY, R. B., 1939. Ein Versuch, den vertikalen und seitlichen Austausch in der Tiefe der Sprungschicht im äquatorialen Atlantischen Ozean zu bestimmen. *Ann. Hydrograph. Maritimen Meteorol.*, 67: 242–246.

MONTGOMERY, R. B., 1958. Water characteristics of Atlantic Ocean and of world ocean. *Deep-Sea Res.*, 5: 134–148.

MONTGOMERY, R. B., 1962. Equatorial Undercurrent observations in review. *Oceanog. Soc. Japan, 20th Anniv. Vol.*, pp.487–498.

MONTGOMERY, R. B. and PALMÉN, E., 1940. Contribution to the question of the Equatorial Counter Current. *J. Marine Res.*, 3: 112–133.

MONTGOMERY, R. B. and STROUP, E. D., 1962. Equatorial waters and currents at 150° W in July–August 1952. *Johns Hopkins Univ., Oceanog. Studies*, 1: 68 pp.

MOORE, D. W., 1963. Rossby waves in ocean circulation. *Deep-Sea Res.*, 10: 735–748.

MORGAN, G. W., 1956. On the wind-driven ocean circulation. *Tellus*, 8: 301–320.

MORSE, R. M., 1957. *The Measurement of Transports and Current Velocities in Tidal Streams by Electromagnetic Method.* M.Sc. Thesis, Univ. Wash., Seattle, Wash.

MOSBY, H., 1946. Experiments on turbulence and friction near the bottom of the sea. *Bergens Museums Årbok, Naturvidenskap. Rekke*, 3: 1–6.

MOSBY, H., 1949. Experiments on bottom friction. *Univ. Bergen Årbok, Naturvidenskap. Rekke*, 19: 1–12.

MOSBY, H., 1954. Oberflächenströmungen in der Meerenge bei Tromsö. *Arch. Meteorol., Geophys. Bioklimatol., Ser. A*, 7: 378–384.

MUNK, W. H., 1950. On the wind driven ocean circulation. *J. Meteorol.*, 7: 79–93.

MUNK, W. H. and ANDERSON, E. R., 1948. Notes on a theory of the thermocline. *J. Marine Res.*, 7: 276–295.

MUNK, W. H. and CARRIER, G. F., 1950. On the wind driven circulation in ocean basins of various shapes. *Tellus*, 2: 158–167.

MUNK, W. H. and PALMÉN, E., 1951. Notes on the dynamics of the Antarctic Circumpolar Current. *Tellus*, 3(1): 53–55.

NAN'NITI, T., AKAMATSU, H. and NAKAI, T., 1964. A further observation of a deep current in the East-North-East Sea of Torishima. *Oceanog. Mag.*, 16(1–2): 11–19.

NANSEN, F., 1902. Oceanography of the North Polar Basin. *Norweg. North Polar Expedition, 1893–1896, Sci. Results*, 3: 427 pp.

NEUMANN, G., 1942. Die absolute Topographie des physikalischen Meeresniveaus und die Oberflächenströmungen des Schwarzen Meeres. *Ann. Hydrograph. Maritimen Meteorol.*, 70: 265–282.

NEUMANN, G., 1947. Über die Entstehung des Äquatorialen Gegenstromes. *Forsch. Fortschr.*, 16–18: 177–179.

NEUMANN, G., 1948a. Über den Tangentialdruck des Windes und die Rauhigkeit der Meeresoberfläche. *Z. Meteorol.*, 2(7/8): 193–203.

NEUMANN, G., 1948b. Bemerkungen zur Zellularkonvektion im Meer und in der Atmosphäre und die Beurteilung des statischen Gleichgewichts. *Ann. Meteorol.*, 1948: 235–244.

NEUMANN, G., 1952a. Some problems concerning the dynamics of the Gulf Stream. *Trans. N.Y. Acad. Sci., Ser.II*, 14(7): 283–291.

NEUMANN, G., 1952b. Über die komplexe Natur des Seeganges. *Deut. Hydrograph. Z.*, 5(2/3): 95–110; 5(5/6): 252–277.

NEUMANN, G., 1952c. On the complex nature of ocean waves and the growth of the sea under the action of wind. In: *Gravity Waves—Natl. Bur. Std. (U.S.), Circ.*, 521: 61–68.

NEUMANN, G., 1953. Ozean und Atmosphäre, Bemerkungen über einige meteorologisch wichtige Wechselbeziehungen. *Naturw. Rundschau*, 6: 405–411.

NEUMANN, G., 1954. Notes on the wind-driven ocean circulation. *N.Y. Univ., Coll. Eng., Res. Div.*, 1954 (May): 54 pp.

NEUMANN, G., 1955. On the dynamics of wind-driven ocean currents. *Meteorol. Papers*, 2(4): 33 pp.

NEUMANN, G., 1956. Zum Problem der dynamischen Bezugsfläche, inbesondere im Golfstromgebiet. *Deut. Hydrograph. Z.*, 9(2): 66–78.

NEUMANN, G., 1958. On the mass transport of wind-driven currents in a baroclinic ocean with application to the North Atlantic. *Z. Meteorol.*, 12(4–6): 138–149.

NEUMANN, G., 1959. Notes on the stress of light winds on the sea. *Bull. Am. Meteorol. Soc.*, 49(3): 146–148.

NEUMANN, G., 1960a. On the dynamic structure of the Gulf Stream as an equivalent-barotropic flow. *J. Geophys. Res.*, 65(1): 239–247.

NEUMANN, G., 1960b. On the effect of bottom topography on ocean currents. *Deut. Hydrograph. Z.*, 13(3): 132–141.

NEUMANN, G., 1960c. Evidence for an equatorial undercurrent in the Atlantic Ocean. *Deep-Sea Res.*, 6: 328–334.

NEUMANN, G., 1965. Equatorial currents in the Atlantic Ocean with special consideration of the Gulf of Guinea during Equalant I and II. *Intern. Conf. Tropical Oceanog., 1965, Miami Beach, Fla, Abstr.*

NEUMANN, G. and PIERSON JR., W. J., 1966. *Principles of Physical Oceanography.* Prentice-Hall, Englewood Cliffs, N.J., 545 pp.

NEUMANN, G. und SCHUMACHER, A., 1944. Oberflächenströmungen und Dichte der Meeresoberfläche vor der Ostküste Nordamerikas. *Ann. Hydrograph. Maritimen Meteorol.*, 72: 277–279.

NEUMANN, G. and WILLIAMS, R. E., 1965. Observations of the Equatorial Undercurrent in the Atlantic Ocean at 15° W during Equalant I. *J. Geophys. Res.*, 70(2): 297–304.

NIILER, P. O., ROBINSON, A. R. and SPIEGEL, S. L., 1965. On thermally maintained circulation in a closed basin. *J. Marine Res.*, 23(3): 222–230.

OKUBO, A., 1962. A review of theoretical models of turbulent diffusion in the sea. *Chesapeake Bay Inst., Johns Hopkins Univ., Tech. Rept.*, 30 (ref. 62–69): 105 pp.

OKUBO, A., 1964a. Notes on dye diffusion experiments. 2. *J. Oceanog. Soc. Japan*, 20(1): 34–36 (Japanese).

OKUBO, A., 1964b. Equations describing the diffusion of an introduced pollutant in a one-dimensional estuary. *Studies Oceanog., Tokyo*, 1964: 216–226.

OLSON, F. C. W., 1951. A plastic envelope substitute for drift bottles. *J. Marine Res.*, 10(2): 190–193.

OSTAPOFF, F., 1957. On the depth of the layer of no motion in the central Pacific Ocean. In: *Contributions to the Study of the Oceanic Circulation—N.Y. Univ., Coll. Eng., Res. Rept.*, 1957 (September): 1–31.

OZMIDOV, R. V., 1959. Extension of Ekman's theory of unsteady purely drift currents to the case of arbitrary winds. *Dokl. Akad. Nauk S.S.S.R.*, 128: 913–916.

PAECH, H., 1926. Die Oberflächenströmungen um Madagaskar in ihrem jährlichen Gang. *Veröffentl. Inst. Meereskunde, Univ. Berlin, Reihe A*, 16: 39 pp.

PAQUETTE, R. G., 1962. Practical problems in the direct measurements of ocean currents. In: R. D. GAUL (Editor): *Marine Sciences Instrumentation.* Plenum Press, New York, N.Y., 2: 135–145.

PARR, A. E., 1936a. On the relationship between dynamic topography and direction of currents under the influence of external (climatic) factors. *J. Conseil, Conseil Perm. Intern. Exploration Mer*, 11: 299–307.

PARR, A. E., 1936b. On the probable relationship between vertical stability and lateral mixing processes. *J. Conseil, Conseil Perm. Intern. Exploration Mer*, 11(4): 308.

PARR, A. E., 1938. Isopycnic analysis of current flow by means of identifying properties. *J. Marine Res.*, 1(2): 133–154.

PETTERSSON, H., 1915. A recording current meter for deep sea work. *Quart. J. Roy. Meteorol. Soc.*, 41: 65–69.

PETTERSSON, H., 1929. Current meter for determination of the direction and velocity of the movement of the water at the bottom of the ocean. *Svenska Hydrograf. Biol. Komm. Skrifter, Ny Ser.: Hydrograf.*, 1(3): 10–11.
PHILLIPS, O. M., 1963. Geostrophic motion. *Rev. Geophys.*, 1(2): 123–176.
PILLSBURY, J. E., 1891. The Gulf Stream. *Rept. Supt., U.S. Coast Geodetic Surv., Year End. June 1890, Appendix*, 19: 459–620.
POCHAPSKY, T. E., 1961. Some measurements with instrumented neutral floats. *Deep-Sea Res.*, 8: 269–275.
POCHAPSKY, T. E., 1962. Measurement of currents below the surface at a location in the western equatorial Atlantic. *Nature*, 195(4843): 767–768.
POCHAPSKY, T. E., 1963. Measurement of small-scale oceanic motions with neutrally-buoyant floats. *Tellus*, 15(4): 352–362.
POCHAPSKY, T. E., 1966. Measurements of deep water movements with instrumented neutrally buoyant floats. *J. Geophys. Res.*, 71(10): 2491–2504.
POISSON, S. D., 1837. Mémoire sur le mouvement des projectiles dans l'air en ayant égard à la rotation de la terre. *J. École Roy. Polytech.*, 16: 1.
PONOMARENKO, G. P., 1963. Deep Lomonosov countercurrent at the equator in the Atlantic Ocean. *Dokl. Earth Sci. Sect. (English Transl.)*, 149(5): 1178–1181.
PRANDTL, L., 1925. Bericht über Untersuchungen zur ausgebildeten Turbulenz. *Z. Angew. Math. Mech.*, 5: 136.
PRANDTL, L., 1942. *Führer durch die Strömungslehre*. Vieweg, Braunschweig, 382 pp.
PRANDTL, L., 1949. *Führer durch die Strömungslehre*, 3. Aufl. Vieweg, Braunschweig.
PRANDTL, L., 1952. *Essentials of Fluid Dynamics*. Hafner, New York, N.Y., 452 pp.
PRITCHARD, D. W. and CARPENTER, J. H., 1960. Measurement of turbulent diffusion in estuarine and inshore waters. *Assoc. Intern. Hydrol. Sci., Bull.*, 20: 37–50.
PROUDMAN, J., 1953. *Dynamical Oceanography*. Methuen, London/Wiley, New York, N.Y., 409 pp.

RADAKOVIC, M., 1914. Zum Einfluss der Erdrotation auf die Bewegungen auf der Erde. *Meteorol. Z.*, 31: 384.
RADAKOVIC, M., 1920. Über Ableitungen der ablenkenden Kraft der Erddrehung. *Meteorol. Z.*, 37: 296.
RAUSCHELBACH, H., 1929. Beschreibung eines bifilar aufgehängten, an Bord elektrisch registrierenden Strommessers. *Ann. Hydrograph. Maritimen Meteorol., Beih.*, 57.
RAUSCHELBACH, H., 1930. Elektrisch registrierender Stromesser nach Rauschelbach zur Stromrichtungs- und Geschwindigkeitsbestimmung. *Werft, Reederei, Hafen*, 11: 456.
REID, J. L., 1959. Evidence of a South Equatorial Countercurrent in the Pacific Ocean. *Nature, Suppl.*, 184(4): 209–210.
REID, J. L., 1964. Evidence of a South Equatorial Countercurrent in the Atlantic Ocean in July 1963. *Nature*, 203 (4941): 182.

REID, J. L., RODEN, G. I. and WYLLIE, J. G., 1958. Studies of the California Current system. *Calif. Coop. Oceanog. Fisheries Invest., Progr. Rept.,* 1 *July 1956–1 January 1958,* pp.27–57.

REID, R. O., 1948a. A model of the vertical structure of mass in the equatorial wind driven currents of a baroclinic ocean. *J. Marine Res.,* 7(3): 304–312.

REID, R. O., 1948b. The equatorial currents of the Pacific as maintained by the stress of the wind. *J. Marine Res.,* 7(2): 74–99.

REYNOLDS, O., 1894. On the dynamical theory of incompressible viscous fluids and the determination of the criterion. *Phil. Trans. Roy. Soc. London, Ser.A,* 186: 123.

RICHARDSON, L. F., 1926. Atmospheric diffusion shown on a distance-neighbor graph. *Proc. Roy. Soc. (London), Ser.A,* 110: 709.

RICHARDSON, W. S. and SCHMITZ, W. J., 1965. A technique for the direct measurement of transport with application to the Straits of Florida. *J. Marine Res.,* 23(2): 172–185.

RICHARDSON, W. S., STIMSON, P. B. and WILKINS, C. H., 1963. Current measurements from moored buoys. *Deep-Sea Res.,* 10: 369–388.

RINKEL, M. O., 1963. *Cruise Report P6302, Equalant II.* Univ. Miami Inst. Marine Sci., Marine Lab., Miami, Fla., pp.1–16.

RINKEL, M. O., SUND, P. and NEUMANN, G., 1966. The location of the termination area of the Equatorial Undercurrent in the Gulf of Guinea based on observations during Equalant III. *J. Geophys. Res.,* 71(16): 3893–3901.

ROBERTS, E. B., 1952. *Roberts' Radio Current Meter Mod. II Operating Manual.* U.S. Dept. Commerce., Coast Geodetic Surv., Washington, D.C., 32 pp.

ROBINSON, A. R., 1960. The general thermal circulation in equatorial regions. *Deep-Sea Res.,* 6: 311–317.

ROBINSON, A. R., 1965. A three-dimensional model of inertial currents in a variable-density ocean. *J. Fluid Dyn.,* 21(1): 211.

ROBINSON, A. R. and STOMMEL, H., 1959. The oceanic thermocline and associated thermohaline circulation. *Tellus,* 11(3): 295–308.

ROBINSON, A. R. and WELANDER, P., 1963. Thermal circulation on a rotating sphere, with application to the oceanic thermocline. *J. Marine Res.,* 21(1): 25–38.

ROLL, H. U., 1965. *Physics of the Marine Atmosphere.* Academic Press, New York, N.Y., 426 pp.

ROSSBY, C. G., 1932. A generalization of the theory of the mixing length with applications to atmospheric and oceanic turbulence. *Papers Phys. Oceanog. Meteorol.,* 1(4): 1–36.

ROSSBY, C. G., 1936. Dynamics of steady ocean currents in the light of experimental fluid dynamics. *Papers Phys. Oceanog. Meteorol.,* 5(1): 1–43.

ROSSBY, C. G. and MONTGOMERY, R. B., 1935. The layer of frictional influence in wind and ocean currents. *Papers Phys. Oceanog. Meteorol.,* 3(3): 101 pp.

SAINT-GUILY, B., 1959. Sur la solution du problème d'Ekman. *Deut. Hydrograph. Z.,* 12: 262–270.

SAINT-GUILY, B., 1962. On the general form of the Ekman problem. *Proc. Symp. Math. Hydrol. Methods Phys. Oceanog.*, *Hamburg, 1961*, pp.61–73.

SAITO, Y., 1951. On the velocity of the vertical flow in the ocean. *J. Inst. Polytech., Osaka City Univ., Ser.B*, 2: 1–4.

SAITO, Y., 1952. On the Oyashio Current. *J. Inst. Polytech., Osaka City Univ., Ser. B*, 3.

SANDSTRÖM, J. W., 1908. Dynamische Versuche mit Meerwasser. *Ann. Hydrograph. Maritimen Meteorol.*, 36: 6–23.

SANDSTRÖM, J. W., 1909. Über die Bewegung der Flüssigkeiten. *Ann. Hydrograph. Maritimen Meteorol.*, 37: 242–255.

SANDSTRÖM, J. W. und HELLAND-HANSEN, BJ., 1903. Über die Berechnung von Meeresströmungen. *Rept. Norweg. Fishery Marine Invest.*, 2(4): 43 pp.

SARKISIAN, A. S., 1954. The calculation of stationary wind currents in an ocean. *Izv. Akad. Nauk S.S.S.R., Ser. Geofiz.*, 6: 554–561.

SARKISIAN, A. S., 1957. On non-stationary wind-driven currents in a homogeneous ocean. *Izv. Akad. Nauk S.S.S.R., Ser. Geofiz.*, 8: 1008–1019.

SCHEMAINDA, R. M., STURM, M. und VOIGT, K., 1964. Vorläufige Resultate der Untersuchungen im Bereich des äquatorialen Unterstroms im Golf von Guinea mit M.S. "Professor Albrecht Penck" in der Zeit von April bis Juli 1964. *Deut. Akad. Wiss. Berlin, Beitr. Meereskunde*, 15: 1–13.

SCHMIDT, W., 1917. Wirkungen der ungeordneten Bewegungen im Wasser der Meere und Seen. *Ann. Hydrograph. Maritimen Meteorol.*, 45: 367–380, 431–445.

SCHMIDT, W., 1925. Der Massenaustausch in freier Luft und verwandte Erscheinungen. *Probl. Kosmischen Physik*, 7: 118 pp.

SCHMITZ, H. P., 1962a. A relation between the vectors of stress, wind, and current at water surfaces and between the shearing stress and velocities at solid boundaries. *Deut. Hydrograph. Z.*, 15(1).

SCHMITZ, H. P., 1962b. Über die Interpretation bodennaher vertikaler Geschwindigkeitsprofile in Ozean und Atmosphäre und die Windschubspannung auf Wasserflächen. *Deut. Hydrograph. Z.*, 15(2): 45–72.

SCHOTT, G., 1898. Weltkarte zur Übersicht der Meeresströmungen und Schiffswege. *Ann. Hydrograph. Maritimen Meteorol.*, 26: 409.

SCHOTT, G., 1915. Die Gewässer des Mittelmeeres. *Ann. Hydrograph. Maritimen Meteorol.*, 43: 63–79.

SCHOTT, G., 1926. *Geographie des Atlantischen Ozeans*. Boysen, Hamburg, 438 pp.

SCHOTT, G., 1942. Grundlagen einer Weltkarte der Meeresströmungen. *Ann. Hydrograph. Maritimen Meteorol.*, 70: 329–340.

SCHUMACHER, A., 1930. Bemerkungen zur Technik der Strommessungen. *Rappt. Proces-Verbaux Réunions, Conseil Perm. Intern. Exploration Mer*, 54: 19–23.

SCHUMACHER, A., 1940. Monatskarten der Oberflächenströmungen im Nordatlantischen Ozean (5°S–50°N). *Ann. Hydrograph. Maritimen Meteorol.*, 68: 109–123.

SCHUMACHER, A., 1943. Monatskarten der Oberflächenströmungen im äquatorialen und südlichen Atlantischen Ozean. *Ann. Hydrograph. Maritimen Meteorol.*, 71: 209–219.

SCRIPPS INSTITUTION OF OCEANOGRAPHY, 1948. The field of mean wind stress over the North Pacific Ocean. *Scripps Inst. Oceanog., Oceanog. Rept.*, 14.

SEIWELL, H. R., 1937. Short period vertical oscillations in the western basin of the North Atlantic. *Papers Phys. Oceanog. Meteorol.*, 5(2): 44 pp.

SHARPY-SCHAFER, J. M., 1952. Observations of tidal streams in deep water. *Intern. Hydrograph. Rev.*, 29(2): 135–136.

SHULEIKIN, V. V., 1945. Convective sea currents in the monsoon field. *Dokl. Akad. Nauk S.S.S.R.*, 46(5).

SIGEMATSU, R., 1933. Some oceanographical investigations of the results of oceanic survey, carried out by "Mansyu' from April 1925 to March 1928. *Bull. Hydrograph. Dept. Imp. Navy*, 1933: 475.

SOULE, F. M., 1939. Consideration of the depth of the motionless surface near the Grand Banks of Newfoundland. *J. Marine Res.*, 2: 169–180.

SPIESS, F., 1928. *Die Meteor-Fahrt. Forschungen und Ergebnisse der Deutschen Atlantischen Expedition.* Meteor-Werk, Berlin, 375 pp.

STALCUP, M. C. and METCALF, W. G., 1966. Direct measurements of the Atlantic Equatorial Undercurrent. *J. Marine Res.*, 24(1): 44–55.

STALCUP, M. C. and PARKER, C. E., 1965. Drogue measurements of shallow currents on the equator in the western Atlantic Ocean. *Deep-Sea Res.*, 12(4): 535–536.

STEWART, R. W., 1958. The problem of diffusion in a stratified fluid. In: *Symposium on Atmospheric Diffusion and Air Pollution, Oxford, August 1958.* Academic Press, New York, N.Y., pp.303–311.

STEWART, R. W., 1959. The natural occurrence of turbulence. *J. Geophys. Res.*, 64(12): 2112–2115.

STEWART, R. W., 1961a. The wave drag of wind over water. *J. Fluid Mech.*, 10(2): 189–194.

STEWART, R. W., 1961b. Wind stress on water. *Proc. Symp. Math. Hydrodynamical Methods Phys. Oceanog.*, Hamburg, pp.399–408.

STEWART, R. W., 1964. The influence of friction on inertial models of oceanic circulation. *Studies Oceanog., Hidaka Vol.*, pp.3–9.

STIMSON, P. B., 1963. The design of anchors for deep sea moorings. *Deep-Sea Res.*, 10.

STOCKMAN, W. B., 1946. Equations for a field of total flow induced by the wind in a non-homogeneous sea. *Dokl. Akad. Nauk S.S.S.R.*, 54(5).

STOMMEL, H., 1948a. The westward intensification of wind-driven ocean currents. *Trans. Am. Geophys. Union*, 29(2): 202–206.

STOMMEL, H., 1948b. The theory of the electric field induced in deep ocean currents. *J. Marine Res.*, 7(3).

STOMMEL, H., 1949. Horizontal diffusion due to oceanic turbulence. *J. Marine Res.*, 8: 199–225.

STOMMEL, H., 1950. An example of thermal convection. *Trans. Am. Geophys. Union*, 31(4): 553–554.

STOMMEL, H., 1952. An elementary explanation of why ocean currents are strongest in the west. *Bull. Am. Meteorol. Soc.*, 32: 21.

STOMMEL, H., 1954. Exploratory measurements of electrical potential differences between widely spaced points in the North Atlantic Ocean. *Arch. Meteorol., Geophys. Bioklimatol., Ser.A,* 7: 292–304.

STOMMEL, H., 1955. Discussion at the Woods Hole Convocation, June 1954. *J. Marine Res.,* 14: 504–510.

STOMMEL, H., 1956. On the determination of the depth of no meridional motion. *Deep-Sea Res.,* 3: 273–278.

STOMMEL, H., 1957a. Florida Straits transports, 1952–1956. *Bull. Marine Sci. Gulf Caribbean,* 7: 252–254.

STOMMEL, H., 1957b. A survey of ocean current theory. *Deep-Sea Res.,* 4: 149–184.

STOMMEL, H., 1958. *The Gulf Stream,* 1st ed. Univ. Calif. Press, Berkeley, Calif.

STOMMEL, H., 1960. Wind-drift near the equator. *Deep-Sea Res.,* 6: 298–302.

STOMMEL, H., 1965. *The Gulf Stream,* 2nd ed. Univ. Calif. Press, Berkeley, Calif.

STOMMEL, H. and ARONS, A. B., 1960a. On the abyssal circulation of the world ocean. I. Stationary planetary flow patterns on a sphere. *Deep-Sea Res.,* 6(2): 140–154.

STOMMEL, H. and ARONS, A. B., 1960b. On the abyssal circulation of the world ocean. II. An idealized model of the circulation pattern and amplitude in oceanic basins. *Deep-Sea Res.,* 6(3): 217–233.

STOMMEL, H. and VERONIS, G., 1957. Steady convective motion in a horizontal layer of fluid heated uniformly from above and cooled non-uniformly from below. *Tellus,* 9: 401–407.

STOMMEL, H. and WEBSTER, J., 1962. Some properties of thermocline equations in a subtropical gyre. *J. Marine Res.,* 20: 42–56.

STURM, M. and VOIGT, K., 1966. Observations on the structure of the Equatorial Undercurrent in the Gulf of Guinea in 1964. *J. Geophys. Res.,* 71(12): 3105–3108.

SUDA, K., 1936. On the dissipation of energy in the density current (2nd paper). *Geophys. Mag. (Tokyo),* 10: 131–243.

SVERDRUP, H. U., 1916a. Druckgradient, Wind und Reibung an der Erdoberfläche. *Ann. Hydrograph. Maritimen Meteorol.,* 44: 413–427.

SVERDRUP, H. U., 1916b. Über Mittelwerte von Vektorpaaren mit Anwendung auf meteorologische Aufgaben. *Meteorol. Z.,* 1917: 411–480.

SVERDRUP, H. U., 1926. Dynamics of tides on the North Siberian shelf. Results from the "Maud" Expedition. *Geofys. Publikasjoner, Norske Videnskapsakad. Oslo,* 4(5): 75 pp.

SVERDRUP, H. U., 1927. Two new recording current meters. *Hydrograph. Rev.,* 4.

SVERDRUP, H. U., 1928. Die Eisdrift im Weddelmeer. *Ann. Hydrograph. Maritimen Meteorol.,* 56: 265–274.

SVERDRUP, H. U., 1929. The waters on the North Siberian shelf. *Norweg. North Polar Expedition "Maud", 1918–1925, Sci. Results,* 4(2): 206 pp.

SVERDRUP, H. U., 1930. Some oceanographic results of the Carnegie's work in the Pacific—The Peruvian Current. *Trans. Am. Geophys. Union*, 11: 257–264.

SVERDRUP, H. U., 1932. Arbeider i luft- og havsforskning. *Chr. Michelsens Inst. Videnskap Aandsfrihet, Beretn.*, 2(5): 20 pp.

SVERDRUP, H. U., 1933. Vereinfachtes Verfahren zur Berechnung der Druck- und Massenverteilung im Meere. *Geofys. Publikasjoner, Norske Videnskapsakad. Oslo*, 10(1): 9 pp.

SVERDRUP, H. U., 1934a. The circulation of the Pacific. *Proc. Pacific Sci. Congr. Pacific Sci. Assoc.*, *5th, Toronto, 1933*, pp.2141–2145.

SVERDRUP, H. U., 1934b. Wie entsteht die Antarktische Konvergenz? *Ann. Hydrograph. Maritimen Meteorol.*, 62(8): 315–317.

SVERDRUP, H. U., 1938. On the process of upwelling. *J. Marine Res.*, 1: 155–164.

SVERDRUP, H. U., 1940a. Hydrology. *Brit. Australian New Zealandic Antarctic Res. Expedition 1929–1931, Ser.A*, 3(2): 88–126.

SVERDRUP, H. U., 1940b. Lateral mixing in the deep water of the south Atlantic Ocean. *J. Marine Res.*, 2(3): 195–207.

SVERDRUP, H. U., 1941. The influence of bottom topography on ocean currents. *Appl. Mechanics*, Th. von Kármán Ann. Vol.: 66–75.

SVERDRUP, H. U., 1947. Wind driven currents in a baroclinic ocean, with application to the equatorial currents of the Eastern Pacific. *Proc. Natl. Acad. Sci. U.S.*, 33(11): 318–326.

SVERDRUP, H. U. and DAHL, O., 1926. Two oceanographic current recorders designed and used on the "Maud" Expedition. *J. Opt. Soc. Am.*, 12:537-545.

SVERDRUP, H. U. and FLEMING, R. H., 1941. The waters off the coast of Southern California, March to July 1937. *Bull. Scripps Inst. Oceanog. Univ. Calif.*, 4(10): 261–378.

SVERDRUP, H. U., JOHNSON, M. W. and FLEMING, R. H., 1942. *The Oceans, their Physics, Chemistry and General Biology*, 1st ed. Prentice-Hall, Englewood Cliffs, N.J., 1087 pp.

SVERDRUP, H. U., JOHNSON, M. W. and FLEMING, R. H., 1946. *The Oceans, their Physics, Chemistry and General Biology*, 2nd ed. Prentice-Hall, Englewood Cliffs, N.J., 1087 pp.

SWALLOW, J. C., 1955. A neutral-buoyancy float for measuring deep currents. *Deep Sea Res.*, 3(1): 74–81.

SWALLOW, J. C., 1957. Some further deep current measurements using neutrally-buoyant floats. *Deep Sea Res.*, 4: 93–104.

SWALLOW, J. C., 1962. Ocean circulation. *Proc. Roy. Soc. (London), Ser.A*, 265: 326–328.

SWALLOW, J. C., 1964. Equatorial Undercurrent in the western Indian Ocean. *Nature*, 204: 436–437.

SWALLOW, J. C., 1965a. The Equatorial Undercurrent in the western Indian Ocean in 1964. *Intern. Conf. Trop. Oceanog., 1965, Miami Beach, Fla., Abstr.*

SWALLOW, J. C., 1965b. The Somali Current. Some observations made aboard R.R.S. "Discovery" during August 1964. *Marine Observer*, 35: 125–130.

SWALLOW, J. C. and HAMON, B. V., 1960. Some measurements of deep currents in the eastern North Atlantic. *Deep-Sea Res.*, 6: 155–168.

SWALLOW, J. C. and WORTHINGTON, L. W., 1957. Measurements of deep currents in the western North Atlantic. *Nature*, 179: 1183–1184.

SWALLOW, J. C. and WORTHINGTON, L. W., 1959. The deep countercurrent of the Gulf Stream off South Carolina. *Intern. Oceanog. Congr., New York, N.Y.*, pp.443–444.

SWALLOW, J. C. and WORTHINGTON, L. W., 1961. An observation of a deep countercurrent in the western North Atlantic. *Deep-Sea Res.*, 8: 1–19.

TAFT, B. A., 1965. Measurement of the Equatorial Undercurrent in the Indian Ocean 1962–1963. *Intern. Conf. Trop. Oceanog., 1965, Miami Beach, Fla., Abstr.*

TAFT, B. A. and KNAUSS, J. A., 1967. The equatorial under-current of the Indian Ocean as observed by the Lusiad Expedition. *Bull. Scripps Inst. Oceanog.*, 9: 163 pp.

TAKANO, K., 1955a. On the Antarctic Circumpolar Current. *Records Oceanog. Works Japan*, 2(1): 71–75.

TAKANO, K., 1955b. An example of thermal convective current. *Records Oceanog. Works Japan*, 2(1).

TAYLOR, G. I., 1915. Eddy motion in the atmosphere. *Phil. Trans. Roy. Soc. London, Ser.A*, 215: 1–26.

TAYLOR, G. I., 1918. Phenomena connected with turbulence in the lower atmosphere. *Proc. Roy. Soc. (London), Ser.A*, 94.

TAYLOR, G. I., 1922. Diffusion by continuous movements. *Proc. Math. Soc., Ser.2*, 20.

TAYLOR, G. I., 1931. Internal waves and turbulence in a fluid of variable density. *Rappt. Proces-Verbaux Réunions, Conseil Perm. Intern. Exploration Mer*, 76: 35–43.

THOMPSON, E. F., 1939a. Chemical and physical investigations. The general hydrography of the Red Sea. *John Murray Expedition 1933–1934, Sci. Rept.*, 2: 83–103.

THOMPSON, E. F., 1939b. Chemical and physical investigations. The exchange of water between the Red Sea and the Gulf of Aden over the "sill". *John Murray Expedition 1933–1934, Sci. Rept.*, 2: 105–119.

THORADE, H., 1909. Über die Kalifornische Meeresströmung. Oberflächentemperaturen und Strömungen an der Westküste Nordamerikas. *Ann. Hydrograph. Maritimen Meteorol.*, 1909: 17–34, 63–76.

THORADE, H., 1914a. Die Geschwindigkeit von Triftströmungen und die Ekman'sche Theorie. *Ann. Hydrograph. Maritimen Meteorol.*, 42: 379.

THORADE, H., 1914b. Die Geschwindigkeit der Triftströmungen. *Wiss. Beil., Realschule Eilbeck*, 1913–1914: 56 pp.

THORADE, H., 1931. Strömung und zungenförmige Ausbreitung des Wassers. *Gerlands Beitr. Geophys.*, 34: 57–76.

THORADE, H., 1933. Methoden zum Studium der Meeresströmungen. In: E. ABDERHALDEN (Herausgeber), *Handbuch der Biologischen Arbeitsmethoden, Abt.2*. Von Urban und Schwarzenberg, Berlin–Wien, 3: 2865–3095.

THORADE, H., 1941. Der äquatoriale Gegenstrom im Atlantischen Ozean und seine Entstehung. Nach neueren Arbeiten. *Ann. Hydrograph. Maritimen Meteorol.*, 69: 201–209.

THORNDIKE, E. M., 1962. A suspended drop current meter. *Deep-Sea Res.*, 10: 263–267.

TOMCZAK, G., 1964. Investigations with drift cards to determine the influence of the wind on surface currents. *Studies Oceanog., Hidaka Vol.*, pp.129–139.

TSUCHIYA, M., 1961. An oceanographic description of the equatorial current system of the Western Pacific. *Oceanog. Mag.*, 13(1): 1–30.

UDA, M., 1930. On some oceanographical researches of the sea water of Korisowo. *Records Oceanog. Works Japan*, 2(2).

UDA, M., 1949. On the correlated fluctuation of the Kuroshio current and the cold water mass. *Oceanog. Mag.*, 1: 1–12.

UDA, M., 1951. On the fluctuation of the main stream axis and its boundary line of Kuroshio. *J. Oceanog. Soc. Japan*, 6.

UDA, M., 1964. On the nature of the Kuroshio, its origin and meanders. *Studies Oceanog., Hidaka Vol.*, pp.89–107.

VERCELLI, F., 1925. Bilancio della Scambio di aqua fra Mar Rosso e Oceano Indiano. *Compendio Idrographico-Sci. Mar Rosso R.N. "Ammiraglio Magnaghi", 1923–1924—Ric. Oceanog. Fis.*, 1, *Correnti Mareo*, pp.178–183.

VERONIS, G., 1963. On inertially-controlled flow patterns in a β-plane ocean. *Tellus*, 15: 59–66.

VERONIS, G. and MORGAN, G. W., 1955. A study of the time-dependent wind-driven ocean circulation in a homogeneous, rectangular ocean. *Tellus*, 7: 232–242.

VERONIS, G. and STOMMEL, H., 1956. The action of variable wind stresses on a stratified ocean. *J. Marine Res.*, 15(1): 43–75.

VINE, A. C., KNAUSS, J. A. and VOLKMANN, G. H., 1954. Current studies of the Eastern Cayman Sea. *Woods Hole Oceanog. Inst., Ref.54–35*, 73 pp., unpublished.

VOIGT, K., 1961. Äquatoriale Unterströmung auch im Atlantik (Ergebnisse von Strömungsmessungen auf einer atlantischen Ankerstation der "Michail Lomonossov" am Äquator im Mai 1959). *Deut. Akad. Wiss. Berlin, Beitr. Meereskunde*, 1: 56–60.

VOLKMANN, G., 1962. Deep-current measurements in the western North Atlantic. *Deep-Sea Res.*, 9: 493–500.

VOLKMANN, G., 1963. Deep-current measurements using neutrally buoyant floats. In: M. N. HILL (General Editor), *The Sea*. Interscience, New York, N.Y., 2: 297–302.

VOLKMANN, G., KNAUSS, J. and VINE, A., 1956. The use of parachute drogues in the measurement of subsurface ocean currents. *Trans. Am. Geophys. Union*, 37(5): 573–577.

VON ARX, W. S., 1950. Some current meters designed for suspension from an anchored ship. *J. Marine Res.*, 9(2): 93–99.

Von Arx, W. S., 1950. An electromagnetic method for measuring the velocities of ocean currents from a ship under way. *Papers Phys. Oceanog. Meteorol.*, 11: 1–62.
Von Arx, W. S., 1952a. A laboratory study of the wind-driven ocean circulation. *Tellus*, 4(4): 311–318.
Von Arx, W. S., 1952b. Notes on the surface velocity profile and horizontal shear across the width of the Gulf Stream. *Tellus*, 4: 211–214.
Von Arx, W. S., 1962. *Introduction to Physical Oceanography.* Addison-Wesley, Reading, Mass, 422 pp.
Von Arx, W. S., Bumpus, D. F. and Richardson, W. S., 1955. On the fine structure of the Gulf Stream front. *Deep-Sea Res.*, 3: 46–65.
Von Helmholtz, H., 1888. Über atmosphärische Bewegungen. *Sitzber. Preuss. Akad. Wiss., Berlin, Math. Physik. Kl.*, p.647.
Von Humboldt., A., 1814. *Reisen in die Äquatorial-Gegenden des Neuen Kontinents.* Also in: H. M. Williams, 1822. *Personal Narrative of Travels to the Equatorial Regions of the New Continent*, 3rd ed. Longman, Hurst Rees, Orme and Brown, London.
Von Kármán, Th., 1930. Mechanische Ähnlichkeit und Turbulenz. *Nachr. Ges. Wiss. Göttingen, Math. Phys. Kl.*, 1930: 58.
Von Schleinitz, Freiherr, 1889. Die Forschungsreise S.M.S. "Gazelle" in den Jahren 1874 bis 1876. *Hydrograph. Amt Reichsmarineamtes, Reiseber.*, 1.

Warren, B. A., 1963. Topographic influences on the path of the Gulf Stream. *Tellus*, 15(2): 167–183.
Warren, B. A., 1965. The Somali Current. *Oceanus*, 12(1): 2–7.
Webster, F., 1951. A description of Gulf Stream meanders off Onslow Bay. *Deep-Sea Res.*, 8(2): 130–143.
Webster, F., 1963. A preliminary analysis of some Richardson current meter records. *Deep-Sea Res.*, 10: 389–396.
Webster, F., 1964. Some perils of measurement from moored ocean buoys. *Trans. Buoy Technol. Symp., 1964*, pp.33–48.
Webster, F., 1967. On the representativeness of direct deep-sea current measurements. *Trans. S.C.O.R. Symp. Oceanic Variability, Rome, 1966, Preprint*, 20 pp.
Welander, P., 1957. Wind action on a shallow sea. Some generalizations of Ekman's theory. *Tellus*, 9(1): 45–52.
Welander, P., 1959a. On the vertically integrated mass transport in the oceans. In: B. Bolin (Editor), *The Atmosphere and the Sea in Motion—Rossby Memorial Volume.* Rockefeller Inst. Press/Oxford Univ. Press, New York, N.Y.
Welander, P., 1959b. An advective model of the ocean thermocline. *Tellus*, 11: 309–318.
Werenskjöld, W., 1922. Mean monthly air transport over the North Pacific Ocean. *Geofys. Publikasjoner, Norske Videnskapsakad. Oslo*, 2(1): 55 pp.
Werenskjöld, W., 1937. Die Berechnung von Meeresströmungen. *Ann. Hydrograph. Maritimen Meteorol.*, 65: 68.

WERTHEIM, G. K., 1954. Studies of the electrical potential between Key West, Florida and Havana, Cuba. *Trans. Am. Geophys. Union*, 35: 872–882.
WILLIAMS, R. G., 1966. An investigation of the intermediate salinity maximum in the equatorial Atlantic during Equalant I. *N.Y. Univ., Dept. Meteorol. Oceanog., Res. Div., Rept.*, GSL-TR-66-4.
WITTE, E., 1878. Über Meeresströmungen, *1.* Pless, 45 pp.
WITTE, E., 1879. Über Meeresströmungen, 2. Die Strömungen der einzelnen Oceane. In: *Sechstes Programm der Fürstenschule zu Pless, Schlesien* Krummer, Pless, pp.27–47.
WITTING, R., 1923. Om en till skeppsbord reporterande strömmätare. In: *Festschrift für Otto Pettersson*. Helsingfors, pp.90–96.
WITTING, R., 1930. Current measurements, direct and indirect. 1. *Rappt. Proces-Verbaux Réunions, Conseil Perm. Intern. Exploration Mer*, 64: 8–18.
WOOSTER, W. S., 1960. El Niño. *Calif. Coop. Oceanog. Fisheries Inst. Rept.*, 7.
WOOSTER, W. S., 1961. Further evidence of a Pacific south equatorial countercurrent. *Deep-Sea Res.*, 8(3/4): 294–297.
WOOSTER, W. S. and CROMWELL, T., 1958. An oceanographic description of the eastern tropical Pacific. *Bull. Scripps Inst. Oceanog. Univ. Calif.*, 7: 169–282.
WOOSTER, W. S. and GILMARTIN, M., 1961. The Peru–Chile Undercurrent. *J. Marine Res.*, 19(3): 97–112.
WOOSTER, W. S. and JENNINGS, F., 1955. Exploratory oceanographic observations in the eastern tropical Pacific, January to March 1953. *Calif. Fish Game*, 41: 79–90.
WOOSTER, W. S. and REID, J. L., 1963. Eastern boundary currents. In: M. N. HILL (General Editor), *The Sea*. Interscience, New York, N.Y., 2: 253–280.
WORTHINGTON, L. V., 1954. Three detailed cross-sections of the Gulf Stream. *Tellus*, 6: 116–123.
WORTHINGTON, L. V., 1965. On the shape of the Gulf Stream system. *Trans. Am. Geophys. Union*, 46(1): 99.
WÜST, G., 1924. Florida- und Antillenstrom. *Veröffentl. Inst. Meereskunde, Univ. Berlin, Reihe A*, 12: 48 pp.
WÜST, G., 1928. Der Ursprung der Atlantischen Tiefenwasser. *Z. Ges. Erdkunde Berlin, Jubiläum Sonderband*, pp.506–534.
WÜST, G., 1936a. Kuroshio und Golfstrom. *Veröffentl. Inst. Meereskunde, Univ. Berlin, Geograph. Naturwiss. Reihe*, 29: 69 pp.
WÜST, G., 1936b. Schichtung und Zirkulation des Atlantischen Ozeans. "*Meteor*"-*Werk. 6. Teil I: Das Bodenwasser und die Stratosphär*, 288 pp.
WÜST, G., 1938. Bodentemperatur und Bodenstrom in der Atlantischen, Indischen und Pazifischen Tiefsee. *Gerlands Beitr. Geophys.*, 54: 1–8.
WÜST, G., 1941. Relief und Bodenwasser im Nordpolarbecken. *Z. Ges. Erdkunde Berlin*, 5/6.
WÜST, G., 1943. Der subarktische Bodenstrom in der westatlantischen Mulde. *Ann. Hydrograph. Maritimen Meteorol.*, 71: 249–255.
WÜST, G., 1949. Die Kreisläufe der atlantischen Wassermassen, ein neuer Versuch räumlicher Darstellung. *Forsch. Fortschr.*, 25(23/24): 285–289.

WÜST, G., 1950. Blockdiagramme der Atlantischen Zirkulation auf Grund der "Meteor"-Ergebnisse. *Kieler Meeresforsch.*, 7: 24–34.

WÜST, G., 1951. Der Wasserhaushalt des Mittelländischen Meeres und der Ostsee in vergleichender Betrachtung. *Geofis. Pura Appl.*, 21.

WÜST, G., 1955. Stromgeschwindigkeiten im Tiefen- und Bodenwasser des Atlantischen Ozeans. In: *Papers in Marine Biology and Oceanography— Deep-Sea Res. (Bigelow Vol.)*, pp.373–397.

WÜST, G., 1957. Stromgeschwindigkeiten und Strommengen in den Tiefen des Atlantischen Ozeans. *"Meteor"-Werk*, 6(2): 261–420.

WÜST, G., 1961. On the vertical circulation of the Mediterranean Sea. *J. Geophys. Res.*, 66(10): 3261–3271.

WÜST, G. und DEFANT, A., 1936. Atlas: Schichtung und Zirkulation des Atlantischen Ozeans. Schnitte und Karten von Temperatur, Salzgehalt und Dichte. *"Meteor"-Werk, 6. Teil A–B: Stratosphäre. Teil C: Troposphäre.*

WYRTKI, K., 1957. Die Zirkulation an der Oberfläche der südostasiatischen Gewässer. *Deut. Hydrograph. Z.*, 10: 1–13.

WYRTKI, K., 1958. The water exchange between the Pacific and the Indian Ocean in relation to upwelling processes. *Proc. Pacific Sci. Congr. Pacific Sci. Assoc., 9th, 1957*, 16: 61–66.

WYRTKI, K., 1960. The Antarctic Circumpolar Current and the Antarctic Polar front. *Deut. Hydrograph. Z.*, 13(4): 153–174.

WYRTKI, K., 1963. The horizontal and vertical field of motion in the Peru Current. *Bull. Scripps Inst. Oceanog. Univ. Calif.*, 8(4): 313–346.

YOSHIDA, K., 1958a. Coastal upwelling, coastal currents and their variations. *Records Oceanog. Works Japan, Spec. No.*, 2: 85–87.

YOSHIDA, K., 1958b. A study on upwelling. *Geophys. Notes*, 11(13).

YOSHIDA, K., 1959. A theory of the Cromwell Current and of the equatorial upwelling—an interpretation in a similarity to a coastal circulation. *J. Oceanog. Soc. Japan*, 15(4): 159–170.

YOSHIDA, K. and MAO, M. L., 1957. A theory of upwelling of large horizontal extent. *J. Marine Res.*, 16: 40–54.

YOSHIDA, K., MAO, H. O. and HOOVER, P. L., 1953. Circulation in the upper mixed layer of the equatorial North Pacific. *J. Marine Res.*, 12(1): 99–120.

YOUNG, F. B., GERRARD, H. and JEVONS, W., 1920. On electrical disturbances due to tides and waves. *Phil. Mag.*, 40(235): 149–159.

ZÖPPRITZ, K., 1878. Hydrodynamische Probleme in Beziehung zur Theorie der Meeresströmungen. *Wied. Ann.*, 3: 582.

INDEX

Absolute current measurements, 4
— — velocities, 38, 140–144
— dynamic topographies, 145–148
— field of pressure, 82
— motion with respect to current meter, 2
— positioning, 2, 10, 25
— pressure differences, 38, 140–141
— vorticity, 112
Abyssal circulation, 46, 48, 303–309
— sediments, 218, 309
Accelerated currents, 91, 145, 149–161, 166
Acceleration of gravity, 81
—, geostrophic, 91
Advection and diffusion, 174–178
— general, 79, 104, 115, 116, 123, 124
— of vorticity, 284
Agulhas Current — *see* currents (geographical)
Air–sea interactions, 44, 45, 61, 78, 79, 197, 222, 271, 290 — *see also* momentum transfer, wind stress, heat transfer, evaporation, precipitation
Air–sea interface, 178, 263, 271
Aliasing, 8, 9, 49, 50
"Altair" — *see* ships
"Altair"-Cone, 210, 211
Analysis of currents (general), 2, 7–9
—, isentropic, 47, 48
Anchored buoys, 3–6, 49
Anchoring gear, 4–6
Anchor stations (general), 3–6
Anemogenic curl effect, 231
Angular velocity of earth rotation, 92
Anisobaric transports, 193–196

Anomalies of dynamic depth, 88, 138–140, 142, 143
— — specific volume, 87, 88, 130, 133, 138
Antarctic bottom water or currents — *see* water masses and currents
— Circumpolar Current — *see* currents (geographical)
— — — dynamic explanation, 299–301
— — — volume transport, 252, 256, 298–301
— convergence, 46, 301, 302, 303, 305
— divergence, 303
— intermediate water — *see* water masses (geographical)
— "Ob"-expedition, 61
— Ocean, 46, 48, 298–303
— Polar Current, 303
Anticyclonic gyres, 201, 208, 209, 268, 271–273, 278
— motion, 151, 153–156, 160, 161, 163
Apparent accelerations, 91–93
Arabian Gulf, 276
"Armauer-Hansen" — *see* ships
Atlantic troughs, 304
"Atlantis" — *see* ships
Average current, 51–54

Bab el Mandeb, 79, 306
Baltic Sea, 118, 153, 154, 167, 179
Bar (unit of pressure), 81
Baroclinic field, 85, 143, 218, 221
Baroclinicity of western boundary currents, 292
Barotropic field, 85
Bathypitometer, 20

Bathyscaph, 20
Benguela Current — see currents (geographical)
"Beta" effect, 229
Bifilar suspension of current meters, 14
"Blake" — see ships
Blake Plateau, 260
"Blanks" at sea surface, 59
Böhnecke current meter, 7, 17
Bottom, critical slope according to Ekman, 236
— current meters, 22, 33, 34
— currents, 193–197, 303–309
— layer of frictional influence, 193–197
— photographs, 218, 309
— topography effects, 168, 227, 231–235, 260, 261, 292
— water — see water masses (geographical)
— water formation, 304, 305
Boundary conditions (general), 89, 110
Boundary currents, 75, 78, 258, 259, 264, 272–275, 284–298
Boundary currents and bottom topography, 232–234
Bouvet divergence, 303
Brazil Current — see currents (geographical)
Breaking of waves and stirring, 96, 97
British–Australian–New Zealand Expedition, 305
"Buccaneer" — see ships
Buoys, anchored, 3–6, 22–24, 49
—, radio, 22, 27

Cabot, multiple ship operation, 12, 295, 296
California Current — see currents (geographical)
Canary Current — see currents (geographical)
Cape Hatteras, 291, 292, 298
Carbon-14, 48, 177

Centrifugal forces, 152, 155, 160, 162, 163, 207–209
"Challenger" — see ships
Charts, current, 2, 10, 12, 13, 49, 52–74
—, dynamic, 85, 145–149
—, isobaric, 86
—, pilot charts, 59
—, transport— see transport
Circle of inertia, 151–153, 163
Circulation acceleration, 107
—, deep sea, 45–48, 261–271, 303–309
—, defined, 105–106
—, general, of oceans, 227–271
—, gyres, 75, 162, 201, 208, 268, 271–273, 285
—, thermohaline, 218, 222–225, 238, 260, 261–271
—, wind-driven, 78, 178–179, 217–222, 238–261, 264, 266–271, 284
Circular motion, 155–163
Circumpolar Water (Antarctic) — see water masses (geographical)
Classical method of dynamic computation, 37–38, 135–149, 263
Classical theory of turbulent diffusion, 120–125
Coastal countercurrents, 287, 295, 298
Coastal upwelling, 56, 78, 288, 289
Coefficient of turbulent mixing, 103, 120, 122, 124, 218, 220, 266
Common Water, 307
Concentration(s) and mixing, 120–125
—, defined, 115
—, local changes, 123
— of dye in experiments, 125, 186
— of sea water properties, 115, 116
Conditioning of water masses, 78, 79, 197, 263
Conservative properties of sea water, 44, 115, 122, 124, 174
Continental margin, 232, 260, 304
— slope, 288, 289, 292, 304

INDEX

Continuity and vertical motion, 112–115
— equation as prediction equation, 114
—, equation of, 88, 103–105
— of incompressible flow, 105
— of salt, 115, 116
Contra solem, defined, 160
Convergence, antarctic 46, 301–303
— in coastal regions, 289
—, hydrodynamic definition, 55, 104
—, lines of, 56, 57–59, 73
— of wind-driven currents, 195, 196, 217, 278, 279
—, subtropical, 69, 70, 302, 303
—, tropical, 303
Cooling and ice formation, 303, 304
Coordinate systems, 89, 159, 241, 251
Core depth, 44
— of high salinity in Equatorial Undercurrent, 281–283
— method, 38, 45–47, 174, 176, 263, 304
Coriolis forces, 92–95
— parameter, variation of, 111, 229, 242, 245–259
Correlation of vectors, 54–55
Countercurrent(s) — *see also* countercurrents, under special names
— east and west of Gulf Stream, 298
— near west coast of continents, 287, 288
— under Gulf Stream, 143, 295
Critical bottom slope (Ekman), 236
Cromwell Current — *see* currents (geographical)
Cum sole, defined, 160
Curl, defined, 108
— of deep current (Ekman), 229–231
— of wind stress, 238, 242–251, 267, 268, 270
Current(s), absolute, 38, 140–144
— and mass distribution, 37, 138, 198–211, 266–271

Current(s) *(continued)*
—, bottom topography effects on, 168, 227, 231–235, 241, 260, 261
—, boundary, 75, 78, 258, 259, 264, 272–275, 284–298
—, frictionless, 91, 127, 199–211
— (geographical), Agulhas C., 61, 75, 144, 272, 284, 291, 299
— —, Alaskan C., 272
— —, Aleutian C., 272
— —, Antarctic Bottom C., 48, 304, 307, 308
— —, Antarctic Circumpolar C., 46, 258, 263, 272, 298–303, 306
— —, Antarctic Intermediate C., 46, 302, 305
— —, Benguela C., 78, 274, 285, 288, 289
— —, Brazil C., 258, 264, 274, 285
— —, California C., 78, 286
— —, Canary C., 78, 257, 288
— —, countercurrents — *see* under special names
— —, countercurrent under Gulf Stream, 143, 295
— —, Cromwell C., 280–281
— —, Davidson C., 26, 288
— —, East Australian C., 75, 274, 285, 287
— —, East Greenland C., 272
— —, El Niño C., 290
— —, equatorial countercurrents, 76, 77, 206, 275–280, 290
— —, Florida C., 33, 257, 291
— —, Guiana, C., 252
— —, Guinea C., 258, 277
— —, Gulf Stream, 12, 31, 36, 75, 78, 130–133, 143, 204, 232, 256, 259, 260, 264, 272, 274, 284, 291–298.
— —, Irminger C., 272
— —, Kamchatka C., 272
— —, Kuroshio, 12, 75, 78, 143, 272, 284, 285, 290–292, 298
— —, Labrador C., 252, 272
— —, Madagaskar C., 273

Currents (geographical) *(continued)*
— —, Mozambique C., 75, 272, 284, 291
— —, North Atlantic C., 78, 252, 260, 272, 291
— —, North Atlantic Equatorial Countercurrent — *see* equatorial countercurrents
— —, North Equatorial C., 76, 206, 275, 277–280, 290
— —, North Pacific C., 78, 285, 292
— —, North Pacific Equatorial Countercurrent — *see* equatorial countercurrents
— —, Oya Shio, 272
— —, Peru C., 78, 274, 287, 288
— —, Peru Countercurrent, 287
— —, Peru Undercurrent, 288
— —, Somali C., 75, 77, 272, 277, 284, 291
— —, South Equatorial C., 76, 206, 275, 277–280, 290
— —, South Pacific Equatorial Countercurrent, 277, 290
— —, Tsushima C., 292
— —, undercurrents — *see* equatorial
— —, West Greenland C., 272
—, geostrophic, 127–145
— gradient, 158–162
— impulse, 104
— inertia, 7, 149–155, 156, 157, 161, 166, 188
—, mean, 1, 7, 13, 51–54, 59, 61–65, 68, 73, 74
—, meandering, 155, 292, 295–298
— measurements — *see* also current meters
— —, absolute, 4, 6
— —, bottom, 18, 20, 33
— —, direct, 9–34
— —, electromagnetic methods, 34–36
— —, Eulerian, 9, 13, 89
— —, evaluation of, 49–55
— —, general, 1–9

Current measurements *(continued)*
— —, indirect methods, 34–48
— —, Lagrangian, 9, 24, 89
— —, representation of, 1, 13, 49, 55, 59–73
— —, vector summation of, 50, 51
— meter(s), anchored, 3–6
— —, Bathypitometer, 20, 21
— —, BBT-Neypric currentograph, 20
— —, bifilar suspension, 14, 15
— —, Böhnecke, 7, 17
— —, Carruthers (various designs), 9, 17, 21, 22
— —, in connection with bottom photography, 18, 20
— —, D.H.I.-Bifilar, 22
— —, D.H.I. paddle wheel, 17, 24
— —, Doodson, 22
— —, Doppler current meter, 21
— —, Ekman, 15–17
— —, Ekman-E.M.W. serial, 17
— —, Ekman-Merz, 17
— —, Geomagnetic Electro-Kinetograph (GEK), 35, 36
— —, Idrac, 18
— —, moored, 4–6
— —, Nylon yarn inclinometer, 20
— —, Pegram, 22
— —, Pettersson, 18
— —, Rauschelbach, 22
— —, recording of direction, 14, 15
— —, records, analysis of, 6–9, 49–54
— —, Richardson, 4, 5, 18, 19
— —, Roberts, 22–24
— —, Sverdrup, 22
— —, telerecording and telemetering, 21–24
— —, ultrasonic devices, 21
— —, Witting, 22
—, non-divergent, 104, 105, 108, 111
—, relative, 38, 138, 139–144
—, representation of, 10–12, 49–52, 55–58, 60–73
—, residual, 1

Current(s) *(continued)*
—, resultant, 2, 52–54
—, rip, 59, 162
— roses, 61–63, 65
—, scalar mean, 1, 52–54
—, seasonal variations, 74–78, 272, 275–279, 299
—, singularities, 55–61, 65
—, stability of, 53, 54, 61, 64, 66–68, 73
—, stationary, 91, 127
—, thermohaline, 79, 218, 222–225, 238, 261–271
—, tidal — *see* tidal currents
—, variability of, 1, 7–9, 49–52, 61
—, vectorial mean, 1, 52–54
— vectors, 50–54, 65
—, vertical component of, 55–58, 112, 288–290
—, wind-driven, 78, 178–197, 227–231, 238–261, 265–271, 278, 284, 298, 299
—, world map of, 69–72
Curvature of streamlines, 155–163
Cyclonic motion, 158–163, 208–211, 268, 271–275 — *see also* gyres
Cyclostrophic motion, 162, 163

Damped inertia currents, 154, 166, 188
Davidson Current — *see* currents (geographical)
Dead reckoning, 12
Decay of a current under friction, 166, 167
Decca, 2
Decibar, defined, 81
Deep current (Ekman), curl of, 229
— sea anchoring, 3, 4, 6
— — currents, 20, 34, 261, 263, 303–309 — *see also* water masses
— — inertia currents, 156, 157
— — laminated stratification, 79, 304
— — mooring, 4
— — sediment structures, 218, 309

Deep *(continued)*
— water masses, 303–309 — *see also* water masses (geographical)
Defant's method of finding layer of no horizontal motion, 141–143, 293
Deflection angle between wind and drift currents, 183–185, 189, 190, 197
Density discontinuity, 197, 198, 201–209
— distribution (vertical), 197, 198, 200, 201, 204–211, 271, 292
— in situ, 82
— of sea water, 37, 39, 47, 81, 82, 85
Depth of frictional influence, 183–185, 189, 193–197, 228, 278
Deuterium, 48
Differential heating and cooling, 78, 79, 263, 266–268, 303–305
Diffusion and advection, 40, 44, 47, 79, 94, 116, 120–125, 164, 174–178, 263 — *see also* mixing and spreading
— and the law of parallel fields, 168–173
— coefficients, 120, 122–125, 174–177
— of dye pattern, 125, 186, 187
— equation, 123–125, 174
—, molecular, 95, 122
Direct current measurements, 9–34
"Discovery II" — *see* ships
Dispersion (turbulent), 120–125, 186, 187 — *see also* diffusion
Dissipative forces and inertia currents, 154, 167
Divergence, Antarctic, 303
—, Bouvet, 303
—, hydrodynamic definition, 104
— in coastal regions, 56, 288, 289
—, lines of, 55–59, 73
— of deep current (Ekman), 229
— of wind driven currents, 195, 196, 241, 244, 278
—, tropical, 278, 279

Doldrums belt, 75–77, 275, 276, 278, 280
Doppler current meter, 21
Doppler shift, 2, 21
Drag coefficient, 179
Drake Passage, 256, 298, 309
Drift, bottles and cards, 25, 26
— measurements, 24–34
— of ships, 2, 6, 12, 13, 25
Drogues, 27–30
Dropsonde, 32–33, 37
Dye concentrations, 125, 186, 187
Dye patches, 26, 186, 187
Dynamic charts, 86, 87, 145–148
— computation, 37, 38, 138–149, 263, 308
— depth, 83, 87, 88, 138–147
— isobaths, 86
— meter, 83, 138–140

Earth's rotation, angular velocity, 92
East Australian Current — see currents (geographical)
Eastern Atlantic Trough, 304
— boundary currents, 78, 273–275, 284, 287–290 — see also boundary currents and currents (geographical)
Eddies in Gulf Stream System, 295–298
Eddy diffusion, 120–125, 169, 174–178, 263, 266
— heat conduction, 120, 122, 266
— viscosity, 95–103, 182, 184, 239, 301
Ekman current meter, 15–17
— layer, 279 — see also layer of frictional influence
— spiral, 184, 185, 196, 228
Ekman's differential equation for currents in a homogeneous ocean, 228–238
— pure drift currents, 180–190, 195–197
— relative currents, 212–216

Electromagnetic methods of current measurement, 34–36
Electronic computing techniques, 227, 251, 266–271
Elementary current system, 195–197, 212–216, 228
El Niño Current — see currents (geographical)
English Channel, 134
Equalant expeditions, 4, 27, 29, 31, 277, 280
Équation(s), Lagrangian and Eulerian, 89
— of continuity, 88, 103–105, 110, 112–119
— of motion, 88–95
Equatorial countercurrents — see currents (geographical)
— currents, 74–77, 205–207, 275–284 — see also currents (geographical)
— undercurrents, 4, 27, 30, 94, 199, 275, 280–284
Equilibrium, geostrophic, 128, 135
—, hydrostatic, 81, 84, 94
—, quasi-static, 85
Equivalent-barotropic flow, 143, 294
Estimates of bottom current speeds, 309
Eulerian equations, 89, 90
— methods, 9–10, 13
Evaporation, 45, 46, 78, 117, 119, 197, 218, 222, 224, 263, 299, 305
Exchange of momentum, 120 — see also wind stress and eddy viscosity

Fetch of wind action, 178
Field(s) acceleration, 91
—, baroclinic, 85, 143, 218
—, barotropic, 85
— of currents, 55, 56, 58, 65, 105, 217
— of density, 37, 81, 82, 148, 216
— of mass, 37, 82, 85, 137
— of pressure, 37, 81–88, 137, 148, 149, 217

INDEX

Forces *(continued)*
Fixed point for current measurements, 2, 3, 6, 14, 25
Floats, neutrally buoyant, 7, 29, 31, 155–157
—, surface, 11, 12, 25
Florida Current — *see* currents (geographical)
Fluid line, 105, 106
Forces — *see also* Guldberg-Mohn friction, momentum exchange, Reynolds stresses, friction
—, Coriolis, 92–94
—, frictional, 95–103
—, gravitational, 81, 83
—, pressure gradient, 84–87
Foucault's pendulum, 153, 188
Friction, 95, 96, 101–103, 145, 154, 164–171, 178, 180–182, 189, 192, 194, 195–197, 212–216, 218–222, 227, 228, 234, 235, 238, 240, 245–251, 266, 300
— and law of parallel fields, 168–173
—, non-isotropic, 227
Frictional influence, bottom layer of, 194–197
— —, surface layer of, 184–186, 189, 194
Fronts, oceanic, 57, 58, 69–72, 201–205, 292, 295, 297, 301–302

"Gazelle" — *see* ships
General ocean circulation
— — —, horizontal wind-driven, 227–231, 238–261
— — —, numerical solutions, 224, 244, 245, 250, 251–258, 266–271
— — — of deep sea, 303–309
— — — of surface waters, 59–79, 271–275
— — —, thermohaline, 222–225, 261–271
Geomagnetic Electro-Kinetograph (GEK), 35, 36, 294, 296
Geopotential, 83
Geostrophic acceleration, 91
— equilibrium, 128, 135, 198–206

Gibraltar — *see* Strait of Gibraltar
Gradient currents, 156–162, 198, 207–212, 220
Grand Banks, 232–234, 291, 292, 305
Gravitational potential, 83
Gravity acceleration, 81
—, specific, 82
Great Belt, 132–134
Greenland–Iceland Ridge, 305
Ground tackle, 4, 5
Guinea Current — *see* currents (geographical)
Guldberg-Mohn friction, 103, 164, 166, 167, 228, 245, 300
Gulf of Guinea, 76, 277, 283
Gulf Stream — *see also* currents (geographical)
— —, countercurrents east and west of, 250, 298
— —, countercurrent under, 143, 295
— — eddies, 291, 295–298
— — inertial models, 259
— —, lower boundary, 142–144, 260, 292–295
— —, main axis of, 12, 293, 297
"Gulf Stream '60", 295
Gulf Stream sections, 130–133, 293
Gyres, anticyclonic, 151, 153–156, 160, 161, 163, 201, 208, 209, 268, 271–273, 278
—, cyclonic, 160–162, 209–211, 249, 268, 271, 272
—, subpolar, 268, 271–273
—, subtropical, 201, 248, 249, 256, 258, 268, 270, 285

Halskov-Rev Lightship, 132, 134
Heat conduction in turbulent flow, 120, 122
Homogeneous ocean, 137, 150, 182–190, 192–197
— — and Ekman's currents, 182–190, 192–197, 228–236, 238
Horse-latitudes, 76
Hot-wire devices in current meters, 14

HUMBOLDT, A. VON, 61
Humboldt Current — *see* currents (geographical), Peru Current
Hydrodynamic roughness of sea surface, 178
Hydrostatic equation, 37, 81, 84, 85, 94

Iceland–Färoer Ridge, 304
Incompressible fluids, 105
Indian Ocean,
— — Antarctic Intermediate Water, 306, 307
— — Countercurrent, 77, 278
— — Deep Water, 48, 306
— —, Monsoon currents, 77, 272, 275, 276, 279
— —, Red Sea Water, 306
— —, surface currents, 75, 276, 278, 279
Indirect methods of current measurement, 34–38
Inertia circle, 151, 156, 163, 167
— currents, 7, 149–157, 161, 167, 188, 211
— period, 152–155, 211
Inertial models of western boundary currents, 259
Initial conditions, 89, 188, 219, 266, 268
Intermediate Water — *see* water masses (geographical)
Internal drift current, 191, 215
Internal waves, 7, 149
International Cooperative Investigations of the Tropical Atlantic (Equalant), 4, 27, 29, 30, 258, 280
— Gulf Stream Survey, 209
— Ice Patrol Service, 233, 234
— Indian Ocean Expedition (IIOE), 77, 279–281, 291
Irish Sea, 118
Irrotational motion, 109, 110
Isentropic analysis, 47, 48
Isobaric charts, 86
— surfaces, 83, 85, 86, 129, 138–140, 142, 145–148, 201, 202

Isobars, curved, 156–162
Isogons (isogonic lines), 65, 68
Isopycnals — *see also* density
— and streamlines, 168–173
—, inclination of, 129, 132, 200, 201, 208, 209
—, slope in Gulf Stream, 130, 132, 204
Isosteric surfaces, 85
Isotachs, 65
Isotherms in North Atlantic computed, 267–271
Isotopes, radioactive, as tracers, 48
Isotopic–salinity relationships, 48
Isotropic turbulence, 121

Kelvin's theorem, 107
Kinematic boundary conditions (general), 110
— eddy diffusion coefficients, 123, 174, 175
— relationships in a current field, 105–110
Knot (unit of speed), 9
Knudsen relationships, 115–119
Kuroshio — *see* currents (geographical)

Lagrangian equations, 89
— methods, 9–11, 25–34
Laminar flow, 95
Laminated stratification of deep sea, 79, 263
Lateral mixing, 45, 47, 121, 125, 177, 227, 266
— shearing stresses, 102, 103, 171, 218, 221, 238–240, 242, 244, 247, 299, 300
— stress vorticity, 238, 247–250, 266, 284
Law of parallel fields, 168
Layer of frictional influence, 184–186, 189, 193–197, 213–215, 228–230, 243
— of no absolute motion, 141

INDEX

Layer *(continued)*
— of no horizontal motion, 136, 137, 140, 141, 143, 145, 148, 217, 293
Lenz's model of thermal circulation, 223
Level surfaces, 83
Leveling, precise, 131, 133, 135, 148
Local changes of concentration, 123–125
Logarithmic spiral (Ekman), 184, 185, 194, 196
Lorac, 25
Loran, 2
Loran-C, 25
Lower Deep Water — *see* water masses (geographical)

Magnetic field, geomagnetic, 34, 35
— — of ship, 14
Margules equation, 201, 204, 205
Mass, field of, 37, 82
— stratification in geostrophic and gradient currents, 198–211
— transport by currents — *see* transport
— turbulent-exchange, 121
MAURY, M. F., 60, 73
Mean arithmetic, 53
— current, 1, 7, 51
— scalar and vectorial, 1, 52–55, 65
Meandering currents, 7, 155, 292, 297, 298
Mean free path, 99
Measurement(s), absolute, 6
— drift, 23–34
— Eulerian, 9, 10, 13, 89, 91
— Lagrangian, 9–12, 24–31, 89
— of currents (in general), 1–9
Mediterranean Sea, 46, 119, 263, 270, 302, 303, 305, 307
— Water — *see* water masses (geographical)
Meridional exchange of water masses, 304–308

"Meteor"-Atlas, 267, 270
Mid-Atlantic Ridge, 260, 304
Middle Deep Water — *see* water masses (geographical)
Mid-water current (Ekman), 194
Mixing coefficients, 120, 122, 174–177
— and diffusion, 79, 120–125, 168–177
—, general, 40–46, 120–125
—, lateral, 45, 47, 121–125, 177, 227
—, length, 99, 101, 121
—, non-isotropic, 121
— of water types, 38–47, 120, 174, 177, 263, 304, 306
—, turbulent, 120–125, 174–178
—, vertical, 45, 47, 121, 124, 174–177
Molecular diffusion, 95, 120, 122
— viscosity, 91, 95, 96, 101
Momentum exchange, 95, 120, 178
— transfer by wind, 178, 251, 263
— *see also* wind stress
Monsoon currents, 77, 275, 276, 278, 279
— winds, 75, 77, 272, 276, 278, 279
Moorings, 4, 5
Motion(s), absolute, computed, 38, 140–149
—, absolute with respect to current meter, 2, 4
—, anticyclonic and cyclonic, 160–162 — *see also* gyres
—, circular, 155–163
—, cyclostrophic, 162, 163
—, equations of, 88–103
—, frictionless, 127–163
—, irrotational, 109
—, layer of no absolute, 141
—, layer of no horizontal, 136, 137, 140, 141, 143, 145, 148, 217, 239
—, net, 1, 52
—, non-accelerated, 91, 127
—, relative, 38, 138–149
—, relative with respect to current meter, 2, 6

Motion *(continued)*
—, ship, 2, 6, 7
—, stray, 2, 7
—, tidal, 1, 7, 8, 52, 99, 115, 153, 155
Motionless layer, 136, 137, 145, 217 – *see also* layer
Mozambique Current — *see* currents (geographical)
Multiple Ship Operation Cabot — *see* Cabot, multiple ship operation

Navier-Stokes equations, 96
Navigational aids, astronomical, 2, 12, 29
— —, electronic, 2, 25
— —, satellites, 2
Net motion, 1, 52
Neutral point, 56, 57
Neutrally buoyant floats, 7, 29–32, 155–157, 294, 309
Newton's first law of mechanics, 88
— formulation of stress, 96
Noise level (aliasing), 9, 51
Non-conservative properties of sea water, 115
Non-divergent currents, 104, 105, 108, 109, 111
Non-homogeneous ocean, 168, 197, 198, 212, 216
Non-isotropic mixing, 121
North Atlantic Current — *see* currents (geographical)
— — Deep Water — *see* water masses (geographical)
North Equatorial Current — *see* currents (geographical)
North Pacific Current — *see* currents (geographical)
Norwegian North Polar Expedition, 180
Norwegian Sea, 168
Numerical solutions, 227, 244, 250–258, 268–271
Nylon-yarn inclinometer, 20

Observation(s), platforms, 2–7

Observation(s) *(continued)*
—, statistical significance of, 8, 9, 49–51
—, synoptic, 12, 149
—, time scale of, 1, 8, 9, 49–51
Oceanic frontal zones, 57, 58, 69–72, 201–205, 292, 295, 297, 301–302
Operation Cabot, 12, 295, 296
Oxygen content, 46, 47, 79, 115
— distribution (max.) in Antarctic Circumpolar Water, 46, 306
— — in Atlantic, 46, 47, 304, 306
— — in North Pacific, 307
— and photosynthesis, 115
Oxygen-18, 48

Pacific Deep and Bottom Water, 307
— *see* also water masses (geographical)
Paddle wheels, 14, 17, 18
Palmer Peninsula, 298
Pendulum day, 153, 187, 188
Peru Coastal Current, 287
— Countercurrent, 287, 288
— Current — *see* currents (geographical)
— Oceanic Current, 287
— undercurrent, 288
Phosphorescence, 59
Photographic recording devices, 17–20
Photosynthesis, 115
Pillsbury's current measurements, 3
Pilot charts, 59
Pingers, 29, 31, 155, 156
Pitot orifices, 14
Planetary curl effect (vorticity), 231, 235–238, 242, 245, 247, 249, 284
Platforms for observation, 2–7
Polar Front, 69, 70, 301, 303
Pollutants, 125
Positioning, absolute, 2, 12, 25, 28, 29
Potential density, 38, 47
— temperature, 47, 122, 304
— vorticity, 112
Prandtl's mixing length, 99, 101, 121

INDEX

Precipitation, 45, 46, 78, 117, 119, 197, 218, 222, 224, 263, 299, 305
Precise leveling, 131, 133–135, 148
Pressure (sea pressure), 37, 78, 81–86
— differences, absolute, 38, 140, 145–148
— —, relative, 38, 82, 138, 141, 145, 216
—, field of, 37, 82, 83–86, 216, 292
—, gradients, 78, 84–86, 135, 136, 140, 222
— recorders, 24
—, static, 37, 81, 84
—, unit of, 81
Propellers in current meters, 14, 17, 22, 23

Quasistatic equilibrium, 85
Quasisynoptic current fields, 10–12
— observations, 149, 295

Radio buoys, 3, 26
Radioactive isotopes as tracers, 48
Radius of curvature (streamlines), 155
Ram pressure, 14
Rauschelbach rotor, 14
Recording devices in current meters, 14–24
Red Sea Water — *see* water masses (geographical)
Reference level, 38, 140, 143, 145–148, 254
Relative currents, 38, 138–144
— dynamic topographies, 86, 87, 145–148
— field of pressure, 82
— pressure differences, 38, 141, 145
— vorticity, 112
Repeating current meter (Ekman), 15
Representation of currents, 1, 12, 49–54, 59–73
Residual current, 1
Resistance (or drag) coefficient, 178–180
Resultant current, 2, 50–54

Reykjanes Ridge, 260
Reynolds stresses, 97–99
Rhodamine-B, 125, 186, 187
Richardson current meter, 3–5, 18, 19, 50
— drop-sonde, 32, 33, 37
— number, 179
Rip currents, 59, 162
Ripple marks in sediments, 218, 309
Roberts current meter, 22–24
Rock outcrops at ocean bottom, 309
Rossby number, 162, 163
Rotors (Rauschelbach and Savonius), 14, 18
Roughness (hydrodynamical) of sea surface, 178

Salinity in core of Equatorial Undercurrent, 282, 283
— and core-method, 45–47
—, defined, 37
— in Gulf Stream section, 131
— in Upper Deep Water, 39, 46, 47, 303–305
—, temperature-salinity diagrams, 39–45
Sampling, discrete, 8
— rate, 8, 9, 49–52
Sandström's experiments, 223
Sargasso Sea, 143, 293
Satellites for navigation, 2
Savonius rotor, 14, 18
Scalar mean, 1, 52–54, 73
Scour marks in sediments, 309
Scylla and Charybdis, 162
Seasonal variations of currents, 1, 74, 76–78, 276–279, 281, 285, 299
Sedimentary structures, 218, 309
Sensitivity of current meters, 7, 8
Serial current meter (Ekman-E.M.W.), 17
Shearing stresses, lateral 98, 102, 103, 171, 218, 221, 238–240, 242, 244, 247, 299, 300
— —, vertical, 98, 103, 178, 182, 191, 192, 212, 214, 218, 228, 239
Ship drifts, 12, 25

Ship motions, 6
Ships, "Altair", 3, 7, 209, 211
—, "Armauer Hansen", 3, 211
—, "Atlantis", 130, 139, 142, 144, 291, 293
—, "Blake", 3
—, "Buccaneer", 3
—, "Challenger", 3, 205
—, "Crawford", 297
—, "Discovery II", 301, 303, 305
—, "Explorer", 28–31
—, "Fram", 180
—, "Gazelle", 3
—, "Meteor", 3, 267
—, "Meteor II", 279
—, "Trieste" (Bathyscaph), 20
Shoran, 2, 25
Siderial Day, 187
Sigma-T, defined, 82
Slack-line moorings, 4
Slicks, 59
Slope currents, 192–196
— of density discontinuity, 201–211
— of isobaric surfaces, 129–131
— of isopycnal surfaces, 130, 132, 136, 217
— — sea surface, 134–136, 216, 217, 278, 280
Somali Current — see currents (geographical)
South Equatorial Current — see currents (geographical)
Soviet Marine Antarctic Expedition, 299
Specific gravity, 81
— volume, 85, 87, 130, 133
Spreading of thermocline at equator, 281, 282
Spreading of water masses, 40, 44–48, 79, 174, 177, 178, 197, 263, 302–308
Stability of currents, 53, 54, 61, 64, 66–68, 73
Standard ocean, 83, 84, 140
Static pressure, 37, 81, 84, 85, 87
— stability, 197

Stationary currents, 91, 127
— vortices, 207–211
Stokes Theorem, 108
Strait(s) of Bab el Mandeb, 79, 306
— of Dover, 134
— of Florida, 3, 33, 36, 257
— of Gibraltar, 46, 79, 119, 263
— of Messina, 162
Stratified Ocean, 85, 168, 212
— —, mass stratification and geostrophic currents, 135–149, 198–201
— —, mass stratification and gradient currents, 207–212
— —, mass stratification, total pressure field, 82, 135
— —, mass stratification, two-layer ocean, 197, 201–206, 214
Stray motions, 2, 6, 8, 99
Stream function, 109
Streamline(s), 10, 55, 58, 61, 65, 68, 109, 110, 113, 234
— and isopycnals, 168–173
—, curved, 155, 158, 163
—, defined, 10, 109
— deflection due to bottom topography, 234, 235, 260
— divergence, 113, 114
— of horizontal mass transport, 245–250, 252–258, 267–270
Streaks of phosphorescence, 59
Stress — see also shearing stresses
—, Newton's formulation, 96
—, Reynolds, 97–99
—, tensor, 98
—, wind, at sea surface, 178–180
Submarine telegraph cables, 34
Subtropical convergence, 69–70, 302, 303
— gyres, 201, 248, 249, 256, 258, 268, 270, 285
Surface floats, 10, 11, 25, 26
—, geopotential, 83
—, isobaric, 83, 85, 86, 129, 138–140, 142, 145–148, 201, 202
—, isopycnal, 130 — see also isopycnals

Surface *(continued)*
—, isosteric, 85
— level, 83
— waves, 6, 7, 178, 179
Suspension, bifilar, 14
— cables for buoy stations, 4, 5, 24
Sverdrup type solution to wind-driven currents, 242–244
Swallow floats — *see* neutrally buoyant floats
Synoptic current fields, 10–12
— observations, 149

"Tagged" water particles, 25
Taut-wire moorings, 4
Telegraph cables (submarine), 34
Telemetering and telerecording, 21–24
Temperature, potential, 47, 122, 304
— structure in deep sea, 39–47, 79, 222–225, 263, 268–271, 303, 304, 307 — *see also* spreading of water masses
Temperature anomaly south of Grand Banks, 232–234
Thermal circulations (Bryan and Cox), 224, 266–271
— circulations (Sandström), 223
— engine (Bjerknes), 223, 224
— equator, 278
Thermistors, 14
Thermocline, 207, 224, 266, 278
Thermo-haline and wind-driven circulations — *see* circulation
Three-dimensional ocean circulation, 218, 227
— — — —, Bryan and Cox's model (1967), 218, 224, 238, 266–271
— — — —, Bryan's model (1963), 218
— — — —, Ekman's results for a homogeneous ocean, 228–238
— — — —, Lineikin's model (1955), 218–222
— — — —, Stommels model (1957), 224, 264–266

Three-dimensional ocean circulation *(continued)*
— — — —, Transient state (qualitative), 216, 217
Tidal currents, 1, 7, 8, 11, 52, 99, 115, 153, 155
Time scale of observation, 1, 8, 9, 49–51
— series measurements, 8, 9
Tongue-like distribution of water properties, 44, 174–178
Topographic effects of bottom on currents — *see* bottom topography effects
— (quasi) curl effect, 231
Topographies (dynamic), absolute and relative, 86–87, 141, 145–149
Total field of pressure, 82, 135, 216, 217
Tracers, 10, 11, 25, 26, 29, 48, 177
Tracing of bottom water, 304
— — Equatorial Undercurrent, 283
Trade Winds, 75, 77, 249, 275, 276, 278, 280
Trajectories, 10, 11, 25, 29, 30, 55
Transfer of momentum, 95–101, 120, 178 — *see* also wind stress
Transformation of relative into absolute dynamic topographies, 145–149
Transient state of bottom currents, 217–218
— of pure wind-driven currents, 187, 188
Transport, anisobaric, 193, 194, 196
— charts, 286
—, horizontal in pure drift currents, 186, 187, 246–250, 255–258
— of water in Antarctic Ocean, 299–303
— of water masses — *see* water masses
— of wind-driven currents in equatorial regions, 278–284
"Trieste" Bathyscaph, 20
Tromsö Sund, 10, 11

Tropical convergence, 303
— ocean circulation, 69–72, 76, 77, 206, 275–284
T–S diagrams, 38–47
Turbulence, general, 95, 120
—, isotropic, 121
—, measurements of, 36
— and mixing, 120, 172–177 — *see also* mixing and diffusion
Two-layer, ocean, 197, 201–209, 214
— — vortex, 208–209

Ultrasonic current meters, 21
— devices for current measurements, 14
Undercurrent(s) beneath Gulf Stream, 143, 295 — *see also* equatorial undercurrents, and currents (geographical)
— near west coasts of continents, 288
Upper Deep Water — *see* water masses (geographical)
Upwelling, 56, 73, 78, 288, 289

Variability of currents, 1, 7–9, 49–52, 61
Variable depth effect on currents, 231–235, 260–261 — *see also* bottom topography effects
Variable wind stress (Ekman), 230
Variation of Coriolis parameter, 229, 231, 245, 247
Variations, high and low frequency, in current meter records, 8, 9, 49–51
—, seasonal, of currents, 1, 74, 76–78, 276–279, 281, 285, 299
—, time dependent, of currents, 187–189, 245, 259, 260, 268
Vector correlation, 54, 55
Vector lines, 56, 57
Vectorial mean, 1, 51–54, 65
— sum, 50, 51, 53
Velocity potential, 109
Vertical component of currents, 55–58, 112, 288–290

Vertical *(continued)*
— density sections, 132, 200, 204, 210
— logs, 27
— mixing, 40–45, 120–125, 174–178
— motion, 112 — *see also* upwelling
— shearing stresses, 98, 103, 178, 182, 191, 192, 212, 214, 218, 228, 239
— structure of pure drift currents, 183–185
Virtual displacements, 50, 51
Viscosity — *see* also friction
—, eddy, coefficients, 95–103, 185–187
—, molecular, 95
Vortex over "Altair" Cone, 209–212
Vortices — *see also* gyres
—, small-scale, 162, 163, 209
—, stationary, 208–212
Vorticity, absolute, 112
—, advection of, 284
—, balance, 112, 230–231, 241, 247, 251
—, defined, 108
—, frictional, 238, 247, 284
—, planetary, 231, 235–238, 247, 251, 284
—, potential, 112
—, relative, 112
— tendency equation, 111, 112, 231, 240–251, 300
—, wind stress, 228, 230, 240–251, 268, 300

Walfisch Ridge, 304
Warm water accumulation in subtropical deep layers, 201, 208, 209, 268–270
Water masses, conditioning of, 78, 79, 197, 263
— — (geographical), An-tarctic Bottom Water, 39, 48, 262, 302, 303, 305, 308
— — —, Antarctic Circumpolar Water, 48, 62, 252, 263, 298, 300–303, 306

Wind stress, 178–180, 230, 242, 247, 248, 251, 267, 300
—— curl — *see* curl
Witte-Margules equation, 205

"Zero"-surface (or layer), 136, 237, 293 — *see also* layer of no horizontal motion and reference level

Zonal circulation, 75, 206, 275, 277–284
— slope of sea surface, 244, 280, 281, 284

Water masses (geographical) *(continued)*
— — —, Antarctic Intermediate Water, 46, 47, 121, 177, 178, 262, 302, 303–307
— — —, Arctic Bottom Water, 262, 304, 305, 307
— — —, Arctic Intermediate Water, 304, 305
— — —, Deep Water, 46, 262, 303–309
— — —, definition of, 40, 41
— — —, general, 39–48, 78, 79, 174, 261, 303
— — —, Indian Antarctic Intermediate Water, 306, 307
— — —, Indian Ocean Bottom Water, 307
— — —, Indian Ocean Deep Water, 48, 305–307
— — —, Intermediate Water — *see* Antarctic and Arctic Intermediate Water
— — —, Lower Deep Water, 39, 303–305
— — —, Mediterranean Water, 46, 79, 119, 263, 270, 304, 305, 307
— — —, Middle Deep Water, 303–307
— — —, North Atlantic Deep Water, 46, 48, 263, 303–307
— — —, Pacific Ocean Bottom Water, 48, 306, 307
— — —, Pacific Ocean Deep Water, 48, 307
— — —, Pacific Ocean Intermediate Water, 305
— — —, Red Sea Water, 79, 306, 307
— — —, Upper Deep Water, 39, 47, 262, 304, 305
— —, meridional exchange of, 262, 303–309 — *see also* water masses (geographical)
— —, spreading of, 40, 44–48, 79, 174, 177, 178, 197, 263, 302–308

Water masses *(continued)*
— —, stratification of, 39–47, 261, 262, 275
— —, transport charts, 286
— transport — *see* transport and water masses
— types and water masses defined, 38–47
Waves, internal, 7, 149
—, surface, 6, 7, 178, 179
—, wind-generated, 103
Weddell Sea, 303
Western Atlantic Trough, 305
— boundary currents, 258–259, 264, 272, 274, 284–287, 290–298
Westward intensification, 247–251
Whirlpools in Strait of Messina, 162
Wind(s), Doldrums, 75, 76, 77, 275, 278, 280
—, Monsoon, 75, 77, 275
—, Polar Easterlies, 248
—, Trade, 75, 196, 243, 248, 275, 278
—, Westerlies, 75, 196, 243, 248, 278
Wind-driven currents
— — — and slope currents, 192, 195–197, 216–221
— — — and thermo-haline circulation, 261–271
— — — and vertical density stratification, 206, 215, 216, 218–222
— — — in deep water, 180–188
— — — equatorial regions, 206, 207, 275–284
— — — of finite depth, 189–192
— — —, general, 78, 178–192, 195–197
— — —, modern approaches, 238–261, 265–271
— — —, modification of Ekman's results, 191
— — —, non-stationary, 187–188, 217, 307
— — —, vertical structure, 184, 189, 195–197